THE OBSERVER'S FIGHTING VEHICLES DIRECTORY WORLD WAR II

RESEARCH AND EDITING BY
BART H. VANDERVEEN
OLYSLAGER ORGANISATION NV

FREDERICK WARNE & CO LTD   London
FREDERICK WARNE & CO INC.   New York

CONTENTS

FOREWORD	3
ACKNOWLEDGEMENTS	3
INTRODUCTION	4
KEY TO TECHNICAL DATA	5
UNITED STATES OF AMERICA	6
GREAT BRITAIN	132
BRITISH COMMONWEALTH OF NATIONS	212
USSR	258
GERMANY	272
ITALY	336
JAPAN	352
ABBREVIATIONS	366
INDEX (BY TYPE)	367
INDEX (BY MAKE)	368

Revised Edition
Published by FREDERICK WARNE & CO. LTD LONDON ENGLAND 1972
© OLYSLAGER ORGANISATION NV and B. H. VANDERVEEN 1972
Printed by BUTLER & TANNER LTD FROME ENGLAND 1058.572

LIBRARY OF CONGRESS CATALOG CARD
No. 72–81146
ISBN 0 7232 1469 7

FOREWORD

by Robert J. Icks, Colonel US Army Reserve (Retired)

So much has been written on the exciting political, strategic and tactical aspects of modern warfare that it is easy for the public to lose sight of the part played by and the importance of supply for fast-moving forces. The late General Eisenhower, Supreme Commander in Europe, referred to the matter in his *Crusade for Victory* when he said (on page 453): 'The contribution of an overwhelming air force and the great mobility provided by the vehicular equipment of the Army enabled us to strike at any chosen point along a front of hundreds of miles.'

The Allied armies not only were well equipped with armored vehicles and self-propelled guns, but also with vast numbers of transport and special-purpose vehicles without which the combat vehicles would have been unable to sustain momentum. The Axis powers likewise relied on the so-called soft-skinned vehicles for similar reasons but lacked the industrial might to provide them in the prodigious quantities possible for the Allies.

Many books have been published which deal with the more glamorous combat vehicles in various countries but, until now, there has been nothing available which dealt with the auxiliary vehicles which made it possible for troops and armored vehicles to keep moving. The well-known Olyslager Organisation has filled this gap in the literature very ably, under the direction of Mr B. H. Vanderveen. Mr Vanderveen is uniquely qualified. In his position as a long-time assistant of Mr Piet Olyslager he has acquired the necessary technical and historical background in motor vehicles of all kinds to enable him to undertake a work of this kind. And, perhaps even more importantly, he possesses the inner driving force of the hobbyist.

Thus, although obviously a 'labor of love' it is also a technically accurate book covering an area long neglected. The Olyslager Organisation, Mr Vanderveen and the publishers are to be commended for their public service in making generally available a book so badly needed.

Elmhurst, Illinois

NOTE This revised edition supersedes the edition published in 1969. It contains numerous factual and pictorial additions and amendments. For post-war use of many of the vehicles covered refer to *The Observer's Military Vehicles Directory—from 1945.*

ACKNOWLEDGEMENTS

Most, if not all, of the historic material used in the compilation of this book has come from the private collection of the editor, who would like to gratefully acknowledge the co-operation of many of his friends with whom, over the years, he has had the pleasure of exchanging information, photographs and other material. Special thanks are due to Colonel Robert J. Icks, the internationally acknowledged, long-time expert and authority on past and present Armoured Fighting Vehicles, for his considerable help and encouragement, closely followed by (in alphabetical order): Brian S. Baxter, Peter Chamberlain, Andrew Mollo, William F. Murray, R. G. Ogorkiewicz, Yasuo Ohtsuka, Walter J. Spielberger, B. T. White and Laurie A. Wright, all of whom have over the years supplied invaluable material. There are many others, private individuals and collectors as well as people in industry and trade, to whom sincere thanks and appreciation are extended, with apologies for not mentioning them by name.

Special notes: Australian War Memorial photographs are reproduced by kind permission, on the following pages: 221, 224, 226, 243, 245, 247–9, 256 (1 each), 239, 252 (2 each), 238, 253 (3 each), 225, 231, 234 (4 each).

Imperial War Museum photographs appear with their reference number, and glossy prints, if desired, can thus be easily acquired by ordering directly from the IWM (Lambeth Road, London, SE1).

INTRODUCTION

This directory is intended to present, in a convenient form, the motor vehicles used by the Allied and Axis fighting forces during the Second World War, 1939–45. The emphasis is on 'soft-skin' vehicles, i.e. the unarmoured vehicles which operated behind the front lines, on the supply routes, at home bases, airfields and similar areas and locations. Of the armoured fighting vehicles only the principal types are included. To deal with the enormous variety of armoured vehicles in detail would have resulted in a book at least twice this size. Furthermore, it would be impossible to incorporate more than the present number of civilian-type and 'pseudo-military' vehicles which were used by the combatant powers, notably the Germans, who employed very large numbers of captured, confiscated and impressed vehicles. Some of these were modified in order to fulfil a specific role, for which the *Wehrmacht* classification system allowed, and such vehicles have been included where possible.

It should be noted that, except in the case of Germany, vehicles have been listed only under the country of origin, even if such vehicles were not used by that country's military forces themselves. This applies, for example, to vehicles which were supplied by the USA, Great Britain and British Commonwealth countries to 'friendly nations' and allies, either on a direct commercial basis or via Lend-Lease.

In the case of Germany this would not be practicable since a considerable portion of *Wehrmacht* vehicles were either taken over from defeated armies, captured *en masse* or produced for them in occupied countries, either to original or to existing German designs. Such vehicles are included in the German section.

The German section, in particular, also includes a number of pre-war vehicle types which were either continued in production after the outbreak of war or were used in large quantities throughout the duration of hostilities.

Countries which were annexed or occupied by Germany, such as Austria, Czechoslovakia, France, Poland, etc., are not included as separate entities. The vehicles which their forces used at the beginning of the war were of pre-war manufacture. Their wartime production is included under Germany. The Free French, Free Poles and similar forces which fought with the Allies used vehicles of US, British and Canadian origin. Such equipment was also used for re-equipping the armed forces of numerous countries after 1945 and much of it remained operational for many years (see *The Observer's Military Vehicles Directory—from 1945*).

In order to present as much information as possible in the amount of space available, it has been necessary to use many abbreviations and to condense the technical and other data generally. The explanatory notes on page 5 should help the reader to use the directory to its best advantage.

A list of abbreviations and an index are at the end of the book.

A work of this kind, especially in its first editions, can never be wholly complete in every detail. In its compilation use has been made of the most reliable sources available, which in many cases included the original vehicle handbooks, parts lists, etc. Reliable information not always being obtainable and authentic sources sometimes being inaccurate or contradictory, it has been necessary in certain cases to approximate such data as dimensions and weights or to omit certain details.

This, however, is believed not to detract from the book's value as a general reference work which has long been overdue, on a subject which, in comparison with nearly every other aspect of the Second World War, had so far been almost completely neglected.

Piet Olyslager, MSIA, MSAE, KIVI

Inasmuch as the value of a work of this kind is determined by the degree of accuracy of its content, the Editor should like to invite readers to call to his attention any inaccuracies and omissions they may detect. Constructive criticisms, suggestions and additional information and material will continue to be welcomed. Please write to the Editorial Productions Division, Olyslager Organisation NV, Book House, Vincent Lane, Dorking, Surrey RH4 3HW, England.

KEY TO TECHNICAL DATA

Abbreviations See page 366.

Vehicle Nomenclature A uniform system has been used throughout, regardless of vehicle's country of origin. Examples: Truck, 2½-ton, 6×6, Cargo (or GS) = 6-wheeled, 6-wheel drive (general service) cargo truck with rated payload capacity of 2½ tons. Car, Medium, 4×2, Radio = 4 wheeled, 2-wheel drive medium-size radio car; etc. The vehicle type is followed by the make and model designation in parentheses.

Engine Cubic capacity in cubic inches (cu in) or cubic centimetres (cc) according to use in country of origin. Letter combination to indicate cylinder and valve arrangement, cooling system and location in chassis. Examples: I-I-A-F = cylinders in-line, valves in head (ohv), air-cooled, front-mounted; V-L-W-R = cylinders in Vee, side valves (L-head), water-cooled, rear-mounted. Fuel is petrol (gasoline) unless stated otherwise.

Transmission Number of gears forward (F) and reverse (R) and number of speeds in transfer case or auxiliary gearbox, if applicable. Example: 5F1R×2 = 5-speed main gearbox with 1 reverse gear and 2-speed transfer, giving overall number of 10 forward and 2 reverse speeds. 4F1R×1 = 4-speed and reverse main gearbox and single-speed transfer.

Tyres (2DT) or (4DT) behind tyre size indicates dual tyres on 2 or 4 rear wheels, depending on number of rear axles. Otherwise dual tyres are considered to be single wheels, i.e. a 6×6 vehicle may have 6, 10 or even 12 tyres.

Wheelbase Distance between front and rear wheel centres or, in the case of tandem axles, the centre of the bogie. The distance between the bogie axle centres is indicated in parentheses, prefixed BC (bogie centres). Thus, on a 6-wheeled vehicle with wb 160 (BC 50) inches the distance between the front and middle axle is (160 inches less ½BC) 135 inches, the distance between middle and rear axle (or foremost and rearmost rear axle) is 50 inches.

Dimensions Overall length, width and height are given where available, in many cases together with the reduced (minimal reducible) figure, in parentheses. All dimensions are given in the units which are generally used in the country of origin, namely inches for US vehicles, feet and inches for British vehicles and millimetres for vehicles from countries where the metric system is used. If only two figures are shown they indicate length and width.

Weights Vehicle weights are net kerb weights (unladen) unless indicated otherwise.

Conversions
Length
1 inch (1″ or 1 in) = 25·40 mm
1 foot (1′ or 1 ft) = 12 in = 304·8 mm = 0·30 m
1 mile = 1760 yards = 1609 m
1 millimetre (mm) = 0·039 in
1 centimetre (cm) = 10 mm = 0·39 in
1 metre (m) = 1000 mm = 39·37 in = 3·28 ft
1 kilometre (km) = 1000 m = 0·62 mile

Weights
1 pound (1 lb) = 0·45 kg
1 quarter = 28 lb = 12·70 kg
1 cwt = 112 lb = 4 quarters = 50·80 kg
1 long ton = 20 cwt = 1·12 short tons = 1016 kg
1 short ton = 2000 lb = 0·89 long ton = 907·2 kg
1 kilogram (kg) = 2·2 lb
1 metric ton = 1000 kg = 0·98 long ton = 1·1 short tons

Capacities
1 cubic inch (1 cu in) = 16·39 cc
1 Imperial gallon = 1·2 US gallons = 4·55 litres
1 US gallon = 0·833 Imp. gal. = 3·79 litres
1 cubic centimetre (1 cc or cm³) = 0·061 cu in
1 litre (1 ltr) = 1000 cc = 61·025 cu in

UNITED STATES OF AMERICA

With their vast industrial resources, the United States, during 1939–45, turned out the impressive total of 3,200,436 military transport vehicles, 88,410 tanks (by comparison, Great Britain built 24,803 and Germany 24,360 tanks), 41,170 half-track vehicles, over 82,000 commercial-type tractors and thousands of other special-purpose vehicles.

The majority of trucks were in the light (¼-ton, ½-ton and ¾-ton) class of which 988,167 were produced, and the light-heavy (2½-ton) class, represented by a total of 812,262. The production figures of medium (1½-ton) and heavy-heavy (over 2½-ton) trucks were 428,196 and 153,686 respectively. Approximately one-third of the total production was procured by the Quartermaster Corps. In August, 1942, the responsibility for army transport vehicles was transferred from the QMC to the Ordnance Department. Peak procurement was in June, 1942, during which month the QMC accepted not fewer than 62,258 trucks of all kinds. This was some 16,000 more than the total production of 1939 and 1940.

Yet the requirements for trucks by the using arms, and not least, by other members of the Allies, were so high that production constantly lagged behind schedule, especially in the very 'popular' 2½-ton class.

To assist other members of the Allies in the supply of military equipment, a programme was set up known as Lend-Lease. This programme was based on the theory that the Allies, by pooling their resources, could win the war more quickly and with less expense than if they attempted to operate separately. Under Lend-Lease the US made available material and services valued at more than 42 billion dollars to 44 countries. The main recipients were Great Britain and the USSR. The Soviet Union alone received well over 400,000 trucks, tractors, scout cars, tanks, etc., together with countless other items. Before the enactment of the Lend-Lease Act in March, 1941, large numbers of commercial-type military vehicles had been supplied directly by the US motor industry to Great Britain, China, France and other countries, through normal commercial channels.

Large shipments of vehicles which were ordered by the French Government in 1939–40, but which could not be delivered before France was occupied by the Germans, were diverted to Great Britain and taken over by the British Army ('ex-French contract' vehicles). Some of these were subsequently passed on to the Soviet Union.

In addition to the huge problems of meeting requirements there were, as always, two other factors which hampered production, supply and maintenance of motor vehicles, namely standardisation of vehicle types and the production and supply of spare parts. Suffice it to say here that the US coped with these problems better than any other nation.

As briefly mentioned before, the Quartermaster Corps was responsible for all the transport needs of the US Army (incl. the Army Air Forces) until August, 1942, when this responsibility was transferred to the Ordnance Department, which was already dealing with all combat vehicles, including combat wheeled and tracked tractors. The using arms or services were responsible for special tractors. In 1943 the responsibility for commercial-type full-track tractors, which operated at a speed of not more than 12 mph, was transferred to the Corps of Engineers, who were the chief users of this type (a total of 82,099 of these tractors were acquired by the Corps of Engineers during the war).

The US Marine Corps and Navy were responsible for the development and procurement of their own motor vehicles. Many of their vehicles closely followed the general design of army types (notably in the light and light-heavy truck class). Certain army types were used (albeit with certain differences in specification; the USMC ¼-ton 4×4 'Jeep', for example, had front towing hooks and lifting rings) and some specific USMC/USN types were later also used by the Army (for example the LVT—Landing Vehicle Tracked).

Vehicle Types

Military motor vehicles can be divided and sub-divided into various groups. A first division can be made between administrative and tactical vehicles.

The former are basically commercial-type vehicles, for use

where stringent military requirements need not be met, e.g. conventional passenger cars, buses, metropolitan-type ambulances, regular production 4×2 trucks, etc. The advantages of such vehicles, employed chiefly at base camps, airfields and for 'domestic' use, are obvious. They are considerably cheaper to procure and operate than purely military vehicles and their maintenance can, if necessary, easily be carried out by the civilian motor trade.

The other main category comprises tactical vehicles. These are specially designed or modified to fulfil military requirements and are intended for use at or immediately behind the front line. Tactical vehicles as used by the US Army were, in turn, sub-divided as follows:

General-purpose Vehicles Wheeled vehicles designed to be used either for the movement of personnel, supplies, ammunition and equipment, or for the towing of guns, trailers or semi-trailers. They can also be used without modification to body or chassis to satisfy general automotive transport needs, by more than one service (this classification was largely made up of trucks equipped with cargo bodies and truck-tractors equipped with a so-called fifth wheel for towing semi-trailers). Example: Truck, 2½-ton, 6×6, Cargo.

Special-equipment Vehicles Wheeled vehicles, the chassis of which are basically identical, except for minor alterations, to those used for general-purpose vehicles but which have a special body or special equipment mounted thereon. Example: Truck, 2½-ton, 6×6, Gasoline Tank.

Special-purpose Vehicles Vehicles where the chassis and body are designed or adapted for special purposes, so differing from those of general-purpose or special-equipment vehicles. They include half-track and full-track vehicles, except AFVs. Example: Truck, 2½-ton, 6×6, Amphibious.

Combat Vehicles Vehicles, with or without armour and/or armament, which are designed for specific fighting functions.

Note: armour protection or armament mounted as supplementary equipment on vehicles of the other classes does not change the classification of such vehicles to combat vehicles.

The following abbreviations came into universal use for various types of combat vehicles:

AFV   Armoured fighting vehicle (all types)
APC   Armoured personnel carrier
SP   Self-propelled artillery carriage

Non-motorised Vehicles
Trailers Vehicles designed to be towed and provided with a suitable drawbar or tongue for attachment to trucks or other prime movers. Trailers, sometimes referred to as full-trailers, have two or more axles and can 'stand on their own feet'.
Semi-trailers Vehicles designed to be towed by, attached to and resting on a truck-tractor by means of a fifth-wheel arrangement. They can be detached at will from the latter and can be used as a full-trailer by using a 'dolly' to replace the rear end of the truck-tractor.

Basic Tactical Vehicle Chassis Types

¼-ton, 4×4
¾-ton, 4×4 (superseded ½-ton, 4×4, in 1942)
1½-ton, 4×4 and 6×6
2½-ton, 4×4, 6×4 and 6×6
4-ton, 4×4, 6×6
4–5-ton, 4×4
5-ton, 4×2 (semi-tactical)
5–6-ton, 4×4
6-ton, 6×6
7½-ton, 6×6
10-ton, 6×4
12-ton, 6×4, 6×6

Colours
US Army vehicles generally were painted a semi-gloss or matt olive drab, unless required to be painted with special finish for camouflage protection. US Marine Corps vehicles were generally painted forest green, US Navy vehicles 'battleship-grey'.

The following exceptions existed for Army vehicles: red for fire-fighting and crash vehicles; yellow for vehicles principally operated on airfields; aluminium with dark blue trim for recruiting vehicles (not mandatory); black for cars used by military intelligence and attachés (not mandatory). Full-gloss olive drab was allowed for use on buses, ambulances (except field ambulances) and passenger sedans.

Registration Numbers
US Army vehicles carried a white or 'blue drab' registration number on both sides of the engine hood (bonnet) and on the

rear of the body or in equivalent positions. The numbering system used was introduced in 1929–30 and continued until the mid-1950s, when it was superseded by a new system.

The first or first two digits of the registration number indicated the vehicle type; the subsequent digits indicated the numerical sequence in which the vehicle was added to the type group. The complete number was prefixed by the letter 'W' which designated War Department vehicles. This letter, however, gradually disappeared towards the end of the war. The following type designation prefix figures were used:

0	Trailers and semi-trailers, all sizes, except kitchen trailers.
00	Trucks, maintenance, all sizes (light repair, small-arms repair; also wreckers).
1	Cars, passenger, 2- to 7-passenger (sedan).
10	Trailers, kitchen, motor-drawn, 1½-ton.
2	Trucks, light, ½- to 1-ton (pickup, panel, etc.).
20	Trucks, reconnaissance, all sizes (incl. carryall, etc.), and buses.
3	Trucks, medium 1½-ton (cargo, dump, truck-tractor).
30	Tanks, all sizes and certain tank-based special vehicles.
4	Trucks, light-heavy, 2½-ton, and heavy-heavy, up to 4–5-ton incl.
40	Vehicles, full- and half-track (cargo carrying, wire-laying, and combat, except tanks).
5	Trucks, heavy-heavy, and prime movers, 5-ton and over, all types.
50	Trucks, fire (fire and crash), all sizes.
6	Motorcycles (solo, w/sidecar, and motor tricycles).
60	Vehicles, special and technical (radio, office, generator, air compressor, searchlight, sterilising, water purification and similar), and armoured cars.
7	Ambulances, field and metropolitan, all sizes.
70	Amphibious vehicles.
8	Tractors, wheeled (light, medium, heavy).
80	Trucks, tank, sprinkler and refuellers, all sizes.
9	Tractors, half- and full-track (light, medium, heavy).

Note: before 1942 truck payload capacity classification was as follows:

2	Utility	¼- to 1-ton
3	Light	1¼- to 2-ton
4	Medium	2½- to 4-ton
5	Heavy	5-ton and over

Vehicles supplied to the US forces in Australia by the Australian motor industry (mainly rhd administrative vehicles: Chevrolet, Ford, IHC, Pontiac) had numbers with a 'U' prefix.

Military Designation Symbols
The following symbols were adopted to designate military vehicles and equipment (if developed by or for US Government):

T	Experimental type	(e.g. Car, Half-Track, *T7*)
M	Standardised type	(e.g. Car, Half-Track, *M2*)
E	Experimental modification	(e.g. Car, Half-Track M2*E*6)
A	Standardised modification	(e.g. Car, Half-Track, M2*A*1)

The US Signal Corps used a number with prefix K to designate special vehicle types. The US Army Air Forces used a similar system but with various prefix letters (A, C, E, F, J, L, N).

SNL Group Numbers
With the exception of experimental and non-classified types, all US Army vehicles, including trailers, can be identified by their SNL number. SNL is the abbreviation for Standard Nomenclature List, which is the Ordnance parts catalogue. There were 19 groups of Ordnance equipment, each of which had a letter. The G group applies to motor vehicles and each vehicle type was allocated a number. This number appears on the vehicle's publication data plate and on the relevant SNL or parts list. The G number is often the easiest and simplest way of indicating a vehicle, especially if it was produced by various manufacturers to a common design. For example: G102 indicates all half-track vehicles produced by Autocar, Diamond T and White; G671 indicates the Truck-tractor, 5-ton, 4×2, which was produced by International, Kenworth, and Marmon-Herrington. Inclusion of a complete summary of G-numbers in consecutive order here is prohibitive for space reasons, but they are shown, where applicable, under the 'Makes and Models' headings.

Who's Who in the US Automotive Industry (1939-45)

In the following pages only the trade names (makes) of the manufacturers are used. Listed below are their full names. Several of them have now disappeared, or been bought out by other manufacturers, so losing their identity. Examples: Autocar, Diamond T, Reo and Sterling were taken over by White; Willys-Overland became Kaiser Jeep and later merged with American Motors, etc.

The locations are those of the companies' head offices at the time. The various manufacturers of locomotives, etc., who, during the war, produced tanks and other AFVs, and the manufacturers of trailers, semi-trailers and special equipment such as fire-fighting equipment, cranes, etc., are not included.

(American) Bantam	American Bantam Car Co., Butler, Pa.
Allis-Chalmers	Allis-Chalmers Mfg Co., Milwaukee, Wis.
Autocar	Autocar Co., Ardmore, Pa.
Available	Available Truck Co., Chicago, Ill.
Bay City	Bay City Shovels, Inc., Bay City, Mich.
Biederman	Biederman Motors Corp., Cincinnati, Ohio.
Brockway	Brockway Motor Co., Cortland, NY.
Buick	Buick Motor Div. of General Motors Corp., Flint, Mich.
Cadillac	Cadillac Motor Car Div. of General Motors Corp., Detroit, Mich.
Caterpillar	Caterpillar Tractor Co., Peoria, Ill.
Chevrolet	Chevrolet Motor Div. of General Motors Corp., Detroit, Mich.
Cletrac	Cletrac Tractor Co. (The Cleveland Tractor), Cleveland, Ohio.
Coleman	Coleman Motors Corp., Littleton, Colo.
Corbitt	The Corbitt Co., Henderson, NC.
Crosley	The Crosley Corp., Cincinnati, Ohio.
Cushman	Cushman Motor Works, Lincoln, Neb.
Dart	Dart Truck Co., Kansas City, Mo.
Diamond T	Diamond T Motor Car Co., Chicago, Ill.
Dodge, Fargo	Dodge Brothers Corp., Div. of Chrysler Corp., Detroit, Mich. (military sales through Fargo Motor Corp., Div. of Chrysler Corp.).
Federal	Federal Motor Truck Co., Detroit, Mich.
FMC	Food Machinery Corp., San Jose, Calif.
Ford	Ford Motor Co., Dearborn, Mich.
FWD	Four Wheel Drive Auto Co., Clintonville, Wis.
GMC	GMC Truck & Coach Div. of General Motors Corp., Pontiac, Mich. (prior to 1943: Yellow Truck & Coach Mfg Co.).
Harley-Davidson	Harley-Davidson Motor Co., Milwaukee, Wis.
Hendrickson	Hendrickson Motor Truck Co., Chicago, Ill.
Indian	Indian Motorcycle Co., Springfield, Mass.
International	International Harvester Co., Chicago, Ill.
John Deere	John Deere & Co., Moline, Ill.
Kenworth	Kenworth Motor Truck Corp., Seattle, Wash.
Linn	Linn Mfg Corp., Morris, NY.
Mack	Mack Mfg Co., New York, NY.
Marmon-Herrington	Marmon-Herrington Co. Inc., Indianapolis, Ind.
Massey-Harris	Massey-Harris Co. (Tank Div), Racine, Wis.
Minneapolis Moline	Minneapolis Moline Power Implement Co., Minneapolis, Minn.
Oshkosh	Oshkosh Motor Truck Inc., Oshkosh, Wis.
Pacific Car & Foundry	Pacific Car & Foundry Co., Seattle, Wash.
Packard	Packard Motor Car Co., Detroit, Mich.
Plymouth	Plymouth Motor Corp., Div. of Chrysler Corp., Detroit, Mich.
Pontiac	Pontiac Motor Div. of General Motors Corp., Pontiac, Mich.
Reo	Reo Motors Inc., Lansing, Mich.
Sterling	Sterling Motors Corp., Milwaukee, Wis.
Studebaker	The Studebaker Corp., South Bend, Ind.
Thew	The Thew Shovel Co., Lorain, Ohio.
Walter	Walter Motor Truck Co., Ridgewood, Long Island, NY.
Ward LaFrance	Ward LaFrance Truck Div. of Great American Industries Inc., Elmira, NY.
White	The White Motor Co., Cleveland, Ohio.
Willys	Willys-Overland Motors Inc., Toledo, Ohio.

USA

MOTOR SCOOTERS

10

Four types of motor scooters were used by the US armed forces, all produced by the Cushman Motor Works of Lincoln, Nebraska. There was one two-wheeler, specially developed for airborne use, namely Model 53 (G683). It was standardised in March, 1944. There were two sidecar models, the Cushman 32 (G551) and 34 (G679). The former was standardised in November, 1943, and superseded by the Model 34 in the following year. Finally there was a three-wheeled type, Model 39 'Package Car' (G672), introduced in 1944. Cushman were well known as producers of golf course three-wheelers, etc.

Scooter, Motor, 2-W, Airborne (Cushman 53) 1-cyl., 4·6 bhp, 2F, wb 57 in, 77×23×38 in, 255 lb. Tyres 6.00–6. GVW 480 lb. Model 16M71 four-stroke 14·89 CID engine with magneto ignition. Fitted with parachute-attaching rings. Small pintle hook at rear. Max. speed 40 mph. 1944.

Scooter, Motor, 3-W, w/Side Car (Cushman 34) 1-cyl., 4·6 bhp, 2F, wb 44 in, 75×44½×36½ in, 350 lb. Tyres 4.00–8. GVW 675 lb. Model 15M70 engine. Used by Army and Navy for light delivery, stock chasing and messenger service. Max. speed 25 mph. 1944.

Scooter, Motor, 3-W, Package Delivery (Cushman 39 'Package Car') 1-cyl., 4 bhp, 2F, wb 62 in, 92×45×46½ in, 414 lb. Tyres 4.00–8. GVW 800 lb. Used for same purposes as sidecar type but chiefly by Air Corps (later USAF). Model 10M70 engine. 1944.

USA
MOTORCYCLES

Solo machines were used by all arms and services and in addition many were supplied to Canada, Britain, etc. Sidecar combinations were less numerous. Principal solo models were the Harley-Davidson WLA (G523, from 1939) and Indian 640B and 644 (G524, from 1941) and 741B (G674, from 1941). These were militarised chain-drive civilian types. Sidecar combinations included the Harley-Davidson 40LE (G680, from 1939) and Indian 340 (G673, from 1939). Lightweight airborne models were supplied by Indian (Model 148, 1944/45) and Simplex ('Servi-Cycle', 1943). The latter had a 1·6-bhp two-stroke engine.

Motorcycle, Solo (Harley-Davidson WLA) V-2-cyl., 23 bhp, 3F, wb 57½ in, 88×36¼×59(41) in, 575 lb. Tyres 4.00–18. 45·12 CID (2⅞×3³⁄₁₆ in) L-head engine. Model WLC was made for Canadian forces, with different lighting equipment. Another solo model was the 74 CID Model UA.

Motorcycle, Solo (Indian 741B) V-2-cyl., 15 bhp, 3F, wb 56¾ in, 88⅛×33½×39¾ in, 460 lb. Tyres 3.50–18. 30·07 CID (2½×3¹⁄₁₆ in) L-head engine. Also used by British Army and RAF. Indian 148 was single-cyl. 13·5 CID airborne type (Motorcycle, Solo, Extra Light, M1), weighing 250 lb.

Motorcycle, Solo (Indian 640) V-2-cyl., 24 bhp, 3F, wb 57 in, 90×32½×46 in, 540 lb. Tyres 4.00–18. 45·44 CID (2⅞×3½ in) L-head engine. This machine differed from the 741B in size and in that the gearshift lever pivoted under the forward end of the fuel tank, rather than at the rear.

12

Motorcycle, Solo (Harley-Davidson XA) HO-2-cyl., 23 bhp, 4F, wb 59½ in, 90×36×60(40¼) in, 525 lb. Tyres 4.00–18. 45·04 CID. Differed fundamentally from other Harleys in having cylinders set at 180°, foot-operated gearshift, shaft-drive and telescopic rear suspension. 1000 produced, 1942 (G585).

Motorcycle, Solo (Indian 841) V-2-cyl., 24 bhp, 4F, wb 59 in, 90¾×36¾×40½ in, 565 lb. Tyres 4.00–18. 45·4 CID L-head engine, mounted across frame. Cylinders set at 90°. Foot-operated (toe and heel pad) gearshift, shaft-drive, rear suspension. Limited production, 1941/42 (G631).

Motorcycle, w/Sidecar (Harley-Davidson UA) V-2-cyl., 33 bhp, 3F, wb 59½ in, 95½×69×42½ in, 850 lb. Tyres 5.50–16 or 4.50–18. 74 CID engine. Max. speed 55 mph. Used by all arms (USMC shown). Model ELA was similar but lighter and had 60·32 CID engine.

Motorcycle, w/Sidecar (Indian 340) V-2-cyl., 30 bhp, 3F, wb 61½ in, 97½×68½×44 in, 845 lb. Tyres 4.50–18. Payload 400 lb. 73·62 CID engine. Sidecar combinations were largely replaced by the ¼-ton 4×4 'Jeep' when it became available. Unit shown was supplied to Britain.

USA

CARS and STATION WAGONS

Although many civilian types were used, few were standardised and/or produced specifically for the armed forces. There were four car types, viz. 5-pass. light, medium and heavy sedans; 7-pass. sedans (limousines); and two types of station wagon, light and medium. Light types were Chevrolet, Ford and Plymouth products, medium types included Buick, Chrysler, Oldsmobile, Packard and Pontiac, heavy types (incl. 7-pass.) Buick, Cadillac, Chrysler and Packard. Most were 1941/42 models.* Production ceased in February, 1942. Late in 1942 the Army bought 29,000 Chevrolet, Ford and Plymouth sedans which had been frozen in dealers' stocks.

*For additional coverage see *American Cars of the 1940s* (Olyslager Auto Library/Warne).

Car, Light Sedan, 5-Passenger, 4×2 (Chevrolet KB Master 85)
6-cyl., 85 bhp, 3F1R, wb 113 in, 193×73×67 in, 3040 lb. 216·5 CID OHV engine. The Army had 306 of these 1940 Chevrolets with sedan and van (Sedan Delivery) bodywork. Sedan shown was used by 501st Parachute Batallion.

Car, Light Sedan, 5-Passenger, 4×2 (Chevrolet BG 1503)
6-cyl., 85 bhp, 3F1R, wb 116 in, 196×72¾×69 in, 3275 lb. 216·5 CID OHV engine. Militarised 1942 Master DeLuxe (Stylemaster) sedan, featuring blackout light equipment and lustreless olive-drab paintwork. US Army acquired 1619 (G520). Also supplied to UK.

Car, Light Sedan, 5-Passenger, 4×4 (Ford/M.-H. 01A-73B)
V-8-cyl., 85 bhp, 3F1R×2, wb 112 in, 192×75×75 in. Four-wheel drive conversion of 1940 Ford DeLuxe Fordor for the army of a South American republic. Many Ford cars and commercial vehicles were thus converted, all by Marmon-Herrington.

14

Station Wagon, Light, 4×2 (Ford 01A-79A Standard)
V-8-cyl., 85 bhp, 3F1R, wb 112 in. Virtually standard 1940 civilian type, used by USMC. Army had 150 Pontiac 1942 medium wagons (Model 42–26, G550). The Pontiac was also used by the USMC. Buick station wagons were also in service.

Car, Light Sedan, 5-Passenger, 4×2 (Ford 21A-73C DeLuxe)
V-8-cyl., 90 bhp, 3F1R, wb 114 in, 194½×73½×68 in, 3395 lb. US Army had 10,317 of these 1942 models (incl. many with 6-cyl. engine, Model 2GA-73C) and 1800 1940/41 01A/11A-73C V8s. See also pages 88 and 218.

Car, Light Sedan, 5-Passenger, 4×2 (Plymouth P11 DeLuxe)
6-cyl., 87 bhp, 3F1R, wb 117½ in, 198½×73½×68 in, 3200 lb. 201·3 CID L-head engine. Militarised 1941 model, one of 2031 used by Army (G521). Army also had 1942 Model P14. 1940 Model P9 was used by Army (60) and Civil Conservation Corps (150).

Car, Medium Sedan, 5-Passenger, 4×2 (Packard 2001 Clipper) 8-cyl., 125 bhp, 3F1R, wb 120 in, 208½×76×63½ in, 3700 lb. Militarised 1942 model (G644). Other medium and heavy Packard 1941–42 sedans and limousines used by the armed forces included 275 of Model 2003, and ambulances on 160-in wheelbase chassis (q.v.).

USA
BUSES
4×2

Buses used by the US armed forces during the war were essentially civilian models, mainly 1½-, 2½- and 5–6-ton class truck chassis with schoolbus-type bodywork. There were also some 15-seater car conversions and a few integral COE highway types (GMC, Mack). For mass transportation of war production workers and other personnel over 4000 'new-car haulaway' semi-trailers were converted into 40–80-pass. buses. Numbers of Dodge 4×4 and GMC 6×6 truck chassis were fitted with bus bodies after the war and so were lengthened White M3A1 scout car chassis. The latter, however, were for civilian use.

Bus, Sedan, Converted, 15-Passenger, 4×2 (Chevrolet BG1503) 6-cyl., 85 bhp, 3F1R, wb 188 in, 268×73×70½ in, 4950 lb. Lengthened 1942 5-pass. sedan. Similar conversions also of Ford 01A, 11A, 21A and 2GA and Packard 2000-6 and 2010-8 sedans. Conversion included reinforced suspension and oversize tyres. Relatively rare.

Bus, 1½-ton, 25-Passenger, 4×2 (Available) Waukesha 6-cyl., 90 bhp, 4F1R, wb 178 in, 286×96×105 in, 9100 lb. Similar schoolbus-type bodies, seating 20–29, on other chassis, incl. Ford 094T/194T (G540), GMC, International K5 and KS5 (G688), etc. Bodywork was made by various manufacturers, incl. Superior and Wayne.

Bus, 2½-ton, 37-Passenger, 4×2 (International K7, KS7) 6-cyl., 87.5 bhp, 5F1R (2-speed rear axle on KS7), wb 230 in, 374×96×112 in, 12,387 lb. SNL G541 and 659. Similar buses, seating 37–40, on other chassis, incl. Available, Diamond T 614HD, Federal, and Mack EH. The latter chassis was rated as a 5–6-ton model.

USA

AMBULANCES
4×2

Most 4×2 ambulances were 'Metropolitan' types and used at base hospitals, etc. Tactical 4×4 type ambulances are dealt with in the relative chassis sections. Most Metropolitan ambulances were civilian 4-litter (-stretcher) types, produced by firms like Henney and Superior on special chassis made by Cadillac (Model 62, 1941, G548; also LaSalle (Model 40-50, 1940–41) and Packard (Model 4294-HDA, 1941–42, G549). The first Linn, introduced in late 1940, was a front-wheel drive large-capacity model. A modified version was later standardised as M423.

Ambulance, Metropolitan, ¾-ton, 4×2 (Packard 4294HDA/Henney) 8-cyl., 125 bhp, 3F1R, wb 160 in, 246×75×76 in, 5125 lb. Tyres 7.50–16. Henney bodywork on special chassis with Model 2001A 282 CID 8-in-line L-head engine. Max. seating capacity 2+10. 92 supplied, 1941–42. Similar: Cadillac 62/Superior.

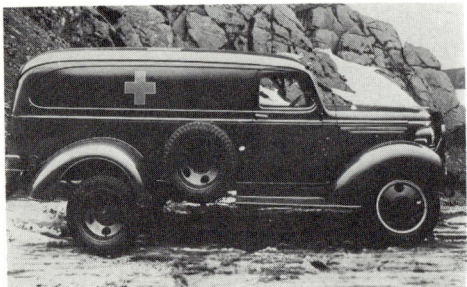

Ambulance, Field, 1½-ton, 4×2 (Chevrolet WA) 6-cyl., 85 bhp, 4F1R, wb 133 in, 288×81¼×84 in, 5500 lb. Conversion of 1940 panel truck. 331 supplied. Similar ambulances on 1941–42 Chevrolet and 1939 Ford chassis. Used also by Civilian Conservation Corps (which was equipped by the Army, without being part of it).

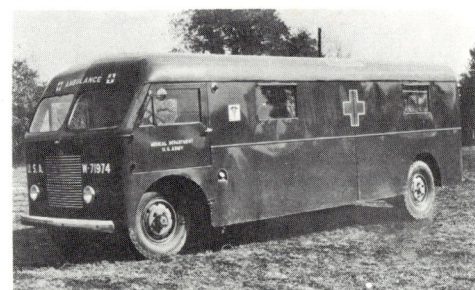

Ambulance, Metropolitan, 1½-ton, 4×2 (Linn) Ford V-8-cyl., 90 bhp, 4F1R×1. Timken F30B driving front axle, Linn trailer-type rear axle. Tyres 8.25–20. A later model was standardised in 1945 as M423 and had a Dodge 6-cyl. 94-bhp engine. It had a 186-in wb and measured 300×88×105 in.

USA

TRUCKS, ¼-TON 4 × 4
COMMAND RECONNAISSANCE

Makes and Models: American Bantam 40 BRC, 1941. Ford GP, 1941. Ford GPW, 1942–45 (G503). Willys MA, 1941. Willys MB, 1941–45 (G503).

General Data: Purpose: to carry personnel and light cargo and to tow 37-mm AT gun or ¼-ton 2-wh. trailer. The first ¼-ton 4 × 4 was produced by Bantam in 1940. It was designed by Karl K. Probst to a US Army requirement. 70 were subsequently made of an improved model. Willys, in late 1940, also produced a similar pilot model, known as 'Quad'. After exhaustive trials both vehicles were almost completely redesigned and 1500 of each were ordered for further tests. By this time Ford had also produced a pilot model and also received an order for 1500 of an improved version. The improved models were designated Bantam 40BRC, Willys MA and Ford GP. Following further tests the Willys, which had the most suitable engine, was chosen for mass production, subject to certain modifications. The resulting model was the MB. It entered production in December, 1941, and Ford became a co-producer early in 1942 (model GPW—General Purpose Willys). Most of the earlier models were released for Lend-Lease and shipped to Britain, Russia, and the Far East. Of the standardised ¼-ton 4 × 4, popularly known as 'the Jeep', a total of 639,245 (277,896 GPWs, 361,349 MBs) were produced until the war ended. They were extensively used for a multitude of purposes by all the United Nations. In the British and Canadian Army they were designated 'Car, 5-cwt, 4 × 4'. Early popular names included 'Peep' and 'Blitz Buggy'. In the interest of standardisation many endeavours were made to adapt the basic 'Jeep' for other purposes, either by using conversion kits or by extensive modifications. The latter included 6 × 6, half-track (snow tractor) and armoured versions but none of these entered series production. After the war many manufacturers, all over the world, started producing 'Jeep'-type vehicles for civilian and military use.

Vehicle shown (typical): Truck, ¼-ton, 4 × 4, Command Reconnaissance (Willys MB) (later re-named: Truck, ¼-ton, 4 × 4, Utility)

Technical Data:
Engine: Willys 442 'Go Devil', 4-cylinder I-L-W-F, 134·2 cu in, 54 bhp @ 4000 rpm (gross: 60 bhp @ 3600 rpm).
Transmission: 3F1R × 2.
Brakes: hydraulic.
Tyres: 6.00–16.
Wheelbase: 80 in.
Overall l × w × h: 132¼ × 62 × 72(52) in.
Weight: 2450 lb.
Note: first 25,808 Willys MBs had slatted radiator grille.

18

Truck, ¼-ton, 4×4, Command Reconnaissance (Bantam 40BRC) Continental 4-cyl., 45 bhp, 3F1R×2, wb 79½ in, 126×54×71½(49) in, 2100 lb. A total of 2675 Bantams were made. Most went to the UK and USSR. Unit price was just over $1166. Was first 'Jeep' to appear in Britain.

Truck, ¼-ton, 4×4, Command Reconnaissance (Willys MA) 4-cyl., 54 bhp, 3F1R×2, wb 80 in, 132¾×62×70(50¾) in, 2160 lb. Immediate predecessor of MB/GPW. Steering column gearshift. Used well-proven engine of Willys 'Americar'. 1500 produced; many of these supplied to USSR.

Truck, ¼-ton, 4×4, Command Reconnaissance (Ford GP) 4-cyl., 45 bhp, 3F1R×2, wb 80 in, 129×62×71(49) in, 2150 lb. 1500 produced in 1941 at $925 each. Used adaptation of Ford tractor engine. GP stood for General Purpose. In all about 3650 were produced. Many went to Britain.

Truck, ¼-ton, 4×4, Command Reconnaissance (Ford GPW) 4-cyl., 54 bhp, 3F1R×2, wb 80 in, 132¼×62×72(52) in, 2450 lb. Produced to Willys MB design. Differed mainly in having inverted 'U'-section chassis front cross member (tubular on MB). Also basis for amphibious GPA.

Bantam Reconnaissance Car, Prototype This was the world's first ¼-ton 4×4 field car, later to be known as the 'Jeep'. Designed by Karl K. Probst and built by Roy S. Evans and his staff of the American Bantam Car Co. Continental engine, Spicer transfer case and axles. Pioneered 'Jeep' type body.

Bantam Reconnaissance Car, Pilot Model The Bantam 'Mark II' featured redesigned front wings and other improvements. 70 were made, including 8 with 4-wheel steering, for further tests in 1940/41. Several components were inherited from Austin/Bantam cars.

Willys 'Quad' Reconnaissance Car, Pilot Model The Willys 'Quad' was produced late in 1940 and submitted for tests at the Camp Holabird proving grounds on 13th November. Appearance of 'Quad' was very similar to Bantam but engine-wise it was superior.

Ford 'Pygmy' Reconnaissance Car, Pilot Model Ford/Ferguson 'Dearborn' tractor engine, Ford Model A 3F1R gearbox, 133×59×59 in, 2100 lb. Tested at Camp Holabird in Nov./Dec., 1940. Pioneered 'Jeep' front end with built-in hinged headlights. With modifications became 1941 Model GP.

Carriage, Motor, 37-mm Gun, T2 (Bantam 40BRC) The $\frac{1}{4}$-ton 4×4 was soon adapted to fulfil other roles such as self-propelled (SP) gun mounts. This 1941 Bantam had its rear body removed and steering column repositioned so that the gun muzzle could pass under the steering wheel.

Carriage, Motor, 37-mm Gun, T2E1 (Bantam 40BRC) This was a later experimental 37-mm gun SP mount, also based on the Bantam but with fewer modifications to the vehicle and with the gun pointing rearward. The British fitted Bantams with machine guns for desert patrol duties.

Truck, $\frac{1}{4}$-ton, 4×4, Ambulance (Ford GPW, modified) The 'Jeep' was frequently converted and employed as a front-line ambulance. These USMC Fords were rather extensively modified for use in the Pacific area and carried three litters. Note equipment locker and vertical windscreen.

Truck, $\frac{1}{4}$-ton, 4×4, Cargo (Willys MB, modified) Many kinds of 'unofficial' modifications were carried out in the field, often using parts of disabled vehicles. This unit had wheelbase and body extended by 30 in. US Coast Guard also used lwb (116-in) conversion, seating 10.

Car, 5-cwt, 4×4 (Willys MB) This was the nomenclature by which the ¼-ton 4×4 'Jeep' was known in Britain. Shown is an early Willys fitted with all-weather equipment, made by Humber Ltd in April, 1944. British fitted sidelights on front wings (fenders).

Truck, ¼-ton, 4×4, Command Reconnaissance (Willys, Ford) Basically standard 'Jeep', fitted with pedestal-mounted machine gun and armour plates to protect crew and radiator (field modification). Used by the US Army on the European Western front. Picture taken in the Ardennes in Dec. 1944.

Truck, ¼-ton, 4×4, with Rotary Wing Australian Autogiro type conversion kit intended for use in New Guinea battles. M.L. Aviation Co. in Britain also converted a 'Jeep' into a rotary-wing glider (1942). It had a 46′ 8″ diameter rotor. Neither was used operationally.

Truck, ¼-ton, 4×4, with Replica 'Kübel' Body A 'Jeep' chassis was fitted with a replica *KdF Typ 82* (Volkswagen *'Kübel'*) body for use by the SAS in North Africa in 1942. The 'driver' was a dummy with *Wehrmacht* helmet, the actual driver sat in the back.

USA

TRUCKS, ¼-TON 4×4
EXTRA LIGHT, LIGHT and MEDIUM WEIGHT

Makes and Models: *Extra Light-weight:* Chevrolet 'Extra Light'. Crosley pilot model. Crosley CT-3 'Pup'. Kaiser 'Midget Jeep'. Willys WAC.
Light-weight: Kaiser 1160. Willys MB-L Special.
Medium-weight: Willys MB-L.

General Data: Purpose: for tactical use by armed forces engaged in mountain, tropic, or arctic warfare and for airborne infantry use. These vehicles were intended to have the same payload capacity as the standard ¼-ton 4×4 but to more nearly conform to the weight originally specified for it so that they could be manhandled when in difficulty. The first pilot model was submitted by Crosley in February, 1943, and given limited service tests at Ft Benning, Ga (Crosley had also produced a lightweight field car in 1940, based on their regular midget car). 36 Crosleys of a modified type (CT-3) were subsequently procured and tested (six went overseas) but failed in several respects. The Chevrolet had an Indian motorcycle power unit but was not successful. The WAC (Willys Air-Cooled) was much like the Crosley pilot model but was powered by a Harley-Davidson engine. It, too, failed but was later modified by the Army into a platform carrier (not unlike the later 'Mechanical Mule'). Willys then produced, at their own expense, the MB-L in several forms. These were in fact stripped and modified 'Jeeps'. Their test results were satisfactory. The Henry J. Kaiser Co. produced two 'Midget Jeeps' and four of the slightly heavier '1160'. None of these was satisfactory. All tests and further experiments were terminated towards the end of 1943. In England a jeep was extensively modified for airborne use by Nuffield Mechanizations Ltd, concurrently with some other 'Jungle Jeep' projects carried out by SS Cars Ltd and the Standard Motor Co.

Vehicle shown (typical): Truck, ¼-ton, 4×4, Light (Willys MB-L Special).

Technical Data:
Engine: Willys 442 4-cylinder, I-L-W-F, 134·2 cu in, 54 bhp @ 4000 rpm.
Transmission: 3F1R×2.
Brakes: hydraulic (front wheels only).
Tyres: 5.00–16.
Wheelbase: 80 in.
Overall l×w×h: 127½×59¼×53½ (39) in.
Weight: 1485 lb.
Note: These stripped 'Jeeps' were colloquially known as Gypsy Rose Lee models.

Truck, ¼-ton, 4×4, Extra Light (Crosley CT3 'Pup') Waukesha–Crosley HO-2-cyl., air-cooled, 38 CID, 13 bhp, 3F1R×2, wb 65 in, 102×49×35 in, 1125 lb. GVW 1625 lb. Tyres 5.00–16. Warner trans., Crosley-Spicer axles. Two seats (one in rear). Canvas mudguards. Speed 45 mph. Mech. brakes. Max. gradeability 40%.

Truck, ¼-ton, 4×4, Extra Light (Chevrolet) Indian V-2-cyl., air-cooled, 45 CID, 20·5 bhp, 3F1R×2, wb 72 in, 106×49×38¾ (cowl) in, 1062 lb. GVW 1562 lb. Tyres 5.00–16. Two seats. Centrally mounted steering gear. Independent suspension. Max. gradeability 60%. Mech. brakes. Two built, 1943. One has survived in a US museum.

Truck, ¼-ton, 4×4, Extra Light (Kaiser 'Midget Jeep') Continental HO-4-cyl., air-cooled, 84 CID, 30 bhp, 3F1R×2, wb 68 in, 104×52×35 in, 1100 lb. GVW 1600 lb. Tyres 5.00–14. Warner trans., Spicer axles. Model 1160 (Light) had full-width four-seater body and 6.00–14 tyres.

Car, 5–cwt, 4×4, Airborne (Willys/Nuffield) Produced experimentally by Nuffield Mechanizations Ltd in England, *c.* 1943, using mechanical components and shortened frame of Willys MB. Body (2-seater) and many other parts were entirely new. Shown with steering column and top removed.

USA

TRUCKS, ½- to 1-TON 4×2

Trucks in this category were militarised 1940–42 commercial panel vans, carryalls, pickups and some special types like bomb service trucks and telephone maintenance trucks. Dodge supplied 1542 ½-ton T112 (G613) units, mostly carryalls and pickups. Chevrolets included 1940–42 ½-tonners (KB, KC, AG, AK, BK; G612), ¾-tonners (KD, KE, AL, BL; G616) and 1-tonners (WA). Ford supplied ½-ton pickups (G615). Bomb service trucks were built on Diamond T, Ford and GMC ½-ton 4×2 chassis until superseded by the Chevrolet 1½-ton 4×4 M6 (q.v.).

Truck, ½-ton, 4×2, Canopy Express (Chevrolet BK) 6-cyl., 85 bhp, 3F1R, wb 115 in, 198½×73×78 in, 3395 lb. 1942. Basically panel van, but open sides and back which could be covered with curtains. Few used by Army. Body styling shown was used on Chevrolet trucks from 1941 until 1946 incl.

Truck, ½-ton, 4×2, Telephone Maintenance (Chevrolet AK) 6-cyl., 85 bhp, 3F1R, wb 115 in. 1941 AK3103 chassis/cab. Similar body on Dodge T112 (WC39, 1941; WC50, 1942). Used by Signal Corps for telephone installation and maintenance. Superseded by Dodge 4×4 models (WC43, WC61).

Truck, 1-ton, 4×2, Cargo and Personnel (Chevrolet WA) 6-cyl., 85 bhp, 4F1R, wb 133 in. Tyres 6.00–20 front, 32×6 (single) rear. Standard production model of 1940 Chevrolet WA 1-ton series, with 'Express' pickup-type body and tilt, used by the US Marine Corps. Could carry 12 passengers.

25

Truck, 1-ton, 4×2, Panel (Dodge T116, WD) 6-cyl., 82 bhp, 3F1R, wb 133 in. Canadian Army, 1941/42. US Army only had ½-ton T112 models, viz. 774 carryalls (WC36, 1941; WC48, 1942), 14 panels (WC37, 49), 752 pickups (WC38, 47) and 2 telephone installation trucks (WC39, 50).

Truck, ½-ton, 4×2, Pickup (Ford 2GC) 6-cyl., 90 bhp, 3F1R, wb 114 in, 194×73×75 in, 3290 lb. Standard body, 1942. Also with V-8-cyl. engine (Model 21C). Cab and front end styling shown was used on Ford trucks from late 1941 until 1947 incl. Similar ½-ton pickups supplied by Chevrolet and Dodge.

Truck, ½-ton, 4×2, Bomb Service, M1 (Ford 19Y/M.-H.) V-8-cyl., 100 bhp, 3F1R, wb 113 in, 203½×73½×83 in, 3960 lb. Commercial 1-ton chassis. Same type bomb-lifting crane on Ford 29Y (1942), Diamond T 201-BS and GMC AC-251 chassis. Non-motorised model supplied by Weaver Mfg Co.

Truck, ½-ton, 4×2, Carryall (GMC AC-101) 6-cyl., 95 bhp, 4F1R, wb 113½ in, 198×78×75 in. Commercial type 1940 'Suburban', GMC body style No. 993 (identical to that mounted on 1940–42 Chevrolets). Used for transport of personnel (six passengers) and cargo.

USA

TRUCKS, ½-TON 4×4
(1940 PATTERN)

Makes and Models: Dodge T202, 1939–40 (G505). Bodies: VC1 Command Reconnaissance; VC2 Radio; VC3 Pickup w/closed cab; VC4 as VC3 but w/o troop seats; VC5 Pickup w/open cab; VC6 Carryall.
Ford/Marmon-Herrington LD2-4, LD3-4, LD4-4, 1938–40.
GMC ACK-101, 1940.
International/Marmon-Herrington (conversion), D2, 1938–40.

General Data: The ½-ton 4×4 was pioneered by the Marmon-Herrington Co. of Indianapolis (military vehicle specialists under Colonel Arthur W. Herrington, since 1931) in July, 1936. The resulting vehicle (Ford) became later known as 'Darling' and has been called 'Granddaddy of the Jeep'. Many Ford ½- and 1½-ton trucks were thus converted and used by various governments, incl. the US and Belgium. Half-ton Internationals were converted by M.-H. for the US Coast Guard, from 1938. Chrysler's Dodge Division became mass-producers of the ½-ton 4×4 in 1939. Their T202 series used many commercial pattern parts, including the front-end sheet metal. 4640 of all types were produced by Dodge, and supplied to the US Army by Chrysler's Fargo Division (which handled fleet and military sales). The bulk of these were pickup trucks (1607 with open cab, 820 with closed cab) and command reconnaissance cars (2155). For 1941 they were replaced by the T207 series, which featured a military pattern front end. The Dodge ½-ton was also known as 'Jeep', when the early ¼-ton was sometimes referred to as 'Peep' (amongst other nicknames), but from early 1942 the name 'Jeep' applied exclusively to the ¼-ton 4×4, even in official US Army literature (TM10-460, May, 1942). Some of the above models were also produced without driven front axle, as conventional 4×2s (see preceding section).

Vehicle shown (typical): Truck, ½-ton, 4×4, Pickup (Dodge T202-VC3)

Technical Data:
Engine: Dodge T202 6-cylinder, I-L-W-F, 201·3 cu in, 79 bhp @ 3000 rpm.
Transmission: 4F1R×1.
Brakes: hydraulic.
Tyres: 7.50–16.
Wheelbase: 116 in.
Overall l×w×h: 186×74×83 in.
Weight: 4620 lb.
Note: One truck was converted into a Multiple Gun Motor Carriage, T1 (cargo box replaced by twin cal. 0·50 turret).

27

Truck, ½-ton, 4×4, Command Reconnaissance (Dodge T202-VC1) 6-cyl., 79 bhp, 4F1R×1, wb 116 in, 188×74×88 in, 4275 lb. Purpose: to provide transportation for staff officers in the field. VC2 Radio truck similar in appearance. A few were built on 4×4 Ford chassis.

Truck, ½-ton, 4×4, Pickup (Dodge T202–VC5) 6-cyl., 79 bhp, 4F1R×1, wb 116 in, 186×74×83 in, 4220 lb. This model was very similar in appearance to Marmon-Herrington's 'Darling' (see 'General Data'). They were the progenitors of the ¾-ton 'Beep'. Also with pedestal-mounted mg.

Truck, ½-ton, 4×4, Ambulance (Ford 1941/Marmon-Herrington) 8-cyl., 100 bhp, 3F1R×1, wb 112 in. Conversion to all-wheel drive by M.-H. Body by Boyertown for USMC. Collapsible body, shown here with roof in lowered position. Same chassis also with pickup/personnel bodywork.

Truck, ½-ton, 4×4, Pickup (GMC ACK-101) 6-cyl., 95 bhp, 4F1R×1, wb 113½, 185×75×84 in. Similar in design to earlier Ford/Marmon-Herrington 'Darling' and Dodge T202-VC5. Only one was built (USA registration No. W22005).

USA

TRUCKS, ½-TON 4×4
(1941 PATTERN)

Makes and Models: (*Note:* figures in brackets are numbers supplied)
Dodge T207, 1941 (G505). Bodies: WC1 Pickup w/Closed Cab (2573); WC3 Weapons Carrier (7808); WC4 ditto w/Winch (5570); WC5 Pickup w/Closed Cab (60); WC6 Command Reconnaissance (9365); WC7 ditto w/Winch (1438); WC8 Radio (548); WC9 Ambulance (2288); WC10 Carryall (1643); WC11 Panel Delivery (353).
Dodge T211, 1941 (G505). Bodies: WC12 Pickup w/Closed Cab (6046); WC13 Pickup w/Open Cab (4039); WC14 Pickup w/Closed Cab (268); WC15 Command Reconnaissance (3980); WC16 Radio (1284); WC17 Carryall (274); WC18 Ambulance (685); WC19 Panel Delivery (103); WC20 Chassis w/Cab (30) (used for Repair and Oil Service trucks).
Dodge T215, 1941–42 (G505). Bodies: WC21 Weapons Carrier (14287); WC22 ditto w/Winch (1900); WC23 Command Reconnaissance (2637); WC24 ditto w/Winch (1412); WC25 Radio (1630); WC26 Carryall (2900); WC27 Ambulance (2579); WC40 Pickup w/Closed Cab (275); WC41 Chassis w/Cab (383) (used for specialist bodies); WC42 Panel Radio (650); WC43 Telephone Installation (370).
International M-1-4, 1941–42. Bodies: Pickup w/Open Cab; Radio; Field Ambulance.

General Data: The Dodge Division of the Chrysler Corporation became the major producer of ½-ton 4×4 trucks for the US Army. During 1942 the ½-ton type, 82,000 of which were delivered, was superseded by the ¾-ton, also produced by Dodge. The International M-1-4 range was rather similar in design but was produced exclusively for the US Marine Corps and Navy. Dodges were supplied through Chrysler's Fargo Division and numbers of them were also released for Lend-Lease. The British Army used the following types: WC21, WC23, WC26, WC27 and WC42. A Pickup version with closed cab was produced by Chrysler in Canada (8-cwt, 4×4, model T212).

Vehicle shown (typical): Truck, ½-ton, 4×4, Panel Radio (Dodge T215-WC42)

Technical Data:
Engine: Dodge T215 6-cylinder, I-L-W-F, 230·2 cu in, 92 bhp @ 3000 rpm.
Transmission: 4F1R×1.
Brakes: hydraulic.
Tyres: 7.50–16.
Wheelbase: 116 in.
Overall l×w×h: 184×76×84 in.
Weight: 4510 lb.
Note: vehicle shown was supplied to British Army. Main body shell was Dodge commercial panel delivery van. Used for special purposes such as telephone and amplifier equipment. Early production had 217·7 CID engine (stroke 4¾ vs 4⅝ in).

Truck, ½-ton, 4×4, Radio (Dodge T207-WC8) 6-cyl., 85 bhp, 4F1R×1, wb 116 in, 179×76×83½ in, 5070 lb. Except radio equipment, similar in appearance to Command Reconnaissance. Body by Edward G. Budd Mfg Co.

Truck, ½-ton, 4×4, Weapons Carrier (Dodge T207-WC3) 6-cyl., 85 bhp, 4F1R×1, wb 116 in, 181×76×88 in, 4440 lb. Similar to WC13 and WC21. Many were fitted with transverse troop seats in rear body. Late 1941 and all 1942 models had T215 engine (92 bhp, 230·2 CID).

Truck, ½-ton, 4×4, Emergency Repair (Dodge T215-WC41) 6-cyl., 92 bhp, 4F1R×1, wb 123 in, 187×91½×81 in, 5870 lb. Purpose: to provide mobile facilities for emergency ordnance repair. Fitted with dual rear tyres (7.50–16). WC40 Pickup had same cab. Wheelbase same as ambulance.

Truck, ½-ton, 4×4, Ambulance (Dodge T215-WC27) 6-cyl., 92 bhp, 4F1R×1, wb 123 in, 195×76×90 in, 5340 lb. Accommodation for four stretchers, or eight men on folding seats. T207-WC9 and T211-WC18 similar except smaller bore engine. In all 6422 were supplied to Army and for Lend-Lease (incl. UK).

Truck, ½-ton, 4×4, Cargo (International) 6-cyl., 85 bhp, 4F1R×1, wb 115 in approx. Produced by International Harvester Co. in 1941. Fitted with commercial type pickup body. Side-valve (L-head) engine. Spare wheel was mounted on right-hand side. Pilot model for USMC/USN Model M-1-4.

Truck, ½-ton, 4×4, Ambulance (International M-1-4) 6-cyl., 85 bhp, 4F1R×1, wb 120 in approx., 190×92×90 in approx. Boyertown collapsible ambulance body (as on Ford/M.-H., see p. 27), for USMC and USN. Also fitted with dual rear tyres and cargo bodies. Five-stud wheels.

Truck, ½-ton, 4×4, Cargo, w/Winch (Ford GC) V-8-cyl., 100 bhp, 4F1R×1, wb 115 in approx., 180×75×90 in approx. Military pattern front end, commercial type cab and pickup body. Few produced, some with 6-cyl. engine. Also with command reconnaissance body (as Dodge, see page 27). 1941.

Truck, ½-ton, 4×4, Cargo (Willys MLW-4) 4-cyl., 54 bhp, 3F1R×2, wb 92 in, 142½×62×72½ in, 2700 lb. Pilot model, produced in 1944. Spicer axles with 7.50–20 tyres, 54-in tread. Another, similar, pilot model had std axles but 5·38 ratio and 7.50–16 tyres. Both had a capstan winch.

USA

TRUCKS, ¾- and 1-TON 4×4

Makes and Models: Dodge T214, 1942–45 (G502, 121). Bodies: WC51 Weapons Carrier; WC52 ditto w/Winch; WC53 Carryall and Command Field Sedan; WC54 Ambulance; WC55 Carriage, Motor, 37-mm Gun, M6 (see page 102); WC56 Command Reconnaissance*; WC57 ditto w/Winch*; WC58 Radio; WC59 Light Maintenance and Installation, K-50; WC60 Emergency Repair, M2; WC61 Telephone Maintenance and Installation, K-50-B (4551 body); WC64 Ambulance KD (Knockdown).
Dodge, 1942–43: various pilot models.
Ford, 1942–43: various pilot models.
Ford/Marmon-Herrington, 1-ton, 1940.
International M-2-4, 1-ton, 1942–45.

General Data: In 1942 the ½-ton 4×4 range of vehicles as first introduced in 1936 was superseded by new ¾-ton models. Prototypes were produced by Dodge and Ford. Dodge became the large-scale producer of this type. International Harvester up-rated the M-1-4 to 1-ton capacity (M-2-4) for the USMC and Navy. The new models were sturdier and wider, and had a lower silhouette. High-floatation combat tyres were fitted. The Dodge Weapons Carrier, which was also produced by Chrysler in Canada, become popularly known as 'the Beep' (possibly derived from 'beefed-up Jeep') and is still in use in many parts of the world. The following types were used by the British Army (Lend-Lease) during the war: WC51/52, 53, 54, 56/57 and 58. Soon after the war Chrysler introduced a commercial version of the Dodge Weapons Carrier which became available as Dodge, DeSoto or Fargo 1-ton 4×4 'Power Wagon' for civilian and military use. The military version had an open cab and was purchased by many Governments. Similar vehicles were subsequently produced in Japan (Nissan, Toyota) and Hungary (Csepel). In the US Army the 'Beep' was superseded in 1950 by the Dodge M37, the ¾-ton Field Ambulance by the Dodge M43.
* Numbers of these were later converted and fitted with S7-MA-51 ambulance body.

Vehicle shown (typical): Truck, ¾-ton, 4×4, Command Reconnaissance (Dodge T214-WC56) and ditto w/Winch (WC57)

Technical Data:
Engine: Dodge T214 6-cylinder, I-L-W-F, 230·2 cu in, 92 bhp @ 3200 rpm.
Transmission: 4F1R×1.
Brakes: hydraulic.
Tyres: 9.00–16.
Wheelbase: 98 in.
Overall l×w×h: 167 (w/winch 175) ×78½×81½(62¼) in.
Weight: 5375 (w/winch 5675) lb.
Note: These models had 12-volt electrical equipment. A pilot model with full-width body was also produced.

32

Truck, ¾-ton, 4×4, Weapons Carrier (Dodge T214-WC51) 6-cyl., 92 bhp, 4F1R×1, wb 98 in, 167×82¾×85(67½) in, 5650 lb. Weapons carriers were used mainly to transport personnel, weapons, tools and equipment, but also for various other purposes. Popularly known as 'the Beep'.

Truck, ¾-ton, 4×4, Weapons Carrier, w/Winch (Dodge T214-WC52) 6-cyl., 92 bhp, 4F1R×1, wb 98 in, 176½×82¾×85½(67½) in, 5940 lb. Produced in Canada, with slightly different engine, as Truck, 15-cwt, 4×4, GS, APT (Model D¾APT T236).

Truck, ¾-ton, 4×4, Command Field Sedan (Dodge T214-WC53 Special) 6-cyl., 92 bhp, 4F1R×1, wb 114 in, 185¾×78¾×80 in, 5750 lb. Modification of WC53 Carryall, provided with rear side doors, map table, special seats, lighting and provision of complete blackout within the body.

Truck, ¾-ton, 4×4, Emergency Repair, M2 (Dodge T214-WC60 6-cyl., 92 bhp, 4F1R×1, wb 121 in, 186 ×81½×88 in, 5580 lb. Purpose: to provide mobile facility for emergency repair. Body of WC61 generally similar but modified for mounting of ladder on top of body.

Truck, ¾-ton, 4×4, Light Maintenance and Installation, K-50 (Dodge T214-WC59) 6-cyl., 92 bhp, 4F1R×1, wb 121 in, 191½×77¼×80½ in, 5400 lb. GVW 6610 lb. Used by Signal Corps for installation of telephones and light maintenance. K-50-B (WC61) had full-width body and carried ladder overhead.

Truck, ¾-ton, 4×4, Ambulance (Dodge T214-WC54) 6-cyl., 92 bhp, 4F1R×1, wb 121 in, 194½×77¾×90 in, 5920 lb. All-steel 4-stretcher panel type body by Wayne. A number were later converted to open-cab configuration, using Weapons Carrier type front end (as KD). Some were used as vans.

Truck, ¾-ton, 4×4, Ambulance, KD (Dodge T214-WC64) As T214-WC54 but open cab and measuring 191½×82¾×90¼ in. Weight 6880 lb. Standardised in March, 1944, to supersede WC54 and ½-ton models. Body was collapsible (KD = knockdown) for ease of air transport. Three rear doors.

Truck, ¾-ton, 4×4, Ambulance, S-7-MA-51 (Dodge T214-WC56) 6-cyl., 76 hhp (net), 4F1R×1. Post-war bodywork by Boyertown Auto Body Works, Inc., on rebuilt and modified wartime chassis. Four stretchers. Two rear doors. Sliding cab doors. Body inside dimensions 104×62½×55 in.

Truck, ¾-ton, 4×4, Weapons Carrier (Dodge, Pilot) 6-cyl., 92 bhp, 4F1R×1, wb 98 in, approx. 176×82×85(67) in, 5890 lb. This was the pilot model of the ¾-ton weapons carrier. A 108-in wb pilot was produced also. The production model (WC52) had different wheels, wings, body, etc.

Truck, ¾-ton, 4×4, Cargo (Dodge, Pilot) 6-cyl., 92 bhp, 4F1R×1, wb 100 in, approx. 180×83×85(62) in. Produced in 1942 this pilot model was, in effect, a weapons carrier with semi-forward control. Another pilot model had normal control but the driver sat outside the frame and lower.

Truck, ¾-ton, 4×4, Utility, T47 (Dodge, Pilot) 6-cyl., 92 bhp, 4F1R×2, wb 105 in, 173¼×69¼×72(58) in, 5760 lb. Converted for command and reconnaissance role. Could carry a crew of eight. Cargo space was 67 in long, 40 in wide. Gradeability over 100% in low transfer ratio.

Truck, 1-ton, 4×4, Cargo (International M-2-4) 6-cyl., 93 bhp, 4F1R×2, wb approx. 127 in, approx. 198×85×88(65) in. Used only by USMC and Navy. Payload capacity 2000 lb. Cargo body also modified for mounting of rocket launcher. Some had capstan winch by driver's seat. Six-stud wheels.

Truck, ¾-ton, 4×4, Cargo (Ford GJL) 6-cyl., 90 bhp, 4F1R×1, wb 92 in. Pilot model for low-silhouette truck, built in 1941. Driver sits alongside engine. Chassis also used for gun motor carriages T33 and T44. 226 CID engine. Tyres 9.00–16/6-ply.

Truck, ¾-ton, 4×4, Cargo (Ford GCA) 6-cyl., 90 bhp, 4F1R×1, wb 103 in, 166×82×84(60) in; approx. 5000 lb. Prototype for weapons carrier, built in 1942. Also with narrower pickup-type body and without winch. Pilots only.

Truck, ¾-ton, 4×4, Cargo (Ford GAJ) 6-cyl., 90 bhp, 4F1R×1, wb 98 in, 161×82×83½(59) in, 5050 lb. Pilot model for 'driver outside frame' low-silhouette ¾-ton 4×4. Troop seat capacity 10 men. All these pilot models had a front-mounted winch of 5000-lb capacity.

Truck, ¾-ton, 4×4, Cargo (Ford GAJ) 6-cyl., 90 bhp, 4F1R×1, wb 92 in, approx. 155×82×83½(59) in, approx. 5100 lb. Forward control version of model GAJ with driver's position alongside engine. Pilots only. Front end similar to model GTB (1½-ton 4×4).

USA

TRUCKS, 1½-TON 4×2

Makes and Models (Chassis/Cab): Chevrolet WA, WB, 1940 (G617); YR, YS, 1941 (G617); MR, MS, 1942 (G617); Diamond T 201S, 404H, 404S, 1940-42. Dodge RE31, T98; T118-WF31, 1940-42 (G618). Ford O1T, O9T, O98T, O18T (V8), 1940 (G540); 11T, 19U, 19T, 198T, 118T (V8), 1941 (G540); 2GT, 2GU, 2G8T (6-cyl.), 1942 (G540); 29T, 218TF (rhd) (V8), 1942 (G540); G8T, G8TA (6-cyl.), 1943-44 (G540). GMC AC-305, ACX-353, 1940 (G655); CC-252, CC-302, CC-303, 1942 (G655). International D-30, 1939-40; K-5, 1943 (G688).

General Data: Between 1930 and 1940 more than half of the United States' truck production was in the 1½–2 ton payload capacity class. Virtually all medium trucks produced by the 'big three' (GM, Ford, Chrysler) were of this size, the bulk of the remainder being ¾-tonners and lighter vehicles. These reliable and relatively inexpensive standard Chevrolet, Ford and Dodge/Fargo/DeSoto trucks were sold all over the world and many governments used them for military transport service. They were also produced in Canada for domestic use and exports, mainly to Commonwealth countries. Many were also shipped out in CKD form and assembled overseas. During the years 1939–43 large numbers were supplied, either direct or under Lend-Lease, to America's allies. Large quantities of truck chassis were purchased by the Chinese Government for use on the notorious Burma Road. They were supplied mainly by Chevrolet, Dodge, Ford, International and Studebaker, assembled at Rangoon and fitted with locally produced cabs and bodywork. On average these trucks lasted about five round trips on the hazardous 1500-mile route. In Britain many Fords were assembled at Dagenham and fitted with an open type cab. The Dodge was widely used by the Soviet Army. Those which were employed by the US forces were mainly for domestic use. Ford supplied a total of 77,604 1½-ton 4×2 trucks during the war. COE types are dealt with in the next section.

Vehicle shown (typical): Truck, 1½-ton, 4×2, Stake and Platform (Ford 2G8T)

Technical Data:
Engine: Ford G8T 6-cylinder, I-L-W-F, 225·78 cu in, 90·7 bhp @ 3400 rpm.
Transmission: 4F1R.
Brakes: hydraulic (hydrovac).
Tyres: 7.50-20 (2DT).
Wheelbase: 158 in.
Overall l×w×h: 256×90×88 in.
Weight: 6500 lb.
Note: One of many types of 1940–42 model Ford commercial trucks supplied for military service. Some had V-8-cyl. engine.

Truck, 1½-ton, 4×2, Pickup (Chevrolet MR) 6-cyl., 85 bhp, 4F1R, wb 134½ in, 224×86×79 in, 4965 lb. Other body styles on Chevrolet 1½-ton chassis included Stake/Platform, Cargo, Canopy Express, Dump, Panel Delivery, Ambulance, Tanker, etc. Some types supplied to India.

Truck, 1½-ton, 4×2, Tractor (Ford G8TA) 6-cyl., 90 bhp, 4F1R, wb 141 in, 202×86×83½ in, 5800 lb. Militarised version of 1942 commercial model, built in 1943. 30-gal. auxiliary fuel tank on lh side of chassis. Other body styles on Ford chassis: Panel Delivery, Dump, Stake, Cargo, Fire.

Truck, 1½-ton, 4×2, Cargo (International D-30) 6-cyl., 81 bhp, 4F1R, wb 155 in, approx. 250×88×115(80) in. Combination cargo/troop-carrying body for USMC by Galion Allsteel Body Co. Galion Allsteel produced large numbers of cargo and other bodies for US military trucks.

Truck, 1½-ton, 4×2, Cargo (GMC ACX-353-6B) 6-cyl., 95 bhp, 4F1R, wb 144 in, approx. 245×87×110(80) in. Budd steel cargo body, model 984 cab. GMC produced large numbers of 6×6 trucks, relatively few of other types. This model also produced as 4×4 (models ACK-353, CCK-353).

USA

TRUCKS, 1½-TON 4×2 (COE)

Makes and Models: Ford 01W86, 11W, 198W, 29W, 291W, 298W, 1940-42 (G540). Bodies: Cargo, Tractor, Motorised Engineer Shops. GMC AFX-312-6C, 1940 (G655). Body: Panel Van. GMC AF-361-6I, 1940 (G655). Body: Signals Van. GMC CF-351-6W/Z, 1941–42 (G620). Body; Signals Van. International K5. Body: Panel Van (USMC).

General Data: Cab-over-Engine type chassis in this class were used mainly for specialist bodies such as mobile repair shops, signals vans (used to house and transport radio equipment), etc. Like the 1½-ton 4×2 trucks with conventional cab mounting, these COE units were suitable vehicles for military 'administrative' roles. Most of the GMCs had bodywork where the cab formed an integral part of the main body. GMC (at that time the Company was still known as Yellow Truck and Coach Mfg Co.) also introduced a 4×4 version, designated AFKX-352, which combined maximum body space with cross-country performance. Earlier types by GMC (and IHC) were used as mobile recruiting offices. The Ford engineer shops were equipped and fitted out by Couse Laboratories of Newark, N.J., who also produced the bodies. They were used by the US Corps of Engineers and the US Navy and carried a very extensive assortment of tools and equipment. Load A (or Couse, Type 'A') was a welding shop, Load B (or Couse, Type 'B') was a machine shop. Just before the war Couse had designed and built a 'mobile airport' on a 6×4 chassis. This completely self-sufficient vehicle which weighed about 12 tons carried some 3000 pieces of equipment including a 10-kW generator with 24 floodlights, radio-telephone and telegraph (with 50-watt transmitter) and enough tools and machinery to rebuild a plane entirely on the spot.

Vehicle shown (typical): Truck, 1½-ton, 4×2, Signals, COE (GMC CF-351-6W and 6Z).

Technical Data:
Engine: GMC 248 6-cylinder, I-I-W-F, 248·5 cu in, 100 bhp @ 3100 rpm.
Transmission: 4F1R.
Brakes: hydraulic (hydrovac).
Tyres: 7.00-20 (2DT).
Wheelbase: 109⅛ in.
Overall l×w×h: 206×87×111 in.
Weight: 6550 lb.
Note: GMC model 1585 cowl; bodywork by Hicks.

Truck, 1½-ton, 4×2, Signals (GMC AF-361-6I) 6-cyl., 100 bhp, 4F1R, wb 107⅝ in, 206×87×112 in, 6512 lb. Bodywork by Luce. 1940 pattern model 1544 cowl and front end. 47 built. Used by Signal Corps as message centre, often in conjunction with large 2-wheel trailer.

Truck, 1½-ton, 4×2, Panel (GMC AFX-312-6C) 6-cyl., 95 bhp, 4F1R, wb 123⅛ in, approx. 210×88×110 in. Bodywork by Proctor-Keefe. 1940 pattern model 1544 cowl and front end. 18 supplied to Quartermaster Corps in 1940. GMC produced only 736 4×2 military vehicles during 1939–45.

Shop, Engineer, Motorised, Aviation Battalion, Welding, Couse, Type 'A' (Ford 29W) V-8-cyl., 100 bhp, 4F1R, wb 134 in, 222×97½×116 in, gross weight approx. 18,000 lb. Built on 1942 Ford COE 1½-ton 4×2 chassis with model 81 closed cab, by Couse Laboratories.

Shop, Engineer, Motorised, Aviation Battalion, Welding, Couse Type 'A' (Ford 29W). This view shows the welding shop opened up and its comprehensive equipment displayed. Type 'B' (machine shop) measured 242×97½×128 in, but was similar in external appearance.

USA
TRUCKS, 1½-TON 4 × 4

Makes and Models: Chevrolet G-4100 Series, 1940–41 (G506). Bodies: Airfield Crash, Cargo, Cargo w/Winch, Cargo lbw, Crash, Dump, Dump w/Winch, Oil Service, Panel (Signal Corps: K-51), Telephone Maintenance (Signal Corps: K-42), ditto w/Winch (Signal Corps: K-43), ditto w/Earth Borer (Signal Corps: K-44), Tractor.
Chevrolet G-7100 Series, 1942–45 (G506). Bodies: Airfield Crash, Bomb Service (G85), Cargo, Cargo w/Winch, Cargo lwb, Dump, Dump w/Winch, Field Lighting (J-3, J-4, J-5), Panel, Telephone Maintenance w/Winch and Earth Borer (Earth Auger, M1, and Earth Borer and Pole Setter, K-44), ditto w/o Earth Borer. Also chassis/cab for special bodies (Army Air Forces: Turret Trainer, E-5, etc.).
Dodge T203, 1939–40 (G621). Bodies: Ambulance, Cargo, Cargo w/Winch, Dump. Also chassis/cab w/Winch for special bodies.
Ford/Marmon-Herrington HH5-4, 00T3-4 (2-ton), LD-4X, LD5-4, 1940-42 (4×4 conversions). Bodies: Cargo, Dump, Earth Borer and Telephone Maintenance.
GMC ACK-353, 1940, and CCK-353, 1941 (G655). Bodies: Earth Borer. Total produced: 100. ACK-352, 1940 (USMC). Bodies: Cargo, Wrecker.
International M-3L-4, 1941–44 (USMC and USN). Bodies: Cargo, Crash, Dump, Refueller, Tank, Wrecker. Total produced: 3742.
International M-3H-4, 1941–44 (USMC and USN). Bodies: Cargo, Crash, Floodlight, and chassis/cab for special bodies. Total produced: 2790.

General Data: The 1½-ton 4×4 had been pioneered during the 1930s by Marmon-Herrington who, since 1934, had converted standard commercial Ford trucks into all-wheel drive units. Dodge started series production in 1939 for the US Army and delivered 6411 units, following earlier small contracts (1934–38). Chevrolet, however, became mass-producer in 1940 and produced very large numbers until the end of the war. GMC only built 100 chassis of this type, to be fitted with Earth Borer equipment (mostly by Buda). For COE types see next section.

Vehicle shown (typical): Truck, 1½-ton, 4×4, Cargo, w/Winch (Chevrolet YP-G-4112)
Technical Data:
Engine: Chevrolet 235 6-cylinder, I-I-W-F, 235·5 cu in, 93 bhp @ 3100 rpm.
Transmission: 4F1R×2.
Brakes: hydraulic (hydrovac).
Tyres: 7.50–20 (2DT).
Wheelbase: 145 in.
Overall l×w×h: 231×86×104½ (87) in.
Weight: 8215 lb.
Note: a similar truck was produced experimentally by Chevrolet using GMC components (see page 53 for additional data). Gar Wood used the Chevrolet chassis, with open cab, for swinging-boom bomb crane trucks for the US Navy.

Truck, 1½-ton, 4×4, Panel (Chevrolet NG-G-7105) 6-cyl., 93 bhp, 4F1R×2, wb 145 in, 221½×87×91 in, 6760 lb. Commercial-type panel delivery body. Used mainly by Signal Corps (K-51) to house and transport radio equipment, incl. SCR-299 set. Sometimes fitted with side extensions.

Truck, 1½-ton, 4×4, Bomb Service, M6 (Chevrolet NQ-G-7128) 6-cyl., 93 bhp, 4F1R×2, wb 125 in, 221×75¾×91½ in, 6325 lb. Superseded earlier ½-ton 4×2 types, M1, for loading, unloading, and towing bomb carrying trailers. Also airborne version (conversion of M6).

Truck, 1½-ton, 4×4, Field Lighting, J-3 (Chevrolet NF-G-7143) 6-cyl., 93 bhp, 4F1R×2, wb 145 in, 233×98½×138 in, 7200 lb. These units (J-3, J-4, J-5) were used to illuminate airfields. J-5 model had four floodlights. J-4 and J-5 became AN2505 and AN2524 resp., in 1943.

Truck, 1½-ton, 4×4, Cargo, 15-ft Body (Chevrolet ZQ-G-4174 and NP-G-7127) 6-cyl., 93 bhp, 4F1R×2, wb 175 in, 296×85¾×106 in, 8150 lb. This truck differed from the regular model mainly in having a divided propeller shaft with support bearing and a longer cargo body.

Truck, 1½-ton, 4×4, Telephone Maintenance, K-42 (Chevrolet YP-G-4112) 6-cyl., 93 bhp, 4F1R×2, wb 145 in, 222½×92×104½ in, 10,215 lb. Used by Signal Corps for telephone maintenance and construction. Also supplied with winch (in rear body) and pole-setting equipment (K-43).

Truck, 1½-ton, 4×4, Telephone Earth Borer (Chevrolet NR-G-7163) 6-cyl., 93 bhp, 4F1R×2, wb 145 in, 268×94×108 in, 7200 lb. Used by Corps of Engineers as Earth Auger, M1, and by Signals Corps as Earth Borer and Pole Setter, K-44. Earth Auger, M1, also on Ford/M.-H. chassis.

Truck, 1½-ton, 4×4, Telephone Construction and Maintenance, K-43 (Chevrolet NS-G-7173) 6-cyl., 93 bhp, 4F1R×2, wb 145 in, 224 (w/o poles)×90×104½ in, 12300 lb (gross). Power winch in forward end of body, produced by American Coach & Body Co. Poles on side usually carried further rearward.

Truck, 1½-ton, 4×4, Dump, w/Winch (Chevrolet YP-G-4112) 6-cyl., 93 bhp, 4F1R×2, wb 145 in, 231×86×108(87) in, 9010 lb. Hyd. operated dump body (Heil and Hercules Steel Products). Used to transport and dump earth, coal, etc., and to carry general cargo. NL-G-7116 (from 1942) similar. Also w/o winch.

Truck, 1½-ton, 4×4, Cargo, Airborne (Chevrolet NJ-G-7107, modified) 6-cyl., 93 bhp, 4F1R×2, wb 145 in. This standard truck was one of several modified in 1944 with a special kit for transport by air. Disassembled it could be transported in a C47A type aircraft. Main modifications on frame and cab.

Truck, 1½-ton, 4×4, Cargo, Low Silhouette (Chevrolet G-7129, Pilot) GMC 6-cyl., 89 bhp, 5F1R×2, wb 137 in, 219×102×99(67) in, 9875 lb. One of several pilot models built for the Ordnance Department (Camp Holabird) in late 1942. The driver's seat was placed outside the chassis frame.

 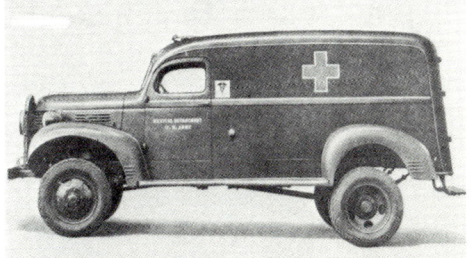

Truck, 1½-ton, 4×4, Cargo, w/Winch (Dodge T203-VF405) 6-cyl., 99 bhp, 4F1R×2, wb 143 in, 233×86×112 in, 8200 lb. Militarised version of regular Dodge model VF truck, produced during 1939-40. Also supplied to Great Britain (ex-French contracts), with different type cargo body.

Truck, 1½-ton, 4×4, Ambulance (Dodge T203-VF407) 6-cyl., 99 bhp, 4F1R×2, wb 143 in. One of only three ambulances (USA Reg. Nos. W71074-76) out of a batch of 6411 1½-ton 4×4 Dodges delivered during 1940. Rest were chassis/cabs, cargo and dump trucks.

Truck, 1½-ton, 4×4, Cargo and Personnel (Ford 09T/Marmon-Herrington) V-8-cyl., 95 bhp, 4F1R×2, wb 134 in. One of many Ford trucks converted to all-wheel drive by Marmon-Herrington. This 1940 model was used by the USMC as Truck, 1-ton, 4×4, Reconnaissance. Similar body on IHC 1-ton 4×4 (q.v.).

Truck, 1½-ton, 4×4, Dump (Ford/Marmon-Herrington) 8-cyl., 95 bhp, 4F1R×2, wb 134 in. Pilot model of a large number of Marmon-Herrington-converted Ford trucks for the US Corps of Engineers. Used on Alcan Highway and Canol Project.

Truck, 1½-ton, 4×4, Earth Borer (GMC CCK-353-8K/L) 6-cyl., 100 bhp, 4F1R×2, wb 159¼ in. Produced in 1941 this model superseded the ACK-353 of 1940. 100 were supplied to the Signal Corps. Buda engine for Earth Borer. Cargo version supplied to France and, later, Britain.

Truck, 1½-ton, 4×4, Cargo (International M-3L-4) 6-cyl., 88 bhp, 5F1R×2, wb 139 in. Produced during 1941–44 to specification of USMC and USN. Model M-3H-4 had 150-in wb. Used mainly in Pacific area, with various body types. Some had soft-top cabs. Total produced of all types: 6532.

USA

TRUCKS, 1½-TON 4×4 (COE)

Makes and Models: Chevrolet YX-G-4103, 1940–41 (G506); NN-G-7123, 1942–45 (G506). Bodies: Stake and Platform.
Chevrolet NX-G-7153, 1942–43 (G506). Bodies: Fire trucks (US Navy).
Ford GTB, GTBS, 1942 (G622). Bodies: Cargo, Bomb Service (USN).
Ford/Marmon-Herrington, 1940–42 (4×4 conversions). Bodies: Air Compressor, Barrage Balloon (model BB-2), Crash (model JJ6-COE-4, American LaFrance and Cardox equipped), Fire.
GMC AFKX-352, 1939–42 (G553, 655). Bodies: Air Compressor, Cargo, Ordnance Repair/Maintenance.
Reo and Studebaker produced pilot models for low-silhouette models, 1942.

General Data: Most of the vehicles in this section used a militarised commercial-type COE (Cab-over-Engine) chassis, modified to have all-wheel drive. Their excessive height made them rather unsuitable for tactical use. The Ford, Reo and Studebaker models, however, were specially designed to have a low silhouette. Only the Ford model reached the production stage. The COE models were used mainly in roles where maximum loading space was required without having excessive wheelbase length. The GMC models were all rated as 1½–3-ton. The GMC Ordnance Maintenance truck was made in the following versions: Small Arms Repair, M1*, Artillery Repair, M1 and M2, Automotive Repair, M1 and M2, Instrument Repair, M1, Light Machine Shop, M3, Machine Shop, M4, Spare Parts, M1 and M2*, Tank Maintenance, M1*, Tool and Bench, M2*, and Welding, M3*. Those types indicated by an asterisk were also used by the British Army. In the US Army all the above Ordnance Maintenance types were superseded by similar types on the standard 2½-ton 6×6 GMC chassis.

Vehicle shown (typical): Truck, 1½–3-ton, 4×4, Ordnance Maintenance (GMC AFKX-352-8G)

Technical Data:
Engine: GMC 248 6-cylinder, I-I-W-F, 248·5 cu in, 100 bhp @ 3100 rpm.
Transmission: 4F1R×2.
Brakes: hydraulic (hydrovac).
Tyres: 7.50–20 (2DT).
Wheelbase: 131 in.
Overall l×w×h: 235×96×120 in.
Weight: 9710 lb.
Note: Body by Wayne.

46

Truck, 1½–3-ton, 4×4, Ordnance Maintenance (GMC AFKX-352-8A) 6-cyl., 100 bhp, 4F1R×2, wb 131 in, 218×88×108 in, 7400 lb. Early model with short bus-type body by Superior. Only 78 produced, then replaced by larger type. Total number produced of all types: 1990.

Truck, 1½–3-ton, 4×4, Cargo (GMC AFKX-352-8D) 6-cyl., 100 bhp, 4F1R×2, wb 131 in, 220×89×105 in, 7500 lb. Cargo body by Heil. Commercial-type closed cab, model 1506. 79 produced in 1939/40. Truck shown was pilot model for early-type Ordnance Repair truck.

Truck, 1½–3-ton, 4×4, Air Compressor (GMC AFKX-352-8J/Q/R/U) 6-cyl., 100 bhp, 4F1R×2, wb 131 in. 820 chassis were supplied during 1940/41 for mounting of air compressor units by LeRoi and Ingersoll Rand. Used by the Corps of Engineers as power supply units for pneumatic tools.

Truck, 1½–3-ton, 4×4, Cargo (GMC AFKX-352–8C) 6-cyl., 100 bhp, 4F1R×2, wb 131 in. Hard-top rear body, produced by Heil, and originally used as mobile repair shop. Lower body sides could be lowered to horizontal position to extend floor area.

47

Truck, 1½-ton, 4×4, Stake and Platform (Chevrolet YX-G-4103 and NN-G-7123) 6-cyl., 93 bhp, 4F1R×2, wb 175 in, 286×96×99 in, 8570 lb. This militarised commercial model retained its basic civilian-type front end. It was in limited production and was used for transport of general cargo.

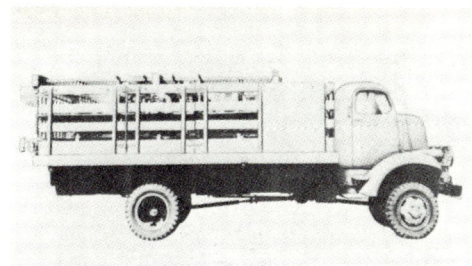

Truck, 1½-ton, 4×4, Stake and Platform, K-33 (Chevrolet NN-G-7123) 6-cyl., 93 bhp, 4F1R, wb 175 in, 322×96×102 in, 8965 lb. The K-33 and the almost identical K-54 were used by the Signal Corps and the AAF to transport components of radio equipment. K-54 was 4 in longer and 15 in higher.

Truck, 1½-ton, 4×4, Air Compressor (Ford 09W/Marmon-Herrington) 8-cyl., 95 bhp, 4F1R×2, wb 134 in, 189×84×99½ in. Produced in 1940 for US Corps of Engineers. Sullivan air compressor and pneumatic tools. Regular Ford 1940 COE chassis, converted to all-wheel drive by M.-H.

Truck, 1½-ton, 4×4, Bomb Service (Ford GTBS) 6-cyl., 90 bhp, 4F1R×2, wb 115 in. This unit was built in limited numbers for the US Navy for loading, unloading and towing bomb-trailers. Like the Chevrolet M6, it was an adaptation of a standard cargo truck, fitted with single rear wheels.

Truck, 1½-ton, 4×4, Cargo, Low Silhouette (Ford GTB) 6-cyl., 90 bhp, 4F1R×2, wb 115 in, 215×88×105 in, 7250 lb. Only low-silhouette type truck to enter series production. Produced chiefly for US Navy, with or without front-mounted winch. Number of chassis produced: 6000.

Truck, 1½-ton, 4×4, Cargo, Low Silhouette (Reo) Another pilot model for a projected low-silhouette cargo truck, produced for the Ordnance Department in 1942. The driver sat alongside the engine. 115-in wheelbase. Front-mounted winch had capacity of 5000 lb.

Truck, 1½-ton, 4×4, Cargo, Low Silhouette (Studebaker LC4×4-125950) Hercules JXD 6-cyl., 109 bhp, 5F1R×2, wb 105 in, 189¾×92×93¾(75) in, 9655 lb. Unusual platform design with driver positioned centrally in front. Front entrance. Body length (inside) 161 in. Pilot only.

Truck, 1½-ton, 4×4, Cargo, Low Silhouette (Studebaker LC4×4-125950) Another view of the same vehicle, with tarpaulin and superstructure removed. Troop seating capacity for 17 men. Front-mounted winch had 5000-lb capacity. Outrigger engine on right-hand side.

USA

TRUCKS, 1½-TON 6×6

Makes and Models: Dodge T223, 1942–45 (G507). Bodies: WC62 Personnel and Cargo; WC63 Personnel and Cargo, w/Winch.
International M-3-6, 1943. Bodies: Cargo w/Winch.

General Data: The Dodge 1½-ton 6×6 cargo truck, which was basically the standard Dodge Weapons Carrier with a tandem rear bogie and lengthened body, was first designed during 1942 for the Ordnance Department. Production started late in 1942 and continued throughout the remainder of the war. Also during the same year experiments were carried out with an armoured version and subsequently with various gun mounts on the standard vehicle. The USMC equivalent of this truck was produced by International Harvester and was, in turn, a conversion of the IHC M-2-4, fitted with a Hendrickson-type rear bogie (featuring conventionally mounted semi-elliptic leaf springs with equaliser beams). This vehicle, however, was not produced in quantity. The Dodge, in the interest of standardisation, had the same engine as the Dodge ¾-ton 4×4 models, but to keep performance at a satisfactory level it was fitted with a two-speed transfer case. Many of these Dodges are still operational, notably in the Middle East. Towards the end of the war a framework rack assembly was designed to carry this vehicle under the fuselage of Douglas C-54A aircraft (fitted with under-fuselage hoist and carrying rack assembly). Similar racks were also made to carry the Dodge ¾-ton 4×4 Weapons Carrier and KD Ambulance and the GMC 2½-ton 6×6 CCKW-352, but these efforts remained in the experimental stage. After the war the French Marmon-Herrington company produced a vehicle which was similar in many respects to the Dodge WC62. It was available with petrol or diesel engine (models FF6.E and FF6.D respectively) and had 9.00×20 tyres.

Vehicle shown (typical): Truck, 1½-ton, 6×6, Cargo and Personnel (Dodge T223-WC62).

Technical Data:
Engine: Dodge T223 6-cylinder, I-L-W-F, 230·2 cu in, 92 bhp @ 3200 rpm.
Transmission: 4F1R×2.
Brakes: hydraulic.
Tyres: 9.00–16.
Wheelbase: 125 in, BC 42 in.
Overall l×w×h: 215×82¾×87(67½) in.
Weight: 7250 lb.

 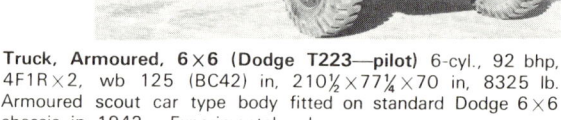

Truck, 1½-ton, 6×6, Personnel and Cargo, w/Winch (Dodge T223-**WC63**) 6-cyl., 92 bhp, 4F1R×2, wb 125 (BC42) in, 224¾×82¾×87(67½) in, 7550 lb. Front-mounted winch with a capacity of 5000 lb. This particular vehicle is part of a private collection of historical military vehicles in England.

Truck, Armoured, 6×6 (Dodge T223—pilot) 6-cyl., 92 bhp, 4F1R×2, wb 125 (BC42) in, 210½×77¼×70 in, 8325 lb. Armoured scout car type body fitted on standard Dodge 6×6 chassis in 1942. Experimental only.

Truck, 1½-ton, 6×6, Multiple Gun Motor Carriage, T74 (Dodge T223) 6-cyl., 92 bhp, 4F1R×2, wb 125 (BC42) in, 224×82¾×76 in, gross weight 11480 lb. This was a gun mount, M33, with twin 0·50 cal. machine guns fitted on the Dodge 6×6 chassis in 1942/43. Experimental only.

Truck, 1½-ton, 6×6, Chassis, w/Winch (International M-3-6) 6-cyl., 93 bhp, 4F1R×2, wb 125 in. Modification of M-2-4 1-ton 4×4 chassis. Walking beam rear suspension. Thornton locking diffs. Tyres 9.00–16. Wt 5870 lb (with exp. Anthony cargo body 6830 lb). GRD-233 L-head engine.

USA

TRUCKS, 2½-TON 4×2

Makes and Models: Diamond T 406H, 406S, 1939–40; 509H, 509HS, 1939–42; 614, 614S, 1939–43 (G554). Bodies: Cargo, Dump, Explosives Van, Telephone Construction and Maintenance.
Federal 2G, 3G, 1943–44 (G539). Bodies: Dump (2G), Telephone Construction and Maintenance (3G). GMC ACX-353, AC-355, AC(X)-453, AF-502, ACX-503, AF-503, AFX-622, 1940–42 (G655). Bodies: Dump, Stake, Van. GMC CC(X)-453, CC(X)-454, 1941–42 (G623). Bodies: Cargo, Dump, Stake, Canopy Express. International K-7, 1941–43 (G541). Bodies: Cargo, Derrick, Dump, Dump Hi-Lift, Stake, also Chassis/cab for special bodies and lwb chassis for bus body.
International KR-8R, 1942 (G541). Bodies: Tractor, w/Winch, for semi-trailer.
Mack EES, EG, EHS, EHDX, 1940–43 (G624). Bodies: Cargo, Dump, Stake, Tank. Reo 21 XHHS, 21 BHHS, 1941 (G625). Bodies: Cargo, Dump.

General Data: With the exception of the Federals, all the above models were regular commercial models, modified only in detail to satisfy military requirements. The Federal 2G and 3G models had a typical American military pattern front end. The latter model had an all-enclosed steel body, similar to that fitted on the Diamond T model 614 chassis. Smaller tactical versions of these telephone construction and maintenance trucks used the Dodge ½-ton and ¾-ton and Chevrolet 1½-ton 4×4 chassis. The dump trucks in this class had a 2-cu yd all-steel body and were not only used to transport and dump earth, sand, gravel, etc., but also for transporting general cargo. Federal also produced a model 29K 3½-ton tractor truck, using the same Hercules engine as in their models 2G and 3G but with civilian type front end (1941–42).

Vehicle shown (typical): Truck, 2½-ton, 4×2, Dump (Federal 2G)

Technical Data:
Engine: Hercules JXD 6-cylinder, I-L-W-F, 320 cu in, 86 bhp @ 2400 rpm.
Transmission: 5F1R.
Brakes: hydraulic (hydrovac).
Tyres: 8.25-20 (2DT).
Wheelbase: 150⅝ in.
Overall l×w×h: 217½×90×89 in.
Weight: 7975 lb.

Truck, 2½-ton, 4×2, Telephone Construction and Maintenance (Diamond T 614) Hercules CB-JXD 6-cyl., 84 bhp, 5F1R, wb 151¾ in, 228×90×105 in, 11,600 lb. Used by US Signal Corps for telephone maintenance, repair and construction. Similar body also on Federal chassis (Model 3G).

Truck, 2½-ton, 4×2, Cargo (GMC CC-453) 6-cyl., 104 bhp, 5F1R, wb 159¼ in, 242×87×95¾ in, 7318 lb. Fitted with regular model 1574 cab and Heil cargo body. Same chassis was also used with Perfection stake body and with Square Deal canopy express body.

Truck, 2½-ton, 4×2, Cargo (International K-7) 6-cyl., 101 bhp, 5F1R, wb 158 in, 315×87×95 in, 8700 lb. All-steel cargo body by Anthony. Model K-7 came in various wheelbase lengths and altogether 7498 were produced. Cargo bodies were also made by Gar Wood.

Truck, 2½-ton, 4×2, Tractor, w/Winch (International KR-8R) 6-cyl., 85 bhp, 5F1R, wb 149 in, 227×92×85 in, 9247 lb. Basically commercial 5-ton truck, fitted with double-reduction rear axle. Winch behind cab. 220 supplied to UK under Lend-Lease, for use with low-loader S/T.

USA
TRUCKS, 2½-TON 4 × 4

Makes and Models: Autocar U-2044, U-4044, U-4144, U-4144T, U-5044, 1940-41 (G626). Bodies: Oil Servicing, Tractor.
GMC AFKX-502, 1939–40 (G627). Bodies: Tractor.

General Data: The 2½-ton 4×4 was basically a commercial-type truck fitted with all-wheel drive to improve cross-country performance. The Autocar tractor versions were used mainly by the Army Air Forces to tow fuel-servicing and other semi-trailers; the GMC mainly by the Cavalry to tow animal and cargo trailers. Their role was taken over by the standardised 4–5-ton 4×4 tractor as produced by Autocar, Federal and White, from 1941. The Autocar model U-2044 oil-servicing truck, L-1, was superseded by the GMC 2½-ton 6×6 oil-servicing truck, L-2. By 1943 all the 1940 pattern 4×4 models in the 2½-ton class were either obsolete or classified as 'limited standard'. In 1943, however, the Ordnance Corps decided to study the possibilities of a compact 2½-ton 4×4 cross-country cargo truck with high-floatation tyres capable of improved operation through mud and swamp areas. The projected vehicle was later designated T23 and quotations for its design and development were asked from Chevrolet, GMC, Studebaker and White. Studebaker and GMC soon had to withdraw their consideration of the project, owing to pressure of other work, and the other manufacturers had to drop the project, too, in 1944. By the end of the war, however, the contract was awarded to Chrysler, who completed two pilot models in 1946, but the vehicle did not go into series production. In order to extend vehicle standardisation the Chevrolet Division of General Motors modified several 2½-ton 6×6 GMCs into 1½-ton 4×4 trucks and tractors in 1943/44. If it had been possible to do this three years earlier it would have had considerable advantages. As it was, it remained experimental.

Vehicle shown (typical): Truck, 2½-ton, 4×4, Tractor, COE (Autocar U-4144T)

Technical Data:
Engine: Autocar D-358 6-cylinder, I-L-W-F, 358 cu in, 100 bhp @ 2600 rpm.
Transmission: 5F1R×2.
Brakes: air.
Tyres: 9.00–20 (2DT).
Wheelbase: 131 in.
Overall l×w×h: 201×92×103 in.
Weight: 10,200 lb.
Note: used mainly in conjunction with fuel-servicing semi-trailers (Air Corps, US Army). Also with oil-servicing tank body (U-4144).

Truck, 2½-ton, 4×4, Oil Servicing, 660-gal., L-1 (Autocar U-2044) Hercules JXD 6-cyl., 110 bhp, 4F1R×2, wb 128 in, 192½×89¼×104 in, 8770 lb. Used by Army Air Forces to provide mobile oil-servicing facilities for airfields. Also produced with Autocar D-358 engine and fifth wheel (model U-4044).

Truck, 2½-ton, 4×4, Tractor, COE (GMC AFKX-502-8E) 6-cyl., 122 bhp, 5F1R×2, wb 136 in, 202×92×105 in, 8567 lb. Used by Cavalry to tow horse-box and cargo semi-trailers. Regular model 1506 cab, but flat section front wings. 81 produced in 1939-40.

Truck, 2½-ton, 4×4, Cargo, w/Winch (Studebaker— project) Hercules JXD 6-cyl., 93 bhp, 5F1R×2, wb 130 in, 223×96×85(69) in, 11,550 lb. One of the original design studies of what became the T23. 16.00×32 tyres. Note similarity with Studebaker 1½-ton low-silhouette truck.

Truck, 2½-ton, 4×4, Cargo, w/Winch, T23 (Chrysler—pilot) 6-cyl., 120 bhp, 5F1R×2, wb 124 in, 229×96×109(84) in, 11,700 lb. 18.00-26 tyres. Central tyre pressure regulation system. Elevated axles with step-down gearboxes and 2·08:1 final drive ratio. Max. speed 43 mph. 5-ton winch.

USA

TRUCKS, 2½-TON 6×4

Makes and Models: Federal 75K-131, 1939. Bodies: Searchlight (COE). GMC AFWX-354, 1939–40 (G655). Bodies: Searchlight/Cargo (COE). GMC CCW-353, 1941–44 (G508). Bodies: Cargo, Platform; also for mounting of special bodies (incl. 1350-gal. Gasoline Tank by Heil and Gar Wood). International M-5-6×4-318, 1942–43 (G637). Bodies: Cargo.
Mack NB, 1939–40 (G629). Bodies: Searchlight (COE).
Studebaker US6×4-U6, U7, U8, 1941–45. Bodies: Cargo, Tractor.

General Data: The vehicles in this class can be divided into two groups, namely the 1939/40-pattern COE searchlight trucks and the 1941–45 conventional types. The former were basically commercial types, fitted with a special searchlight body and an extended cab to accommodate the searchlight crew (Federal, GMC, Mack). The latter were based on the GMC, IHC and Studebaker 6×6 tactical trucks and were similar, with the exception of the front-wheel drive which was omitted. These vehicles were produced mainly for long-distance haulage of supplies, where all-wheel drive was not required. From the point of view of spare parts this proved an attractive solution. Also, as these vehicles were unsuitable for use off the road, the payload capacity rating could be increased to 5 tons, which was the capacity of the 2½-ton 6×6 on made roads. In fact, the 6×4 versions were usually referred to as 2½-5-ton or sometimes as 5-ton trucks. GMC produced a total of 23,649 6×4 trucks for the US Government during 1940–42. The Studebakers were almost exclusively produced for international aid (Lend-Lease). The Duplex Truck Co. of Lansing, Mich., supplied Sperry-Duplex power-plant trucks for use in conjunction with AA Searchlights (1940-41; no further details). Reo, in 1941, supplied 5-ton 6×4 commercial-type trucks, model 23XHHRS and a Ford COE 6×4 type tanker was used by the US Army in North Africa.

Vehicle shown (typical): Truck, 2½-5-ton, 6×4, Cargo (GMC CCW-353)

Technical Data:
Engine: GMC 270 6-cylinder, I-I-W-F, 269·5 cu in, 104 bhp @ 3000 rpm.
Transmission: 5F1R×1.
Brakes: hydraulic (hydrovac).
Tyres: 7.50–20 (4DT).
Wheelbase: 164 in, BC 44 in.
Overall l×w×h: 256¼×88×93 in.
Weight: 10,050 lb.
Note: transfer case low-speed range blocked out, hence no transfer case shift levers. Some chassis were later fitted with Gar Wood and Heil refuse collection bodies.

56

Truck, 2½–5-ton, 6×4, Cargo (International M-5-6×4-318)
6-cyl., 85 bhp, 5F1R×2, wb 169 in, approx. 250×88×106 in, 9700 lb. 6×4 version of IHC M-5-6. 3000 supplied to US Army Quartermaster Corps in 1942, probably for Lend-Lease. Budd all-steel cargo body. Commercial-type closed cab.

Truck, 2½–5-ton, 6×4, Cargo (Studebaker US6×4-U7)
Hercules JXD 6-cyl., 5F1R×1 (later 5F1R×2), wb 162 (BC44) in, 258×88×109 in, 9555 lb. 6×4 version of Studebaker US6. Open or closed cab. Model US6×4-U8 similar but fitted with winch. US6×4-U6 was tractor with 148-in wb.

Truck, 2½-ton, 6×4, Searchlight, COE (Mack NB) Continental FO 6-cyl., 78 bhp, 4F1R×2, wb 172 in, 288×96×139 in, 10,630 lb. Special 'sleeper'-type cab with rear extension and accommodation for crew of five. Similar type supplied by Federal. Limited production.

Truck, 2½-ton, 6×4, Searchlight, COE (GMC AFWX-354-10A)
6-cyl., 91 bhp, 4F1R×2, wb 167 (BC44) in, 288×96×142 in, 10,820 lb. 149 supplied to US Army, 1939/40. Also ordered by French Government (with smaller engine) but shipments diverted to Great Britain in 1940.

USA

TRUCKS, 2½-TON 6×6

Makes and Models: GMC ACKW(X)-353, 1939–41, CCKWX-353, 1941 (G508); CCKW-352, CCKW-353, 1941–45 (G508). Bodies: Bomb Service, Cargo, Cargo w/Winch, Dump, Dump w/Winch, Stock Rack, Tank (Gasoline and Water); also Chassis/cab for special bodies (mobile workshops, etc.).
International M-5-6, M-5H-6, 1941–45 (G651, 659). Bodies: Cargo, Cargo w/Winch, Dump w/Winch, Refueller, Tractor, Wrecker; also Chassis/cab for special bodies.
Reo US6-U3, 1943–45 (G630). Bodies: Cargo.
Studebaker US6-U1 to U5, U9 to U13, 1941–45 (G630). Bodies: Cargo, Cargo w/Winch, Rear Dump, Side Dump, Tank; also Chassis/cab for special bodies.
GMC, International and Studebaker also produced several pilot models for low-silhouette cargo trucks.

General Data: The 2½-ton 6×6 'light-heavy' truck was the most widely used tactical transport vehicle of the US forces during World War II. Over 800,000 were produced, the majority of them by GMC, namely 562,750. The first GMC 2½-ton 6×6 was basically a six-wheel drive commercial type, the ACKWX-353, which was also ordered by the French Government. In 1941 the familiar military pattern front end was introduced and mass-production started. The resulting model became popularly known as 'Jimmy' and saw service in all theatres of war. There were two basic chassis models: the CCKW-352 with 145-in wheelbase and the CCKW-353 with 164-in wheelbase (for AFKWX COE type see next section). These were fitted with either Chevrolet banjo-type or Timken split-type axles, closed or open cab, and with or without winch. The open cab later became standard equipment and many closed-cab trucks were thus converted. Since GMC could not produce sufficient numbers of these 'Workhorses of the Army', other truck manufacturers were also engaged in the production of this type, namely International Harvester (for the USMC and USN) and Studebaker (mainly for Lend-Lease). Reo produced one of the models in the Studebaker range. The 2½-ton 6×6 as a type had been arrived at as being the heaviest truck that could be produced in large quantities.

Vehicle shown (typical): Truck, 2½-ton, 6×6, Cargo (GMC CCKWX-353-120)

Technical Data:
Engine: GMC 270 6-cylinder, I-I-W-F, 269·5 cu in, 104 bhp @ 3000 rpm.
Transmission: 5F1R×2.
Brakes: hydraulic (hydrovac).
Tyres: 7.50–20 (4DT).
Wheelbase: 164 in, BC44 in.
Overall l×w×h: 256¼×88×110 in.
Weight: 10,350 lb.
Note: cab was inherited from Chevrolet and light GMC civilian type trucks but featured hinged windshield and military pattern instrumentation, etc. Truck shown fitted with AA mg on ring mount.

Truck, 2½-ton, 6×6, Cargo (GMC CCKW-352) 6 cyl., 104 bhp, 5F1R×2, wb 145 (BC 44) in, 231×88×108(76) in, 10,500 lb approx. Two spare wheels and tranverse fuel tank (with filler neck at each side) behind closed cab. Steel body with tarpaulin in 'ventilating position'.

Truck, 2½-ton, 6×6, Water Tank (GMC CCKW-353-12CC) 6-cyl., 104 bhp, 5F1R×2, wb 164 (BC44) in, 253×90×92 in, 10,185 lb. Body by Columbian. 700-gallon capacity. Model 1608 cab. Later models had open-type cab. There were several other types of tankers on GMC chassis.

Truck, 2½-ton, 6×6, Cargo, w/Winch (GMC CCKW-352-12CI) 6-cyl., 104 bhp, 5F1R×2, wb 145 (BC44) in, 245×88×108 (76) in, 11,050 lb. Short-wheelbase version, usually supplied with cargo body but a number were later reworked and fitted with dump body. Model 1619 open cab.

Truck, 2½–3-ton, 6×6, Cargo, Low Silhouette, w/Winch (GMC DCKW-3) 6-cyl., 104 bhp, 5F1R×2, wb 151 in, 243½×88×110 (74) in, 11,540 lb. One of several prototypes experimentally built by GMC in 1942/43. Other similar GMCs were designated DAKW-602 and DOKW-602.

Truck, 2½-ton, 6×6, Cargo, w/Winch (GMC CCKW-353-12AU) Only one of its type, this 164-in wheelbase chassis was fitted with a centre winch and the cargo body normally fitted on the 145-in wheelbase model. It was delivered in 1942 and had USA registration number W466455.

Truck, 2½-ton, 6×6, Ordnance Maintenance (GMC CCKW-353) Standard shop van body, used with various types of equipment as Truck, Artillery Repair, M9, Truck, Instrument Repair, M10, etc. Superseded similar model on GMC 1½-ton 4×4 chassis (q.v.). Later models had open cab. 228×95×108 in.

Truck, 2½-ton, 6×6, Bomb Service, w/Winch, M27B1 (GMC CCKW-353) Modification of cargo body (both steel and wood bodies were used). Employed for lifting and transporting bombs from depots to airplanes and airfields. One of many body types on the GMC 'deuce-and-a-half' chassis. Dimensions 352×88×129 in.

Truck, 2½-ton, 6×6, High-Lift, w/Van Type Body (GMC CCKW-353) Used by Army Air Forces to expedite loading and unloading of high-door cargo airplanes. 'Jack-Knife' high-lift hoist mechanism produced by Heil. Similar but not identical type supplied by Gar Wood. Dimensions 256×88×113 in. Weight 12,100 lb.

Truck, 2½-ton, 6×6, Cargo, w/Winch (International) 6-cyl., 85 bhp, 5F1R×2, wb 149 in. First International 2½-ton 6×6, produced in 1940. Developed into M-5-6 (Military, 5000-lb payload, 6-wheel drive) for US Marine Corps and Navy, over 35,000 of which were built until 24th September, 1945.

Truck, 2½-ton, 6×6, Cargo, w/Winch (International M-5H-6) 6-cyl., 111 bhp, 5F1R×2, wb 169 in, approx. 270×88×92(76) in, 11,250 lb. Also produced with steel body and with closed cab (as shown in section on 2½-ton 6×4). Widely used by USMC and USN in Pacific area.

Truck, 2½-ton, 6×6, Cargo, w/Winch (Studebaker US6-U2) Hercules JXD 6-cyl., 87 bhp, 5F1R×2, wb 148 (BC44) in, 244×88×110 in, 10,000 lb. Studebaker commercial-type closed cab or open type. Also on this 148-in wb chassis: rear dump and side dump (1943).

Truck, 2½-ton, 6×6, Gasoline Tank, 750 gal. (Studebaker US6-U5) Hercules JXD 6-cyl., 87 bhp, 5F1R×2, wb 162 (BC44) in, 10,585 lb. Body by Columbian or Heil, also supplied on GMC chassis. Studebaker produced over 197,000 2½-ton trucks, more than half of which went to the USSR.

Truck, 2½-ton, 6×6, Cargo (Studebaker US6-U3) Hercules JXD 6-cyl., 87 bhp, 5F1R×2, wb 162 (BC 44) in, 257½×88 ×109 in, 10,140 lb approx. Model US6-U4 had winch. Both were supplied with wooden (shown) and steel body, open (shown) and closed cab. Model U3 also produced by Reo.

Truck, 3-ton, 6×6, Cargo, Low Silhouette, w/Winch (International M-6-6) 6-cyl., 111 bhp, 5F1R×2, wb 146 in. Another experimental model for a projected family of low-silhouette tactical cargo trucks produced for the Ordnance Department in 1942/43. The project was later cancelled.

Truck, 3-ton, 6×6, Cargo, Low Silhouette, w/Winch (Studebaker LD6×6) Hercules JXD 6-cyl., 109 bhp, 5F1R×2, wb 148 in, 250×92¾×92(75) in. 10.00–20 tyres, single rear. Produced in two versions. The other version had longer (162-in) wheelbase and 8.25–20 (4DT) tyres.

Truck, 2½–3-ton, 6×6, Cargo, Low Silhouette, w/Winch (Studebaker LB6×6-125949) Hercules JXD 6-cyl., 109 bhp, 5F1R×2, wb 154 in, 270½×90×92¾(76) in. On this model the driver's seat was placed outside the frame. Mechanically very similar to model LD.

USA

TRUCKS, 2½-TON 6×6 (COE)

Makes and Models: GMC AFKWX-352, AFKWX-353, 1940–45 (G508, 655). International and Studebaker: pilot models for low-silhouette cargo trucks.

General Data: Sole producer of this type (excepting some low-silhouette pilot models) was the GM Truck & Coach Division of the Yellow Truck & Coach Manufacturing Company, which in 1943 was taken over by the General Motors Corporation to become General Motors' GMC Truck & Coach Division. (Since 1925 GM had had a controlling interest in Yellow Truck & Coach, one of the pioneer manufacturers of taxicabs and coaches.) The GMC AFKWX models were basically commercial types, fitted with three driving axles, military-type cargo body, radiator guard and other military hardware. The civilian-type front-end sheet metal remained, but an open-type cab was introduced and fitted on later models. The body of the AFKWX-353 models was exceptionally long and was produced in two lengths: 15 and 17 ft. The 15-ft body was equipped with troop seats. Many were fitted with truck mount, M36, for anti-aircraft machine gun. Axles were split or banjo-type, as on the models with conventional cab mounting. In fact, the AFKWX-352 and -353 combined main features of the GMC AFKX-352 (1½–3-ton 4×4 COE) and the ACKWX-353 (2½-ton 6×6). Of the swb AFKWX-352 only a few were made, in 1940, all with Timken-Detroit split-type axles and 256 cu in engine. Series production of the lwb AFKWX-353 commenced in 1942, during which year 671 units were delivered, all with split-type axles and 270 cu in engine. This model was still in production (with 17-ft cargo body) when the war ended. The pilot model of the 'Duck' (GMC DUKW-353) was built on an AFKWX-353 chassis.

Vehicle shown (typical): Truck, 2½-ton, 6×6, Cargo, 15-ft Body, COE (GMC AFKWX-353).

Technical Data:
Engine: GMC 270 6-cylinder, I-I-W-F, 269·5 cu in, 104 bhp @ 3000 rpm.
Transmission: 5F1R×2.
Brakes: hydraulic (hydrovac).
Tyres: 7.50–20 (4DT).
Wheelbase: 164 in, BC 44 in.
Overall l×w×h: 266½×88×106 (84) in.
Weight: 10,800 lb.

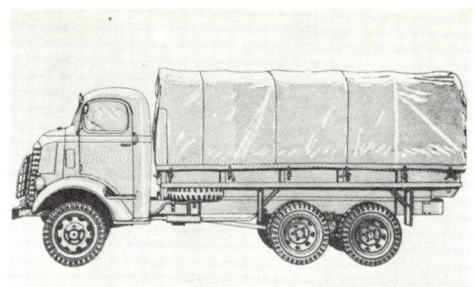

Truck, 2½-ton, 6×6, Cargo, M427, 17-ft, COE (GMC AFKWX-353) 6-cyl., 104 bhp, 5F1R×2, wb 164 (BC 44) in, 289½×88×106 in, 11,950 lb. Early model with closed cab. Steel cargo body. Also built as 4×4 (model AFKX-352).

Truck, 2½-ton, 6×6, Dump, COE (GMC AFKWX-352) 6-cyl., 77 bhp, 5F1R×2, wb 143 (BC 44) in. Rebuilt, fitted with combat wheels and 'hi-floatation' tyres. Originally produced in 1940 with cargo body. At one time designated Truck, 1½–3-ton, 6×6. Also with winch and soft-top cab.

Truck, 2½–3-ton, 6×6, Cargo, Low Silhouette, w/Winch (International M-6-6) 6-cyl., 111 bhp, 5F1R×2, wb 135 in. Pilot model produced for Ordnance Department in 1942. Driver-beside-engine type. Note 'underslung' front springs and machine-gun ring mount.

Truck, 2½–3-ton, 6×6, Cargo, Low Silhouette, w/Winch (Studebaker LA6×6-125948) Hercules JXD 6-cyl., 109 bhp, 5F1R×2, wb 126 in, 242½×90×92¾(76) in, 11,330 lb. Like other Studebaker pilots had WC Lipe clutch, Warner transmission, Timken transfer and axles. Ross steering gear.

USA

TRUCKS, 3- and 4-TON 4×4

Makes and Models: 3-ton: Corbitt, pilot models, 1943. GMC DAK-602 and DOK-602, pilot models, 1942–43. International DOB-M-6-4, pilot models, 1942. 4-ton: FWD HAR-1, 1942–44 (G531). Bodies: Cargo. GMC AFKX-804, 1941–42 (G632, 655). Bodies: Van.

General Data: The majority of 4-ton trucks of the US Army were of the 6×6 type (Autocar, Diamond T, White). The only 4×4 types were the FWD HAR-1 and the GMC AFKX-804 which were basically commercial-type all-wheel drive chassis, fitted with militarised front end and bodywork. The 3-ton 4×4 class (backbone of British and Commonwealth motor transport) was non-existent in the USA, except for a few pilot models produced by Corbitt, GMC and International Harvester in 1942–43, the former for the Research and Development Division of Detroit Arsenal in Centerline, Michigan. These were of the low-silhouette type. The FWD HAR-1 had an open cab and was also supplied to the British, who used it for smoke generating and other purposes. The RAF had the HAR-1 chassis with snow-fighting equipment. These units had a Bros rotary snow-plough fitted to the front of the chassis, driven by a Climax R6 engine mounted on the chassis frame behind the cab. For Canada FWD produced the HAR-01, which was mechanically similar to the HAR-1, but fitted with civilian-type front end, closed cab, wheel-arch body and steel spoke wheels. These FWDs were among the very few 4×4 vehicles which had permanent four-wheel drive, i.e. the front-wheel drive could not be disengaged but, like the rear axle, was permanently driven from the transfer case. The transfer case featured chain drive and incorporated a (third) differential with manual locking arrangement.

Vehicle shown: Truck, 4-ton, 4×4, Cargo, w/Winch (FWD HAR-1)

Technical Data:
Engine: Waukesha 6BZ 6-cylinder, I-L-W-F, 320 cu in, 88 bhp @ 2800 rpm.
Transmission: 5F1R×1.
Brakes: hydraulic (hydrovac).
Tyres: 9.00–20 (2DT).
Wheelbase: 156 in.
Overall l×w×h: 266¾×96½×111 in.
Weight: 11,425 lb.
Note: early models were classed as '2½–4-ton, 4×4'.

Truck, 4-ton, 4×4, Chassis/Cab (FWD HAR-1) Most of these FWDs were supplied to other nations, incl. Britain. The body-less truck shown here was one of a number obtained by the Dutch Government after the war and fitted with a coach-built cab (with roof hatch). The winch was removed.

Truck, 4-ton, 4×4, Van, COE (GMC AFKX-804-8F, 8M) 6-cyl., 177 bhp, 5F1R×2, wb 165 in, approx. 300×98×130 in, 19,000 lb. GMC 426 engine. Body by York-Hoover. 316 supplied in 1941–42. Superseded by Autocar U-8144 which had similar body on standardised 5–6-ton chassis.

Truck, 3-ton, 4×4, Cargo, w/Winch (GMC DAK-602) 6-cyl., 160 bhp, 5F1R×2. Prototype, built in 1942/43. It was similar to the 3-ton 6×6 GMC Model DAKW-602 except for rear bogie, body, larger tyres (11.00–20 vs 8.25–20) and cab mounted further forward. Gar Wood 5000-lb winch.

Truck, 3-ton, 4×4, Cargo, w/Winch (GMC DOK-602) 6-cyl., 160 bhp, 5F1R×2. Another experimental low-silhouette cargo/personnel/prime mover truck. Similar to 3-ton 6×6 GMC DOKW-602 except rear bogie, body and tyres (12.00–20 vs 8.25–20). Note twin spare wheels, carried on right of (off-set) cab.

Truck, 3-ton, 4×4, Cargo, w/Winch (Corbitt, pilot) 6-cyl., 200 bhp, 5F1R×2, wb 130 in. Tyres 12.00–20. Low silhouette cargo/prime mover truck with rear-mounted engine, produced in 1943 and delivered to Detroit Arsenal's Research and Development Division early in 1944.

Truck, 3-ton, 4×4, Cargo, w/Winch (International DOB-M-6-4) 6-cyl., 111 bhp, 5F1R×2, wb 124 in., 204¾×99¾×104½(68) in, 13,500 lb. GVW 19,500 lb. Tyres 12.00–20. Prototype low-silhouette truck with integral cab/body. Troop seat capacity 10. Body inside length 73 in, width 92 in.

Truck, 3-ton, 4×4, Cargo, w/Winch (International M-6-4) 6-cyl., 111 bhp, 5F1R×2, wb 124 in., 204×99×102(66) in approx. Basically similar to Model DOB-M-6-4 but driver positioned alongside engine and consequently longer and roomier body (not unlike Ford GTB 1½-ton 4×4; see page 48).

Truck, 3-ton, 4×4, Cargo, w/Winch (International M-6-4) All International (IHC) 3-ton 4×4 low-silhouette prototypes had IHC 361 CID engine, Rockford clutch, Fuller gearbox, IHC transfer case, Eaton/IHC axles, Wagner brakes, Ross steering gear, Delco-Remy electrics (6 volt), Gar Wood winch.

USA

TRUCKS, 4-TON 6×6

Makes and Models: Autocar C-7066, 1939–40. Bodies: Cargo and Prime Mover.
Canadian-American 'Jumbo Jeep', 1943 (pilot models only).
Diamond T 967, 968, 968A and B, 969, 969A and B, 970, 970A, 972, 975 and 975A, 1941–45 (G509). Bodies: Cargo and Prime Mover, Crane, Dump, Flat Bed, Pontoon, Tank, Torpedo Air Compressor, Tractor, Water Distributor, Wrecker, etc.
Ward LaFrance 106, 1940. Bodies: Wrecker.
White 950×6, 1939–41 (G633). Bodies: Cargo and Prime Mover.

General Data: The 4-ton 6×6 cargo and prime-mover trucks produced during 1939–41 by Autocar and White were superseded in 1941 by the well-known Diamond T model which was the standard type for the remainder of the war. In March, 1943, the Canadian-American Truck Co. of New York City produced a 4-ton 6×6 cargo truck with interchangeable engines (either the Continental 22R, the Hercules RXC or the Waukesha SRKR unit could be used) as a private venture. Two pilot models were submitted to the Ordnance Department for test and evaluation at Aberdeen Proving Ground, but after extensive tests the vehicles proved unable to meet the Army's requirements. The project was closed in September, 1944. The Diamond T was widely used in all theatres of war and was also supplied to other nations, under Lend-Lease. The majority had a front-mounted winch and many had an open cab. The famous wrecker version, which is dealt with in a separate section, became very popular after the war as a civilian vehicle; many are still in use by garages all over the world. The British RAF used the 151-in wb Diamond T chassis/cab with a Bros rotary snow-plough. The Canadian Army acquired over 700 151-in wb models (incl. Wreckers) and more than 1100 special long-wheelbase (201 in, models 975, 975A) chassis/cabs for mounting revolving cranes, workshop and other special bodies.

Vehicle shown (typical): Truck, 4-ton, 6×6, Cargo, w/Winch (Diamond T 968A)

Technical Data:
Engine: Hercules RXC 6-cylinder, I-L-W-F, 529 cu in, 106 bhp @ 2300 rpm.
Transmission: 5F1R×2 (OD top)
Brakes: air.
Tyres: 9.00–20 (4DT).
Wheelbase: 151 in, BC 52 in.
Overall l×w×h: 268½×96×118½(84) in.
Weight: 18,050 lb.
Note: Model 970 had 172-in wheelbase and bridging pontoon body. Also produced with steel body. Gross (SAE) bhp was 119 @ 2200 rpm.

Truck, 4-ton, 6×6, Cargo, w/Winch (Autocar C-7066) 6-cyl., 106 bhp, 5F1R×2, wb 156 in, 257×96×119 in, 17,060 lb. Truck featured soft-top cab with non-folding windshield and Autocar NB-447 engine. Eleven produced, incl. three wreckers.

Truck, 4-ton, 6×6, Cargo, w/Winch (White 950×6) 6-cyl., 115 bhp, 5F1R×2, wb 186 (BC 52) in, 270×96×116 in, 15,240 lb. White 20A engine. Gar Wood winch. Basically commercial model with six-wheel drive. 79 produced, including six wreckers.

Truck, 4-ton, 6×6, Cargo, w/Winch (Diamond T 967) Hercules RXB 6-cyl., 98 bhp, 5F1R×2, wb 151 (BC 52) in, 268½×96×118½ in, 18,400 lb. approx. First Diamond T 4-ton 6×6, featuring RXB 501 CID engine and separate radiator/headlights guard. 998 produced in 1941, incl. 22 wreckers.

Truck, 4-ton, 6×6, Dump (Diamond T 972) Hercules RXC 6-cyl., 106 bhp, 5F1R×2, wb 151 (BC 52) in, 254½ (w/Winch 264½)×94×106 in, 17,200 (w/Winch 18,050) lb. From September 1943 these vehicles were fitted with an open cab. The 3½-cu yd dump body was operated hydraulically.

Truck, 4-ton, 6×6, Map Reproduction Equipment (Diamond T 970A) Hercules RXC 6-cyl., 106 bhp, 5F1R×2, wb 172 (BC 52) in, 300×96×136 in, gross wt 24,000 lb. Van-type body with various types of equipment, used by Engineers.

Truck, 4-ton, 6×6, Crane, Swinging Boom, M1 (Diamond T 968A) Hercules RXC 6-cyl., 106 bhp, 5F1R×2, wb 151 (BC 52) in, 341×96×130 in, 18,400 lb. Used by Chemical Warfare Service to lift and transport heavy chemical containers. Load (normal) 5500 lb at 13-ft radius.

Truck, 4-ton, 6×6, Machinery 'RE 25-KW' (Diamond T 975) Hercules RXC 6-cyl., 131 bhp (gross), 5F1R×2, wb 201 (BC 52) in, 323×98×118 in, 25,400 lb. Model 975 only for Canada. No headlight brush guards. This model was equipped for metal working, wood sawing, etc. Lathe, etc., could be lowered by chain block for ground operation.

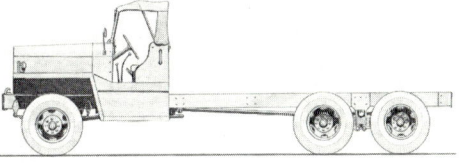

Truck, 4-ton, 6×6, Chassis (Canadian-American 'Jumbo Jeep') Continental 22R 6-cyl., 126 bhp (or Hercules RXC, 111 bhp, or Waukesha SRKR, 120 bhp; see 'General Data'), 5F1R×2, wb 200 in, 296⅛×96×114 in, GVW 25,135 lb. Clark gearbox, Timken transfer and axles, Ross steering. Air brakes. LWB and SWB pilot models, 1943.

USA

TRUCKS, 4- to 5-TON 4×4

Makes and Models: Autocar U-7144T, 1941–45 (G510, 691). Canadian-American 'Jumbo Jeep', 1943 (pilot only). Federal 94×43A, B and C, 1941–45 (G513). Kenworth and Marmon-Herrington, 1945 (produced to Autocar design). White 444T, 1944–45 (G691).

General Data: The 4–5-ton 4×4 truck-tractor was first produced by Autocar and superseded the 2½-ton 4×4 type (Autocar, GMC). Federal built a similar, but not identical, version. White, and towards the end of the war also Kenworth and Marmon-Herrington, supplemented production of the Autocar design. Large numbers were produced. Mechanically this model could be called the 4×4 version of the Diamond T-built 4-ton 6×6. Together these vehicles formed an important part of the US Army's class of 'heavy-heavy' vehicles. In 1943 the Canadian-American Truck Co. produced a pilot model for a proposed replacement for the 4–5-ton tractor. It was a short-wheelbase 4×4 version of the 4-ton 6×6 mentioned in the preceding chapter. Both were designated 'Jumbo Jeep'. Their tests were discontinued in 1944. In the Signal Corps the 4–5-ton 4×4 tractor was designated K-32. Early vehicles had a closed cab; later production had an open-type cab. They were mostly used in conjunction with a 10- or 11-ton cargo or 2000-gallon gasoline semi-trailer, but there were also various other types. A similar open-cab military truck-tractor was produced after the war by Hino Motors in Japan for the Japanese Army.

Vehicle shown (typical): Truck, 4–5-ton, 4×4, Tractor (Autocar U-7144T, White 444T)

Technical Data:
Engine: Hercules RXC 6-cylinder, I-L-W-F, 529 cu in, 112 bhp @ 2200 rpm.
Transmission: 5F1R×2.
Brakes: air.
Tyres: 9.00–20 (2DT).
Wheelbase: 134½ in.
Overall l×w×h: 203×95×112(92) in.
Weight: 11,660 lb.
Note: also supplied with closed cab (early models).

Truck, 4–5-ton, 4×4, Tractor (Autocar U-7144T, White 444T) with Semi-trailer, 12-ton Gross, 2-W, Van, K-78 (Fruehauf) Typical truck-tractor and semi-trailer combination. Tractor shown with closed cab. This particular semi-trailer was used by Anti-aircraft Artillery and Army Air Forces to house and transport radio equipment. It measured 245½ × 97¼ × 125¼ in. A dolly (K-83) was used when a full-track tractor was used as prime mover; this dolly could be towed behind the semi-trailer when not in use.

Truck, 4–5-ton, 4×4, Tractor (Federal 94×43A, B, C) with Semi-trailer, 10-ton Gross, 2-W, Refrigerator Body. Shown are the Federal tractors with open and closed cab. Except for the rounded front end and a different steering linkage, the Federal was very similar to the Autocar design. Semi-trailer was produced by various manufacturers and used for items needing temperature of 10°F or 32°F. Dimensions of the semi-trailer: 253½ × 96 × 126 in. Dimensions of tractor: 203 × 95½ × 109(92) in, wheelbase 134⅛, weight 11,950 lb.

USA

TRUCKS, 5-TON 4×2

Makes and Models: Autocar C50, C70, U70, C90T, 1939–45. Bodies: Cargo, Dump, Tank, Tractor.
Diamond T 806-W-DR, 1940–42. Bodies: Cargo.
Federal 55L, 1941–42 (G555). Bodies: Cargo, Dump, Stake.
GMC AC-723, AC-725, AC-803 (G655), AC-805, 1940–41. Bodies: Dump, Stake.
International KR-11, 1941–45 (G542). Bodies: Cargo, Dump, Tractor. K-10 and KR-10 Tractors, 1941.
International H-542-9, 1944, and H-542-11, 1944–45 (G671). Bodies: Tractor COE.
Kenworth and Marmon-Herrington, H-542-11 (to IHC design).
Mack EH series, 1941–44 (G533) and FPD (diesel), 1941 (G640). Bodies: Cargo, Tractor.
Reo 23BHRS, 1941. Body: Dump.

Vehicle shown (typical): Truck, 5-ton, 4×2, Tractor, COE, M426 (IHC H-542-11, also produced by Kenworth and Marmon-Herrington)

Technical Data:
Engine: IHC RED-450-D ('Red Diamond') 6-cylinder, I-I-W-F, 451 cu in, 124 bhp @ 2600 rpm.
Transmission: 5F1R.
Brakes: air.
Tyres: 11.00–20 (2DT).
Wheelbase: 120 in.
Overall l×w×h: 200×97½×106¾(83¼) in.
Weight: 12,100 lb.
Note: for differences between M425 and M426 see 'General Data'.

General Data: Most of the above models were civilian types with military bodywork and slight modifications to make them acceptable for military use. The Mack EH came in several versions, with either conventional (closed or open) or COE cab mounting, various wheelbases, rear axles, etc. Many of these Macks were supplied to Lend-Lease countries (mainly the UK). The open-cab type Mack EH (cargo and tractor) had a militarised front end. The International H-542-9 and H-542-11 were the only specific military types in this class. The former was used originally by the Transportation Corps and by Ordnance to haul 5-ton semi-trailers; the latter was an improved and up-rated tractor for 10-ton semi-trailers and was classified standard. These semi-tactical tractors were used mainly for long-distance haulage on supply routes in Europe. Their availability for peace-time use after 1945 was welcomed by many West European haulage firms. Many are still in use, now fitted with coach-built cabs and usually diesel engines. The M425 model differed from the M426 (shown here) mainly in the following respects: tyres (9.00–20), wheelbase (117 in), overall dimensions (200×94½×102½(79) in), weight (11,400 lb). One external distinguishing feature was the position of the fuel tank filler cap, which was at the forward end of each tank.

Truck, 5-ton, 4×2, Cargo (Mack EH) 6-cyl., 110 bhp, 5F1R, wb 170 in, 268×96×114 in, 10,500 lb. 12-ft cargo body. 2400 supplied to Britain in 1943. Also produced as truck-tractor with 146-in wheelbase. For closed-cab types see section on trucks supplied to other nations.

Truck, 5-ton, 4×2, Dump (International KR-11) 6-cyl., 133 bhp, 5F1R, wb 161 in, 251½×95½×87 in, 12,815 lb. Dump bodies were produced by Galion, Gar Wood, and Daybrook. Also supplied with cargo body (197- and 179-in wb) and as truck-tractor (161-in wb).

Truck, 5-ton, 4×2, Cargo (Federal 55L) Hercules WXLC3 6-cyl., 118 bhp, 4F1R, wb 184 in, 292×96×90 in, 10,450 lb. Basically a commercial model, fitted with Galion Allsteel cargo body. Similar trucks built by Diamond T, GMC, IHC, and Mack. Federal also built 3½-ton 4×2 model 29K, 1941–42.

Truck, Fuel or Oil Servicing, 1600-gal., 4×2 (Autocar U70) 6-cyl., 106 bhp. 4F1R, wb 163 in, GVW 26000 lb. Body by Heil. One of several types built for the Bureau of Aeronautics of the US Navy, in 1939/40. 11.00–20 tyres. Similar tanker truck also supplied with dual-drive rear bogie (6×4).

USA

TRUCKS, 5- to 6-TON 4×4

Makes and Models: Autocar U-8144, U-8144T, 1941–45 (G511). Bodies: Tractor, Van.
Autocar U-5044 (5-ton), 1940–41 (G635).
FWD CU, 1942 (G649); SU, 1942–45 (G638); YU, 1941–42; HST, 1940.
Mack NJU-1, NJU-2, 1940–41 (G639); NJU-9D, 1941.

General Data: The 5–6 ton 4×4 cab-over-engine type was used mainly as truck-tractor for pontoon semi-trailers. As such it was first manufactured by Mack, using a civilian-type closed cab with military modifications. In 1941 a new type was introduced and standardised, using components which were also used on other standardised army trucks. This was the model U-8144T, produced by Autocar. It was basically an up-rated and longer version of their 4–5-ton tractor. This truck superseded the Mack NJU and was also used with a large van body, similar to that of the earlier GMC 4-ton 4×4, which it replaced. The truck-tractor was used mainly by the US Corps of Engineers for transportation of pontoon-bridging equipment. Later Autocar models were fitted with an open-type cab; all had a front-mounted winch. The FWDs in this category were basically commercial models, modified to meet military requirements. From 1942 they were used mainly for Lend-Lease requirements. The FWD SU-COE was widely used by the British army in various forms (cargo, medium artillery tractor, bomb disposal, snow plough; see also section on trucks produced for other nations) and in the Canadian Army. After the war the SU-COE remained in military use in several countries, either in original or modified form. In the Netherlands, for example, numbers were fitted with a double cab and a fifth wheel for towing aircraft refueller semi-trailers. In the civilian sector they also became popular, especially with fairground operators, just like the FWD Model B after World War I.

Vehicle shown (typical): Truck, 5–6-ton, 4×4, Tractor (Mack NJU-1)

Technical Data:
Engine: Mack EN532 6-cylinder, I-I-W-F, 532 cu in, 136 bhp @ 2500 rpm.
Transmission: 5F1R×2.
Brakes: air.
Tyres: 9.75–20 (2DT) (later changed to 12.00–20).
Wheelbase: 148 in.
Overall l×w×h: 237×96×114 in.
Weight: 16,230 lb.
Note: Model NJU-2 had no box behind cab and was used for topographical semi-trailers. Total production (both models) 692.

Truck, 5–6-ton, 4×4, Pontoon Tractor, w/Winch (Autocar U-8144T) Hercules RXC 6-cyl., 112 bhp, 5F1R×2, wb 163½ in, 246½×98×114¾ in, 16,660 lb. Early production had closed cab. Winch was std equipment. Box behind cab carried bridge installation tools. Tyres 12.00–20. Also wrecker version (q.v.).

Truck, 5–6-ton, 4×4, Van (Autocar U-8144) Hercules RXC 6-cyl., 112 bhp, 5F1R×2, wb 163½ in, 300×98×130½ in, 19,280 lb. Body by York-Hoover. Similar vehicles, designated K-30, K-31 and K-62, used by Army Air Forces to house and transport radio equipment. Gross weight 27,700 lb. No winch.

Truck, 6-ton, 4×4, Pontoon Bridge Construction, w/Winch (FWD YU) Waukesha 140 GK, 6-cyl., 5F1R×1. Tyres 11.00–20. GVW 27,000 lb. This 'Erecting and Transporting Unit' was used by the US Corps of Engineers in 1941/42, until superseded by the 6-ton 6×6 bridging truck.

Truck, 5–6-ton, 4×4, Utility Line (FWD HST-COE) Used by US Coast Guard to build and maintain electric power and telephone communication pole lines. Many of these units saw service with USN 'Seabees' (CBs—Construction Battalions) in their island-hopping activities in the South Pacific.

USA

TRUCKS, 6-TON 6×6

Makes and Models (Chassis): Brockway B-666, C-666, 1942–45 (G514, 547). Corbitt 50SD6, C-666, 1940–45 (G512, 514). FWD B-666, C-666, 1945 (G514, 547). Mack NM-1, 1940, NM-3, -5, -6, 1941–45 (G535). Ward LaFrance B-666, 1945 (G547). White 666, 666CE, ECT, 1941–45 (G514, 526, 690). Bodies: see General Data.

General Data: With the exception of the Mack NM series, which was developed from a commercial model in 1939/40, all the above models were built to a common standardised military design. First in production were Corbitt and White, followed by Brockway. Towards the end of the war FWD and Ward LaFrance joined in to boost production. The most common model was the Cargo and Prime Mover (Corbitt 50SD6, White 666). First built with closed cab, later with open type, these heavy trucks were used in most theatres of war and after 1945 they remained in use for many years by various governments. The following derivations were produced: Crane carrier (half-cab, 197-in wb), Bridge erector ('Treadway', 220-in wb), Gasoline tank (2000-gal., 185-in wb), Fire truck (197-in wb, class 155), Power van (185-in wb), Tractor (185-in wb). Most of these trucks were fitted with a winch, either behind the cab (on the cargo and prime movers) or at front. The Mack 6-ton 6×6 was used extensively for Lend-Lease requirements. With the winch removed and an extended chassis and cargo body, the Canadian Army used the Mack NM as a 10-ton load carrier. All Cargo and Prime Mover models were fitted with controls for electric trailer (gun) brakes. Early Whites and Corbitts (with closed cab) had a steel cargo body, later production (with open cab) had a wooden body. After the war numbers of these Corbitts, Whites and Macks were converted into tractor trucks with fifth wheel. In 1944 the Heil Co. developed a modification of the crane-carrier chassis (Corbitt C-666) by fitting a full-width cab and a 2500-gal. gasoline tank. A 6-ton 4×4 was produced by Sterling in 1942–43 (Model DDS115).

Vehicle shown (typical): Truck, 6-ton, 6×6, Cargo and Prime Mover, w/Winch (Corbitt 50SD6, White 666)

Technical Data:
Engine: Hercules HXD 6-cylinder, I-L-W-F, 855 cu in, 202 bhp @ 2150 rpm (early models: Hercules HXC, 799 cu in, 165 @ 2000 rpm).
Transmission: 4F1R×2.
Brakes: air (plus controls for electric trailer brakes).
Tyres: 10.00–22 (4DT) (early models: 9.75–22).
Wheelbase: 185 in, BC 52 in.
Overall l×w×h: 289×96×114 in.
Weight: 22,900 lb (Corbitt 22,020 lb).
Note: winch located behind cab.

Truck, 6-ton, 6×6, Cargo, w/Winch (Mack NM3) 6-cyl., 159 bhp, 5F1R×2, wb 177 (BC 52) in, 282×96×119 in, 22,659 lb. This is an early model with closed cab and louvres in hood side panels. The NM series was first introduced in 1939 and had a Mack EY engine.

Truck, 6-ton, 6×6, Cargo, w/Winch (Mack NM6) 6-cyl., 159 bhp, 5F1R×2, wb 177 (BC 52) in, 282×96×119(95) in, 22,960 lb. In general design the Mack was similar to the standardised type; transfer case and driving axles were identical (Timken).

Truck, 6-ton, 6×6, Bridge Erection (Treadway) (Brockway B666) Hercules HXD 6-cyl., 202 bhp, 4F1R×2, wb 220 in (BC 52) in, 374×100×110 in, 26,500 lb. Also built by other mfrs. Used by the Corps of Engineers to transport and erect bridging equipment. Open cab on later models.

Truck, 6-ton, 6×6, Prime Mover (White 666) Hercules HXD 6-cyl., 202 bhp, 4F1R×2, wb 185 (BC 52) in, 288×96×131 in, 21,200 lb. Used by Army Air Forces as prime mover and power van; classified standard when part of radio set SCR-545-A. Bodywork by Superior and P. A. Thomas.

USA
TRUCKS, 7½-TON 6 × 6

Makes and Models: Biederman C-2, F-1, P-1 (AAF), 1942–44 (G692). Corbitt 50SD6, 1940. Federal 605, 606, 1942–44 (G692). Hug 50-6, 1939–40. Mack NO-1, -2, -3, -6, -7, 1940–45 (G532). Reo 27XFS, 29XS, 1942–45 (G692).

General Data: During the 1930s military trucks in the 7½-ton 6×6 class had been produced in various shapes and forms by some of the smaller specialist firms like Clydesdale, Corbitt, Hug, Marmon-Herrington and White. Clydesdale and Hug went out of business in 1937 and 1940 respectively. Mack became the major manufacturer of this type of vehicle for the US Army and the United Nations. The war-time Biederman, Federal and Reo models in this class were all built to a basic Army Air Forces specification and were very similar. The latest Federal and Reo, in fact, looked identical. All three had Cleveland Pneumatic model YMB air springs at front (resting on the forward ends of the conventional front springs). These AAF models were built in two basic versions: C-2 Wrecking Truck (with fifth wheel for semi-trailer) and F-1 Tractor for Refuelling Semi-trailer (4000-gal. capacity). In addition there was a special version for mounting a revolving crane (model P-1, built by Biederman) and the Reo 27XFS was also used with Cardox fire-fighting equipment. For these special equipment vehicles see the relevant separate sections. The Mack NO was interesting in that it had a double gear reduction in the steering ends of the driven front axle. The steering ends or wheel drive units were of a type in which bevel gears concentric with the king pins drove the wheels and allowed for steering, thereby dispensing with universal joints. As a result the axle housing was also higher than the wheel hubs, providing extra ground clearance. The Belgian forces, after the war, equipped a series of these Macks with coach-built closed cabs. For Mack Models NO4 and NO5 see section on Wreckers.

Vehicle shown (typical): Truck, 7½-ton, 6×6, Prime Mover (Mack NO2)

Technical Data:
Engine: Mack EY 6-cylinder, I-I-W-F, 707 cu in, 159 bhp @ 2100 rpm.
Transmission: 5F1R×2.
Brakes: air (plus controls for electric trailer brakes).
Tyres: 12.00–24 (4DT).
Wheelbase: 156 in, BC 58 in.
Overall l×w×h: 297×103×124(94) in.
Weight: 29,103 lb.
Note: fitted with chain hoist at rear for use when attaching special trail clamp for 155-mm gun. Also towed 8-in howitzer.

Truck, 7½-ton, 6×6, Prime Mover (Mack NQ) 6 cyl., 159 bhp, 5F1R×2, wb 155 (BC 56) in, 276½×101¾×113 in, 27,700 lb. This was the pilot model for the NO series. Note differences in cab (hard-top), hood, higher front winch mounting with two capstan pulleys. Three were built, one as a truck tractor.

Truck, 7½-ton, 6×6, Tractor (Biederman F1) Hercules HXC 6-cyl., 165 bhp, 5F1R×2, wb 196 (BC 48) in, 287×94½×102 in, 18,960 lb approx. Tractor for Air Corps' Type F-1 Fuel Servicing Semi-Trailer. Similar to Federal and Reo but more rounded cab and front end. Also with 180-bhp Hercules HXD engine.

Truck, 7½-ton, 6×6, Tractor, Type F-1 (Reo 29XS) with Semi-trailer, Fuel Servicing, 4000-gal., Type F-1A (Standard Steel Works) *Tractor:* Hercules HXD 6-cyl., 180 bhp, 5F1R×2, wb 196 (BC 48) in, 283×96×104 in 18,960 lb. *Semi-trailer:* 360×95½×112 in, 19,676 lb. Gross combination weight: 77,636 lb. Used by Army Air Forces to transport aircraft engine fuel from distributing points to airfields, and to service aircraft. Semi-trailer equipped with two engine-driven pumps having a rated total capacity of 320 gpm.

USA

TRUCKS, 8- and 10-TON 6×4

Makes and Models: *8-ton:* Corbitt 40SD6, 1942 (G556). GMC ACW-853, 1940–41.
10-ton: Autocar C-9064, 1940. Corbitt 50SD4, 1942. Dart 200/353, 1941. Federal 604, 1942 (G648). International K8F (1940–41) and H-542-9 (modified). Mack NR-4 to -14, 1941–45 (G528). Ward LaFrance 204, 1940–41. White 1064, 1942–45 (G642).

General Data: The most common and best-known vehicles in this class were the diesel-engined Macks and Whites. During the early years of the war these Macks and Whites were supplied mainly to Great Britain and Russia, under Lend-Lease. Following D-Day, they were used extensively by such transport organisations as the 'Red Ball Express' to haul supplies from Cherbourg in Normandy and other ports to the ever-moving front line in Western Europe. Early Macks had a closed cab. In the Canadian Army the Mack NR was used by the RCASC, which also used a 10-ton conversion of the Mack 6×6 model NM. The Federal 604 10-tonner was a fuel tanker using the chassis/cab of the 20-ton tank hauler. These units saw service on the Alaska Highway in 1943–44. The 10-ton cargo truck, T48, was produced during the latter part of the war. It was developed from the 5-ton 4×2 semi-tactical truck-tractor, M425, probably in the interest of further standardisation. The US Marine Corps used a 6×4 K-series International as an artillery tractor (AA Bn) in 1940/41. It had an all-steel cargo body and overall chains could be used over the rear wheels. The Autocars had a Cummins Diesel engine and were used by the US Navy, equipped with stake bodies. A similar Autocar, but fitted with a power winch behind the cab, was produced as a tank transporter, probably for the Soviet Army. An 8-ton 4×4 was produced in 1942 by Sterling (Model DD105). For transport of partly assembled aircraft between Michigan, Texas, and California, the US Army used a fleet of 6×4 tractor trucks, powered by twin Ford-Mercury engines, with 60-ft semi-trailers.

Vehicle shown (typical): Truck, 10-ton, 6×4, Cargo (Mack NR9)
Technical Data:
Engine: Mack-Lanova ED Diesel 6-cylinder, I-I-W-F, 518·6 cu in, 131 bhp @ 2000 rpm.
Transmission: 5F1R×2.
Brakes: air.
Tyres: front 10.50 or 11.00–24, rear 13.50 or 14.00–20 singles (from vehicle serial no. 8567 (NR14): 11.00–24 all round (4DT)).
Wheelbase: 200½ in, BC 55 in.
Overall l×w×h: 322×103×101 in.
Weight: 21,600 lb.
Note: early production had closed cab and was used mainly by British in Middle East.

Truck, 8-ton, 6×4, Cargo and Prime Mover (International K-8F) 6-cyl., 85 bhp, 5F1R×2. Basically 1940–41 commercial truck, fitted with military type all-steel cargo body, radiator brush guard, rear bumperettes, etc., for USMC. Hendrickson-type rear bogie with walking beams. Overall chains fitted over rear wheels.

Truck, 10-ton, 6×4, Cargo (White 1064) Cummins HB.600 6-cyl., diesel, 150 bhp, 5F1R, wb 200 (BC 52) in, 327×97×122 in, 20,000 lb. Front end and cab similar to 6-ton 6×6. Tarpaulin box on cab roof. Top speed 39 mph, cruising range 750 miles.

Truck, 8-ton, 6×4, Tractor (Corbitt 40SD6) Continental 22R 6-cyl., 133 bhp, 5F1R, wb 183½ (BC 52) in, 285×86×104 in, 14,300 lb. Used in conjunction with heavy semi-trailer for long-distance haulage. Timken front axle and rear bogie. Basically commercial design.

Truck, 10-ton, 6×4, Cargo, T48 (IHC H-542-9, modified) 6-cyl., 124·5 bhp, 5F1R. Developed and tested towards the end of the war, based on IHC 5-ton truck-tractor M425. A long-wheelbase 4×2 cargo version was also produced. All had IHC 'Red Diamond' engine.

USA

TRUCKS, 8-TON, 6×6 and 12-TON, 6×4 and 6×6

Makes and Models: Diamond T 980, 981, 1941–45 (G159) (12-ton, 6×4). Kenworth T30 (8-ton, 6×6, pilots only). Pacific Car & Foundry TR-1 M26, M26A1, 1943–45 (G160) (12-ton, 6×6). Sterling T28, T28E2 (8-ton, 6×6, pilots only).

General Data: The Diamond T 12-ton prime mover, M20, was originally designed to meet a British requirement for tank transporting and recovery. It was introduced in 1940/41 and first saw active service with the British Army in North Africa. Later it was also used by the US Army. It featured a 20-ton winch and a ballast body behind the cab. From August, 1943, these vehicles were fitted with an open-type cab and provision for AA mg ring mount, M36. There were two models: 980 and 981. The 980 had a 300-ft winch cable and two winch-cable roller sheaves at rear; the 981 had a 500-ft winch cable, three cable sheaves at rear and a roller assembly in the front bumper, allowing the winch cable to be paid out from the front of the vehicle. The equally impressive 'Pacific' M26 went into service in June, 1943, as a truck-tractor. It featured six-wheel drive (chain-drive at rear) and an armoured cab. The heavy armoured cab was later found to be of limited tactical use and a modified model with soft-skin cab, the M26A1, was introduced in 1944. These trucks had three winches, one at front with a capacity of 35,000 lb and two at rear with a capacity of 60,000 lb each. The 8-ton, 6×6, T28 and T30 were produced as prototypes in 1944/45. The idea was to develop a heavy-duty truck with performance characteristics superior to standard vehicles in the 6- and 7½-ton 6×6 class, capable of successfully fulfilling any of the following roles: a large cargo-carrying vehicle, a prime mover for the 155-mm gun, and a truck-tractor for semi-trailer. The cessation of the war caused the termination of such projects. For tank transporters see also special section. After the war both the Diamond T M20 and the Pacific M26/M26A1 were extensively used by civilian contractors for heavy haulage, especially in Western Europe.

Vehicle shown (typical): Truck, 12-ton, 6×4, Prime Mover, M20 (Diamond T 981)

Technical Data:
Engine: Hercules DFXE diesel, 6-cylinder, I-I-W-F, 895 cu in, 185 bhp @ 1600 rpm.
Transmission: 4F1R×3.
Brakes: air.
Tyres: 12.00–20 (4DT).
Wheelbase: 179¼ in, BC 52 in.
Overall l×w×h: 280×101×101 in.
Weight: 26,650 lb.
Note: Prime Mover, M20, was used in conjunction with Trailer, 45-ton, 12-wheel, M9, to form Truck-Trailer, 45-ton, Tank Transporter, M19.

Truck, 12-ton, 6×4, Prime Mover, M20 (Diamond T 980)
Hercules DFXE diesel, 6-cyl., 185 bhp, 4F1R×3, wb 179¼ (BC 52) in, 280×101×100 in, 26,950 lb. Used for recovery and transport of damaged tanks and material in conjunction with 45-ton trailer. Max. towed load 115,000 lb.

Truck, 12-ton, 6×6, Tractor, M26 (Pacific Car & Foundry TR-1) Hall Scott 440 6-cyl., 240 bhp, 4F1R×3×1, wb 172 (BC 63) in, 307×130½×118 (w/ringmount 127) in, 48,895 lb. Used to tow 40-ton semi-trailers M15 and M15A1 in recovery and transportation of tanks and material weighing up to 80,000 lb.

Truck, 12-ton, 6×6, Tractor, M26A1 (Pacific Car & Foundry TR-1) Hall Scott 440 6-cyl., 240 bhp, 4F1R×3×1, wb 172 (BC 63) in, 307×130½×118(108) in, 45,000 lb. Similar to M26 but cab not armoured. Collapsible hoist could be erected on rear of chassis for lifting purposes.

Truck, 8-ton, 6×6, Cargo, T30 (Kenworth) Hercules HXLD 6-cyl., 275 bhp, 4F1R×3, wb 185 in, 338×115×129 in, 27,000 lb. 14.00–24 tyres. Also projected as truck-tractor (overall dimensions 310×115×116 in, max. weight on fifth wheel 20 tons). Pilots only.

USA

TRUCKS, 8- and 12-TON 8×8

Makes and Models: 8-ton, 8×8, T20 (Cook Bros). 8-ton, 8×8, T20E1 (Corbitt). 12-ton, 8×8, T26 and T26E1-E4 (Sterling). 12-ton, 8×8, T33 and T33E1 (Corbitt).

General Data: The above models were designed to achieve higher mobility for tactical cross-country cargo and tractor trucks and produced in prototype form for the Ordnance Department during 1943–45, together with a number of other high-mobility vehicles. The end of the war in mid-1945 put an end to the development of some of them, although others were used for further experiments until the early 1950s. The 8-ton series came in two configurations, 6×6 and 8×8. The 6×6 type is dealt with in the preceding section. Of the 8×8s, the first was designated T20 and built by Cook Brothers in California and had chain drive on all wheels. The pilot models were ordered in 1943 and completed in December, 1944. Tests took place in 1945. There were cargo and tractor versions. An improved model, T20E1, was built by Corbitt in Henderson, N.C., featuring conventional axles and platform steering. The first 12-tonner 8×8 truck was projected in 1943 and designated T26. It was first planned to have a straight frame, but in order to reduce overall height the pilot model, produced by Sterling in Milwaukee, Wis., had a 'swan-neck' frame. Like the T20 it was produced in cargo and tractor form. Featuring two chain-driven Sterling bogies and platform steering, they could be fitted with dual tyres all round. It was followed by the T26E1 (cargo and tractor), also built by Sterling. The T26E1 had a Ford V8 engine which was normally used in certain versions of the Sherman tank. Developments of the T26E1 were the T26E2 (wrecker), T26E3 (cargo and tractor with double cab) and T26E4 (cargo and tractor with Fuller torque converter transmission). Finally there were the T33 and T33E1, both built by Corbitt in cargo form. These, too, had a tank engine, namely the 9-cylinder radial Continental R975, but were not delivered until 1947 and 1946 resp.

Vehicle shown (typical): Truck, 8-ton, 8×8, Cargo, w/Winch, T20 (Cook Bros.)

Technical Data:
Engine: Continental R6602 6-cylinder, I-I-W-F, 602 cu in, 240 bhp @ 2600 rpm.
Transmission: 5F1R×2.
Brakes: air.
Tyres: 14.00-20 (4DT).
Wheelbase: 130 in (between bogies).
Overall l×w×h: 328×124×128 in.
Weight: 37,090 lb.
Note: also built as truck-tractor, T20.

Truck, 8-ton, 8×8, Cargo, w/Winch, T20E1 (Corbitt)
Continental 6-cyl., approx. 250 bhp, 5F1R×2, wb approx. 130 in (between bogies). This truck was a further design exercise in the 8-ton 8×8 project. It had turntable steering, conventional axles with hub reduction gears and a 'swan neck' chassis frame.

Truck, 12-ton, 8×8, Cargo, w/Winch, T26 (Sterling)
American LaFrance V-12-cyl., 275 bhp, 5F1R (OD top)×2×2, wb 150 (BC 60) in, 353×126×132(124) in, 51,000 lb. Tyres 14.00–24 (8DT). 'Swan neck' frame with hydraulic turntable steering. Development began in 1943, tests in 1945.

Truck, 12-ton, 8×8, Tractor, w/Winch, T26 (Sterling)
American LaFrance V-12-cyl., 275 bhp, 5F1R (OD top)×2×2, wb 150 (BC 60) in, 350×126×125 in, 50,000 lb. Max. load on fifth wheel 60,000 lb. Intended for Semi-Trailer, 20-ton, Machinery Hauling, which was also under development (1944).

Truck, 12-ton, 8×8, Cargo, w/Winch, T33 and T33E1 (Corbitt)
Continental R975-C4 radial 9-cyl., 450 bhp, 5F1R (OD top)×3, wb 180 (BC 58) in, 365×111×134(110) in, 51,300 lb. Tyres 14.00–24. Knuckle-type steering. T33E1 had Spicer 961 torque converter trans. (3F1R). Both had 3-speed transfer.

USA

CARS AND TRUCKS, MISCELLANEOUS
SUPPLIED TO OTHER NATIONS

During the years 1939–41 (prior to Pearl Harbor) the US motor industry supplied vast quantities of commercial-type cars and truck chassis to friendly nations, such as Britain, China, France, India, etc., through normal commercial channels (usually via purchasing commissions set up in New York). A selection of such types is presented here, in alphabetical order (by make). The bodywork in many cases was fitted in the country of destination. Many were also supplied by the Canadian factories of GM, Ford, and Chrysler. In March, 1941, the Lend-Lease Act was passed and US military-type vehicles became also available. These, however, are included in the preceding and following sections.

Truck, 1½-ton, 4×2, Water Tank (Chevrolet VA) 6-cyl., 90 bhp, 4F1R, wb 133 in. This chassis was particularly popular in the Middle East. They were often fitted with oversize desert tyres. Most had open cab. 1939 model shown was used by British Commonwealth troops in Egypt. 1940–42 models were also used.

Truck, 1½-ton, 4×2, Cargo (Chevrolet MS 4403) 6-cyl., 90 bhp, 4F1R, wb 160 in. 1942 model long-wheelbase truck supplied to China for use on Burma Road. Also supplied with stake body (MS 4409). Note black-out driving lights on top of headlamps. 1941 models (Series YS) similar in appearance.

Truck, 3-ton, 6×4, Cargo (Chevrolet YS4103/Thornton) Basically as 4×2 but fitted with Thornton tandem-drive bogie incorporating 2-speed aux. gearbox. Originally ordered by French Government but shipments diverted to UK. Dimensions 276×94×119 in, wb 174¼ (BC 42) in, weight 8672 lb.

Truck, 3-ton, 4×2, Cargo (Dodge VH-48) 6-cyl., 99 bhp, 5F1R, wb 178 in, GVW 15,000 lb. 1500 of these trucks were supplied to France just before the outbreak of war. Thousands of similar Dodges (chassis and chassis/cabs) were supplied to Russia (via Persia) and to China (for use on Burma Road).

Truck, 3-ton, 4×2, Personnel (Dodge VK-62B) 6-cyl., 100 bhp, 5F1R, wb 188 in, 312×93×111 in, 10,304 lb. Used by British RAF as Crew Coach. Also with cargo body. Most of these heavy-duty Dodges in British service were 'ex-French contract' vehicles (diverted shipments, originally ordered by the French).

Truck, 1½-ton, 4×4, Cargo (Dodge T203) 6-cyl., 99 bhp, 4F1R×2, wb 160 in, 259×91×117 in, 8010 lb. 'Ex-French contract' truck, used by British Army. All-steel GS body with flat floor, 16-in fixed sides and removable folding troop seats. Vacuum servo-assisted hyd. brakes. (US Army version page 43.)

Truck, 3-ton, 6×4, Breakdown, w/Winch (Dodge WK-60) 6-cyl., 100 bhp, 5F1R×2, wb 164 (BC 48) in, 260×94×135 in, 13,680 lb. Specially modified to British requirements. Thornton bogie. RHD. 10.50–20 tyres on split-type wheels. Gar Wood winch. British body with overhead gantry.

88

Car, 5-passenger, 4×2 (Ford 01A-73B) V-8-cyl., 85 bhp, 3F1R, wb 112 in, 192×75×72 in, 3140 lb. 1940 Fordor sedan, assembled by Ford Motor Co. in Dagenham, England. 9.00-13 tyres and modified wings. 1942 Fords were also thus modified. All had LHD.

Truck, ½-ton, 4×2, Machine Gun (Ford 19C/Marmon-Herrington) V-8-cyl., 85 bhp, 3F1R, wb 112 in. Basically standard 1941 Ford commercial chassis, modified by M.-H. for Netherlands Government use in Surinam. Similar trucks used by Indian and Irish armed forces. Also as 4×4 (M.-H. Model LLDMG5-4).

Truck, 1½-ton, 4×2, Cargo (Ford E018T) V-8-cyl., 85 bhp, 4F1R, wb 158 in, 231×87×114(90) in, 5180 lb. Assembled in England. British GS body. Some had closed cab. Similar-looking 15-cwt version (E01Y) also produced. Many militarised Ford 4×2 trucks saw service in North Africa.

Truck, 2½-ton, 6×6, Searchlight and Sound Ranging (Ford 198T/Marmon-Herrington) V-8-cyl., 95 bhp, 4F1R×2. A large number of these trucks was ordered for the Netherlands East Indies but diverted before delivery to US contracting agencies when this country was overrun by the Japanese. 1941.

89

Snow Plough, Rotary, Truck Mounted (Ford/Snogo) V-8-cyl., 95 bhp, 4F1R×2, wb 132 in, 264×81×108 in. 4×4 (M.-H.) chassis with Snogo Model LTR plough, driven by separate engine (IHC U-21). Used from 1942 by British RAF, which also had American FWD SU/Snogo, FWD HAR/Bros and Diamond T 968/Bros units.

Truck, 5–6-ton, 4×4, Tractor, Timber Haulage, w/Winch (FWD CU) Cummins AA-600 diesel, 6-cyl., 100 bhp, 5F1R×1, 15,200 lb. Tyres 10.00–20. GVW 22,000 lb. Towed load 24,000 lb. Gar Wood winch behind cab. 85 supplied to Britain, under Lend-Lease, in 1942. 377 CID engine.

Truck, 5–6-ton, 4×4, Tractor Chassis, w/Winch (FWD SU-COE) Waukesha SRKR 6-cyl., 126 bhp, 4F1R×1, wb 144 in. Many supplied to UK and fitted with British Medium Artillery Tractor body (as AEC Matador). Similar chassis were equipped with augers for bomb digging; these had smaller tyres (dual rear).

Truck, 5–6-ton, 4×4, Cargo (FWD SU-COE) Waukesha SRKR 6-cyl., 126 bhp, 5F1R×1, wb 144 in, 257×96×113 in, 13,235 lb. Produced principally for Lend-Lease (UK) as GS load carrier. Also used as tractor truck (smaller tyres, dual rear), snow-plough, etc. FWDs had permanent all-wheel drive.

Truck, 3-ton, 4×2, Cargo (GMC ACX-504) 6-cyl., 85 bhp (net), 5F1R, wb 168 in, 284×95×122(89) in, 9968 lb. Diverted to Britain, following cancellation of French contract, in 1939–40. Shown with non-original body (original was all-steel fixed-side type). GMC Model 1500 cab. 4·53-litre engine.

Truck, 1½-ton, 4×4, Cargo (GMC ACK-353) 6-cyl., 77 bhp (net), 4F1R×2, wb 157¾ in, 256×90×117(89) in, 7532 lb. Ex-French contract vehicle, used by British Army. Steel GS body with fixed sides and troop seats. 4·07-litre engine. 7.50–20 tyres. Front hubs designed to take dual wheels.

Truck, 3-ton, 6×4, Cargo/Searchlight (GMC AFWX-354) 6-cyl., 77 bhp net), 4F1R×2, wb 167 (BC 44) in, 287×95×135(100) in, 11,004 lb. Ex-French contract vehicle. 'Sleeper' type extension to cab for searchlight crew. 4·07-litre engine (GMC Model 248). 7.00–20 tyres. See also page 56.

Truck, 3-ton, 6×6, Cargo (GMC ACKWX-353) 6-cyl., 77 bhp (net), 4F1R×2, wb 175¾ (BC 44) in, 264×94½×118(90) in, 9856 lb. Another 'ex-French' type, diverted to Britain in 1940. US Army had similar trucks with 162-in wb and US standard type cargo body. IWM photo H24419.

Truck, 5-ton, 4×2, Cargo (International K-8) 6-cyl., 100 bhp, 5F1R, wb 179 in. One of many types of International K-line trucks supplied for military service, mainly during 1940–42. Truck shown was assembled in Britain and used by Royal Navy. Also supplied to civilian operators, under special licence arrangement.

Truck, 5-ton, 4×2, Tractor (International KR-8) 6-cyl., 85 bhp (net), 5F1R, wb 137 in. Used with Fruehauf Model 220 10-ton flat-bed semi-trailer. Overall length (w/S-T) 372 in, weight 16,800 lb. Tractor similar to KR-8R (see page 52) but no winch and different gearbox and rear axle. Tyres 9.00–20 (36×8).

Truck, 5-ton, 4×2, Cargo, COE (Mack EHU) 6-cyl., 100 bhp, 5F1R (OD top), wb 162 in, 264×84×117 in, 10,080 lb. 70 supplied to Britain in 1942, together with 510 Model EH trucks which had conventional cab mounting and 158-in wb. Mack EN310 engine. (Some EHs had 115-bhp EN354 engine.)

Truck, 5-ton, 4×2, Tractor (Mack EHT) 6-cyl., 115 bhp, 5F1R, wb 141 in, 211×91×84 in, 6750 lb. Coupled to Mack Model ST20 semi-trailer. British nomenclature: 6-ton, 4×2-2 Semi-trailer GS. Overall dimensions 372×96×120 in, weight 13,552 lb. Also supplied as COE tractor (Model EHUT).

Truck, 5-ton, 6×4, Heavy Breakdown (Mack LM-SW)
6-cyl., 160 bhp, 5F1R×2, wb 166½ (BC 55) in, 294×99×118 in, 22,000 lb. 14.00–20 tyres. Gar Wood body and equipment. Produced for British Army, equivalent to Scammell SV/2S. Two power winches. Power divider in rear bogie.

Truck, 5-ton, 6×4, Heavy Breakdown (Mack LM-SW)
6-cyl., 160 bhp, 5F1R×2, wb 166½ (BC 55) in, 294×99×118 in, 24,640 lb. This type was supplied to the Canadian Army (RCEME and RCASC) and had double swinging booms. Maximum lifting capacity 16 tons. Some had soft-top cab.

Truck, 18-ton, 6×4, Tank Transporter (Mack EXBX)
6-cyl., 131 bhp, 5F1R, wb 209¼ (BC 52½) in, 370½×121×99 (cab) in, 22,112 lb. Supplied to British Army for use in North Africa. Some were ex-French contract. Either Mack or Timken rear bogie. Loading ramps later changed to detachable type.

Truck, 13-ton, 6×4, Tank Transporter (Mack NR 4) 6-cyl., diesel, 131 bhp, 5F1R×2, wb 200½ (BC 55) in, 327×102×102 in, 22,000 lb. Supplied to British Army, 1940/41. One-piece loading ramps, handled manually. No winch. 12-ton hydraulic support jacks at rear. Originally had cargo body.

Truck, 5-ton, 4×2 Cargo (Studebaker K30) Several hundreds of these heavy-duty Studebakers were supplied to the French Army in 1939. Many of them subsequently fell into German hands and were used by the *Wehrmacht*. Hercules WXC3 6-cyl. engine. Weight 7995 lb.

Truck, 10-ton, 4×2, Cargo (White 760) 6-cyl., 115 bhp, 5F1R, wb 195½, 328½×102×96 (cab) in, 13,340 lb. 11.25–24 tyres. One of the heaviest four-wheelers used by the British (RASC) in the Middle East. Similar Whites used by RAF in Britain for transporting aircraft.

Truck, 18-ton, 6×4, Tank Transporter (White 920) 6-cyl., 134 bhp, 5F1R, wb 212 (BC 53) in, 383×123×94½(cab) in, 22,150 lb. Supplied to British, some ex-French contract. British body. Hinged folding-type loading ramps, operated by hand winches, later replaced by separate man-handled ramps.

Truck, 18-ton, 6×4, Tank Transporter (White-Ruxtall 922) 6-cyl., 134 bhp, 5F1R×3, wb 226 (BC 52) in, 369½×117½×98½ in, 25,765 lb. Ex-French contract. Body built in Britain. 10-ton winch. Detachable loading ramps. All these types were later utilised as heavy cargo trucks.

USA
WRECKERS

Makes and Models: Autocar U90, 1940, U-8144T, 1941 (G511). Available, 1943. Biederman C2, 1940–44 (G692). Diamond T 969, 969A, 969B, 1941–45 (G509). Federal 605, 606, 1940–44 (G692). GMC ACK-352, 1940, CCKW-353, 1942. Kenworth 570, 571, 572, 573, 1941–45 (G116); 580, 1943 (pilots). Mack NO-4, NO-5, 1943 (pilots), LM-SW (Lend-Lease). Oshkosh W709, 1944. Reo 29XS, 1941–44 (G692). Sterling DDS 150, DDS 225, DDS 235, HCS 330, 1942–44. Ward LaFrance 106, 1940; 1000 Series 1, 2, 3, 4, 5, 1941–45 (G116), etc.

General Data: Purpose: to lift, tow, salvage and recover disabled vehicles and equipment. Air Force wreckers were used to salvage crashed planes and to perform general towing. The Diamond T 969 Wrecker was introduced in 1941 and was based on the standard 4-ton 6×6 truck chassis. It superseded several earlier types of 6×6 wreckers and remained in production throughout the war. The heavy wrecker, M1, was first produced by Ward LaFrance in 1940, at which time it was classed as a 4-ton 6×6. By 1943 this type was officially designated Truck, Heavy Wrecker, 10-ton, 6×6, M1, and soon afterwards Truck, Wrecking, Heavy, M1. After the war it was redesignated again and became the Truck, Heavy Wrecker, 6-ton, 6×6, M1. The M1, in slightly different form, was also produced by Kenworth. In 1943 a modified type was introduced, featuring an open cab and military type front end. Designated M1A1, it was produced to the same specifications by Kenworth and Ward LaFrance. The USAAF had their own type of wrecker. Built on a special version of their standard 7½-ton 6×6 chassis, it had a swinging boom crane and a fifth wheel so that semi-trailers for the conveyance of wrecked planes could be towed. The Navy used similar types. A range of new heavy wreckers was developed experimentally in 1943–44 by the Ordnance Department, including the Mack NO-4 (with fifth wheel, for Air Corps) and NO-5 (for Field Artillery), which were developed from the Mack 7½-ton 6×6. Other pilot models were produced by Kenworth, Sterling and Ward LaFrance, with 'soft skin' and armoured cabs, but none of these entered series production.

Vehicle shown (typical): Truck, 4-ton, 6×6, Wrecker (Diamond T 969A, 969B)

Technical Data:
Engine: Hercules RXC 6-cylinder, I-L-W-F, 529 cu in, 106 bhp @ 2300 rpm.
Transmission: 5F1R×2.
Brakes: air.
Tyres: 9.00–20 (4DT).
Wheelbase: 151 in, BC 52 in.
Overall l×w×h: 291½×99½×116 in.
Weight: 21,350 lb.
Note: these models also supplied with closed cab. Holmes W-45 wrecker equipment rated at 10 ton (5 ton per boom).

Truck, 4-ton, 6×6, Wrecker (Diamond T 969) Hercules RXC 6-cyl., 106 bhp, 5F1R×2, wb 151(BC 52) in, 291½×99½×116 in. Twin-boom wrecker equipment by Ernest Holmes Co. Picture shows booms swung out and brace leg in position. Model 969 only supplied with closed cab.

Truck, Heavy Wrecker, 6-ton, 6×6, M1 (Kenworth 570, 571) Continental 22R 6-cyl., 145 bhp, 5F1R×2, wb 181 (BC 52) in, 276×101×122 in, 27,330 lb. Model 572 differed from the above mainly in that the rear body, crane and front bumper were similar to that of the heavy wrecker, M1A1.

Truck, Heavy Wrecker, 6-ton, 6×6, M1 (Ward LaFrance Model 1000, Series 2, 3) Continental 22R 6-cyl., 145 bhp, 5F1R×2, wb 181 (BC 52) in, 318×100×122 in, 27,130 lb. Front and rear winch. Series 1 had front winch only. Series 4 had curved boom and front bumper as M1A1. Gar Wood crane.

Truck, Heavy Wrecker, 6-ton, 6×6, M1A1 (Kenworth 573, Ward LaFrance Model 1000, Series 5) Continental 22R 6-cyl., 145 bhp, 5F1R×2, wb 181 (BC 52) in, 348×99½ ×106½ in, 30,000 lb. Introduced in 1943, superseding the various models in the M1 series. 5-ton swinging-boom crane.

Truck, 1½-ton, 4×4, Wrecker (GMC ACK-352) 6-cyl., 100 bhp, 4F1R×2, wb 135 in approx. Militarised commercial GMC 1940 chassis/cab, equipped with four-wheel drive, radiator brush guard, etc. Wrecking equipment consisted of simple tubular crane structure with power-driven winch. Used by USMC.

Truck, 2½-ton, 6×6, Light Wrecker (GMC CCKW-353) These light wreckers consisted of a standard GMC 2½-ton 6×6 cargo truck (q.v.), fitted with a wrecker equipment set No. 7. Purpose: to provide mobile facilities for maintenance and repair. Overall dimensions: 310×88×127 in, net weight 11,165 lb.

Truck, Heavy Wrecker, 6×6 (Ward LaFrance Model B) Waukesha 6-cyl., 226 bhp, 5F1R×2, wb 195 in, 377×111×126 in, 45,110 lb. Pilot model with armoured cab produced in 1943. Project discontinued in 1944 and vehicle assigned for facility use at Ordnance Desert PG, Camp Seeley, California.

Truck, Heavy Wrecker, 6×6, COE (Kenworth 580) Hercules 6-cyl., 202·5 bhp, 5F1R×2, wb 200 (BC 58) in, 358½×110 ×126 in, 47,886 lb. Pilot models built in 1943 with open (shown) and armoured cab. Armoured-cab version weighed 50,203 lb. Holmes wrecking gear.

Truck, Medium Wrecker, 6×4 (Available) Waukesha 6-cyl., 90 bhp, 5F1R. Holmes Model W-45 wrecking equipment, mounted on militarised commercial chassis for USN Naval Ammunition Depot in 1943. Holmes supplied over 10,000 of these wreckers but mostly on the Diamond T 4-ton 6×6 chassis.

Truck, 7½-ton, 6×6, Tractor, Wrecking, Type C-2 (Biederman) Hercules HXD 6-cyl., 180 bhp, 5F1R×2, wb 189½ (BC 48) in, 405½×96½×126 in, 28,000 lb. Used by AAF to salvage crashed planes and to perform general towing. Payload on fifth wheel 13,000 lb. Early-style front end and cab.

Truck, 7½-ton, 6×6, Tractor, Wrecking, Type C-2 (Federal 605) Spec. as Biederman. Early C-2 models had Gar Wood 10-ton (at 10-ft radius) capacity gooseneck boom crane, 15 ft long. All AAF C-2 wreckers had underslung winch at rear, air-assisted steering, inertia-type starter, air springs at front.

Truck, 7½-ton, 6×6, Tractor, Wrecking, Type C-2 (Federal 606, Reo 29XS) Spec. as Biederman. Late model with Gar Wood 5-ton (at 8-ft radius) telescoping boom, having lengths of 18, 22 and 26 ft. All had 110-volt DC electric power plant of 3000-watt capacity and flood lamps.

Truck, 5-ton, 4×2, Tractor, Wrecking (Autocar U90)
Hercules RXB 6-cyl., 98 bhp, 5F1R, wb 163 in, GVW 30,000 lb. Heavy-duty wrecker produced in 1940 for the US Navy, Bureau of Aeronautics. Note fifth wheel for semi-trailer. Gar Wood crane and equipment.

Truck, 5-6-ton, 4×4, Tractor, Wrecking (Autocar U-8144T)
Hercules RXC 6-cyl., 112 bhp, 5F1R×2, wb 163½ in. Basically 5-6-ton tractor truck (q.v.), with Gar Wood swinging-boom crane. Only one supplied to the Army (0032253). Picture taken on Tarawa, Central Pacific, 1944. Similar crane on Oshkosh W709 4×4.

Truck, 7½-ton, 6×6, Wrecker (Sterling DDS 235) Waukesha 6-cyl., 5F1R×2. Produced in limited numbers for US Navy, Bureau of Aeronautics, for salvage of crashed planes. Gar Wood power-operated swinging-boom crane of 10-ton capacity. Sterling also produced 12-ton 6×6 model.

Truck, 15-ton, 6×4, Heavy Wrecker (Sterling HCS 330) Waukesha 145 GK, 6-cyl. Dual chain-drive rear bogie, used by US Navy, Bureau of Aeronautics. Gar Wood model US7-T-28 wrecker equipment. Note ballast weight on front bumper, winch underneath. Produced 1942-3. GVW 80,000 lb.

USA

TRANSPORTERS
TANK and HEAVY EQUIPMENT

(See also pages 82–83)

Makes and Models (prime movers): Dart 300AWD, 1942 (pilots only). Diamond T 980, 981, 1941–45 (G159). Federal 604, 1942–45 (G648). Mack, 1945 (pilot only). Pacific Car & Foundry TR-1, M26, M26A1, 1943–45 (G160). Reo 28XS, 1944 (G645).

General Data: Purpose: to recover and transport tanks and other heavy equipment and material. The first American tank transporter to go into production during World War II was the Diamond T tractor, M20, with the 45-ton, 12-wheel (12DT) trailer, M9. The trailer was a standard design, produced by various manufacturers. This combination, designed Truck-Trailer, 45-ton, Tank Transporter, M19, was supplied mainly to Great Britain under Lend-Lease. The tank transporter which the US Army themselves procured in quantity was the M25, which consisted of the armoured 'Pacific' tractor, M26, and the Fruehauf 40-ton, 8-wheel semi-trailer, M15. This M25, popularly known as the 'Dragon Wagon', was designed in 1942 and officially adopted in 1943. Fruehauf Trailer Co., holder of the prime contract, subcontracted production of the tractor unit first to the Knuckey Truck Co., later to Pacific Car & Foundry Co., which had greater capacity. During 1944 the M15 semi-trailer was modified to take heavier loads (45 tons) and redesignated M15A1. The M26 tractor was modified to reduce weight; it was fitted with a 'soft skin' cab and redesignated M26A1 in 1944. The M25 was unsuitable for the increasing weights of the tanks which were developed as the war went on and it was often operated at gross weights of up to 155,000 lb with a load of up to 38,000 lb on each of the tractor rear axles. Officially the payload on the fifth wheel was 55,000 lb and the overloading was disastrous not only for the tractor but also for the existing road networks where these vehicles were operated. Experiments with tank transporters with lower axle/wheel loading were carried out throughout the war and some examples are shown here. The first of these (not shown) was the T1, an 8×8 4-engined unit produced by Cook Bros. The British FVPE equipped a 'Pacific' M26 with 21.00–24 tyres and ballast box.

Vehicle shown (typical): Truck-Trailer, 40-ton, Tank Transporter, M25 (Pacific TR-1 Tractor, M26, and Fruehauf Semi-Trailer, M15)

Technical Data:
Engine: Hall Scott 440 6-cylinder, I-I-W-F, 1090 cu in, 240 bhp @ 2000 rpm.
Transmission: 4F1R×3×1.
Brakes: air (rear bogie and S-T only).
Tyres: 14.00–24.
Wheelbase: tractor 172 (BC 63) in, S-T 372½ (BC 64) in.
Overall l×w×h: 684×150×137 in.
Weight: 84,400 lb approx.
Note: Max. fifth wheel load 55,000 lb. Semi-trailer measured about 462 (ramps up)×150 (reducible to 124)×105 in and weighed 36,100 lb. Tank tracks passed between rear wheels. Known as 'Dragon Wagon'. Tractor rear wheel brakes at either side could be operated individually. Hyd. steering assistance.

Truck-Trailer, 45-ton, Tank Transporter, M25 (Pacific TR-1 Tractor, M26, and Fruehauf Semi-Trailer, M15A1) M15A1 semi-trailer was modified M15 model, capable of carrying 90,000 lb. Note ramps for loading of wider tanks over rear tyres (shown in propped-up position).

Truck-Trailer, 45-ton, Tank Transporter, M25 (Pacific TR-1 Tractor, M26A1, and Fruehauf Semi-Trailer, M15A1) In 1944 the M26A1 tractor truck was introduced. It was similar to the M26 except for the cab, which was unarmoured. Both had Knuckey chain-drive rear bogie.

Tractor, 6×6, Tank Transporter (Pacific TR-1, modified) British (FVPE) modification, used for exp. purposes in connection with development of heavy drawbar tractors. 21.00 (vs. 14.00)–24 tyres. The British had many US tank transporters, mostly Diamond T tractors with full-trailers.

Truck, 6×4, Tractor (Federal 604) with Semi-Trailer, 20-ton, 4-W (4DT) (Trailmobile) Cummins HB600 diesel, 6-cyl., 150 bhp, 4F1R×2, wb 167 (BC 52) in, 265×95¼×97 in, 20,000 lb (tractor). Tractor co-produced by Reo (Model 28XS). Mainly for Lend-Lease (UK, Canada). (See also page 80.)

Truck-Trailer, 45-ton, Tank Transporter, M19 Consisting of tractor, M20 (see 12-ton 6×4), and trailer, M9. Trailer designed and built by Rogers Bros (also produced by four others), measured 360×114×62 in and had 24 8.25–15 tyres. Unit shown was modified for transport of ammunition.

Truck-Trailer, 40-ton, Tank Transporter, T3 Consisting of tractor, 6×6, T13 (Dart), and semi-trailer, M15. Tractor, built by Dart in 1942, had Waukesha 6-cyl. 250-bhp engine, 4F1R×2 transmission, 14.00–24 tyres, 230¼ (BC 58½) in wb, measured 340×125½×112 in and weighed 42,100 lb.

Tank Transporter, Double Ended, 4×4, T4 Built by the Ordnance Dept. in 1942 as a test rig for a new concept in tank transporter design. Basically two LeTourneau 'Tournapull' tractor units, each powered by twin Cadillac engines. 614×126×154 in, 63,000 lb. Tyres 24.00–32.

Tank Transporter, Double Ended, 45-ton, 8×8, T8 Experimental model with two 4×4 tractor units, produced by Mack. Two 6-cyl. engines, 240 bhp each, wb 502 (BC 76) in, 713×124×132 in. Tyres 21.00–28. Later modified (T8E1, 40-ton) and fitted with two 500-bhp Ford GAA V8 tank engines.

USA

TRACTORS, WHEELED
(PRIME MOVERS)

Makes and Models: Austin-Western A-W4-44, 1940 (pilot). FWD SU-COE (USMC), 1940–43. Minneapolis-Moline GTX, 1940–43 (G641). Oshkosh TR, 1940. Walter ADUM, 1940–41.

General Data: Wheeled tractors used solely for towing of artillery pieces were rare in the US Army and were built only in limited numbers. The main reason for this was that these tractors had very limited space for crew, ammunition and equipment. 6×6 trucks were therefore used for this role (2½-ton, 6-ton, 7½-ton types, depending on size required). The Austin-Western unit listed above was an all-wheel-steer 4×4 model with dual 9.00–24 tyres all round, handlebar steering and diesel engine. The Oshkosh and the Walter were specially built as prime movers for the 8-in howitzer and the 155-mm gun. The FWD was a special short-wheelbase tractor version of the firm's SU-COE model and was built only for the Marine Corps. After the war the USMC modified some of these FWDs to have smaller tyres (2DT) and 5-cu yd Heil dump bodies. The Minneapolis-Moline was basically a much-modified farm tractor. In 1938 this company started converting a farm tractor into a 4×4 artillery tractor and in 1940 several of these experimental tractors were placed in the Army manœuvres at Camp Ripley, Minnesota. It is claimed that National Guardsmen there called these vehicles 'Jeeps' ('after the Popeye cartoon figure called Jeep, which was neither fowl nor beast, but knew all the answers and could do most anything'), and the maker was quick to coin this name for his multi-purpose tractor. When later the name 'Jeep' became synonymous with the ¼-ton 4×4, Minneapolis-Moline resented and started advertising their claim (whole-page advertisements appeared in magazines) but to no avail. The Company produced several models, including the GTX 6×6 and a 1½-ton 4×4 aircraft prime mover for the US Navy. Conventional 4×2 tractors (industrial type) were produced by Allis-Chalmers and Case. Some John Deere tractors were experimentally fitted with armour plates in 1940/41. Clark produced small tractors for warehouse and airfield use.

Vehicle shown (typical): Truck, 5-ton, 4×4, Prime Mover (Walter ADUM).

Technical Data:
Engine: Hercules diesel 6-cylinder, I-I-W-F, 672 cu in, 150 bhp @ 1800 rpm (or Hercules DHXB 707 cu in, 150 bhp @ 1400 rpm).
Transmission: 6F2R. 'Four-Point Positive Drive.'
Brakes: air.
Tyres: 15.00–24.
Wheelbase: 126 in.
Overall l×w×h: 232×96×109 in.
Weight: 19,000 lb.
Note: shown towing 155-mm gun.

Truck, 5-ton, 4×4, Prime Mover (Oshkosh TR) Built in 1940 and seen here with the 8-in howitzer. Like the Walter ADUM it had a special goose-neck type gun carriage connection attachment, designed to eliminate excessive weight transfer from front to rear axle under heavy pulling. Four-wheel steering.

Truck, 4×4, Prime Mover (FWD SU-COE Special) Waukesha 6-cyl., 561 cu in. Designed to meet quick boarding and un-boarding requirements of USMC. Based on SU-COE truck, with high-floatation tyres (14.00-20), open cab, winch behind cab. Some had longer cargo body, no winch.

Truck, 1½-ton, 4×4, Prime Mover (Minneapolis-Moline) Aircraft prime mover produced for the US Navy. 201 CID L-head engine, wb 101 in, gross weight 6800 lb. Timken front driving axle. One of the vehicles which Minneapolis-Moline persistently called 'Jeep' (see General Data).

Truck, 7½-ton, 6×6, Prime Mover (Minneapolis-Moline GTX) 6-cyl., 604.8 cu in. Timken front axle. MM rear axles with 14.00-24 high-cleat tractor-tread tyres. Latest MM 6×6 model, featuring Gar Wood winch above front rollers. Used to tow 90-mm anti-aircraft gun.

USA

CRANES
TRUCK-MOUNTED

Makes and Models: Available CS-600(L)-SW, C(S)-700(D)-SW, 1942–45 (6×4; Harnischfeger or Link-Belt crane). Bay City CraneMobile T40, T50, T60, T61, T66, 1940–45 (6×6, 6×4). Biederman P-1, 1944–45 (special version of std 6×6 AAF chassis; Michigan crane). Brockway C-666, 1942–45 (G514) (6×6; Quick-Way crane). Coleman G55A, 1941–44 (G684) (4×4; Quick-Way crane). Corbitt C-666, 1944 (G514) (6×6; Quick-Way crane). Dart 200/353, 1941 (6×4; Browning crane), 200/454AWD, 1941 (6×6), 300AWD, 1942 (6×6). FWD C-666, 1944–45 (G514) (6×6; Quick-Way crane). Hendrickson 500B, 1944–45 (6×4; Browning crane). Thew (Lorain Moto-Crane) MC-2, -3, -4, 1944–45 (G172) (6×6, 6×4).

General Data: Purpose: to provide mobile facilities for crane, shovel, dragline, clamshell and pile driver operations. Used mainly by the US Corps of Engineers and produced by several specialist manufacturers. There were relatively many different types and standardisation was difficult. Types to receive a 'standard' classification were the Crane, Truck Mounted, M2 (Thew Shovel Model MC 6×6) and the Crane, Revolving, Truck Mounted, 4- to 8-ton, Class X (Quick-Way crane on Truck, Crane Chassis, 6-ton, 6×6, Brockway/Corbitt/FWD C-666). The former was used in conjunction with Trailer, 3-Ton, 2-Wheel, Clamshell, M16 (Gramm). This type of revolving crane on 3-axle truck chassis became very popular after the war for military and civil engineering use. The chassis were fitted with outriggers. A separate engine for the crane was mounted in the turntable. The maximum lifting capacity of the M2, with the outriggers extended, was 40,000 lb at 11-ft boom radius (21,850 lb with outriggers retracted), down to 21,500 (9,950 lb) at 20-ft boom radius. In addition to the above truck-mounted models the US forces used several other types of mobile cranes, such as tractor-operated one-axle types of 10- to 20-ton lifting capacity. These were towed by a crawler-type tractor, using the winch of the tractor for the hoisting operation. There were also crawler tractor-mounted and crawler-mounted revolving cranes. See also section on wreckers.

Vehicle shown (typical): Crane, Truck Mounted, Gasoline Engine Driven, ⅜ cu yd (Quick-Way E-55 on Truck, 4-ton, 4×4, Coleman G55A), with Trailer.

Technical Data:
Engine: Buda LO-525 6-cylinder, I-L-W-F, 522 cu in, 125 bhp @ 2100 rpm.
Transmission: 5F1R×2.
Brakes: air.
Tyres: 10.00–20 (2DT).
Wheelbase: 147 in.
Overall l×w×h: 408 (w/trailer 736)×96 ×128 in.
Weight: 25,200 (w/trailer 26,410) lb gross.
Note: turntable engine: International U-9, 4-cyl., 35 bhp.

Crane, Truck Mounted (Michigan TMCT 16; 10-ton, 6×4) This type of crane was used by the Army Air Forces to lift loads of up to 20,000 lb. It was usually mounted on a variation of the standard AAF 7½-ton 6×6 chassis, as produced by Biederman with model designation P-1. The unit shown here was in use by the RAF on British airfields. The RAF also used the Bay City T50 and Thew Lorain MC3, both 6×4.

Crane, Truck Mounted, M2 (Thew Shovel MC 6×6) Hercules HXC 6-cyl., 179 bhp, 4F1R×2, wb 168½ (BC 52) in, 414½ × 108 × 130 in, 53,500 lb gross. Turntable engine Waukesha MZR, 6-cyl., 70 bhp. Actual truck chassis was 299½ in long. Boom hoisting time, from horizontal to 10-ft radius, 45 sec. Swing speed of turntable 4·19 rpm. During the war the M2 was used principally in conjunction with heavy artillery (240-mm howitzer and 8-in gun), handling components when changing from travelling to firing position and vice versa.

USA

TRUCKS
CRASH, FIRE and SNOW-FIGHTERS

Makes and Models:
(a) *Truck, Airfield, Crash* and *Fire, Crash, COE* (on Chevrolet and Ford/M.-H. 4×4 chassis).
(b) *Truck, Fire, Crash, High Pressure Fog Foam, Class 125* and *Class 135* (on Dodge, Ford, International, Mack, and Ford 4×2, and Chevrolet 4×4 chassis). *Class 155* (on Brockway C-666 and Kenworth 572 6×6 chassis).
(c) *Truck, Fire, Crash, Low Pressure Carbon Dioxide, Class 150* (on Mack NO, Reo 27XFS and 29FF, and Sterling DDS-235 6×6 chassis).
(d) *Truck, Fire, Brush, Class 300* (on Chevrolet and GMC 4×2 and Chevrolet 4×4 chassis).
(e) *Truck, Fire, Pumper, Class 325, 300 GPM* (on Chevrolet, Dodge, and Ford 4×2, and Chevrolet 4×4 chassis). *Class 500, 500 GPM* (on Chevrolet and Ford 4×2 chassis). *Class 525, 500 GPM* (on Chevrolet, Dodge, and Ford 4×2, and Chevrolet 4×4 chassis). *Class 530, 500 GPM, Overseas Type* (on GMC CCKW-353 6×6 chassis). *Class 750, 750 GPM* (on American LaFrance, Buffalo, GMC, Mack, Maxim, Pirsch, and Seagrave 4×2 chassis).
Note: (a), (b), (c) used by Army Air Forces for combating fires resulting from airplane crashes; (d), (e) used by Corps of Engineers for combating fire in brush and wooded areas and structural fires. The USMC and USN had fire-fighting trucks on Ford/M.-H., GMC and International and other chassis (4×2, 4×4, 6×6).
Snow-Fighters: FWD, Oshkosh, and Walter 4×4 chassis.

General Data: Some of the above types were basically commercial-type appliances, adapted where necessary to fulfil military requirements. Fire-fighting equipment and bodywork were produced by more than 20 different firms, including American LaFrance, Boyer, Buffalo, Cardox, Darley, General Detroit, Hahn, Howe, Mack, Maxim, Oren, Peter Pirsch, Seagrave, and Ward LaFrance. The British RAF used the following US equipment: Chevrolet 4×4 and Federal 4×2 fire-fighters, Ford/M.-H. and FWD (HAR, SU and SU-COE) 4×4 and Diamond T 6×6 snow-fighters.

Vehicle shown (typical): Truck, 1½-ton, 4×4, Fire Crash, COE (Chevrolet NX-G-7153)

Technical Data:
Engine: Chevrolet 235 6-cylinder, I-I-W-F, 235·5 cu in, 93 bhp @ 3100 rpm.
Transmission: 4F1R×2.
Brakes: hydraulic (hydrovac).
Tyres: 7.50–20 (2DT).
Wheelbase: 175 in.
Overall l×w×h: approx. 280×87×99 in.
Note: built for US Navy, 1942–43; carried carbon dioxide and 300-gal. water tank with pump.

Truck, Fire, Pumper, Class 325, 4×2, 300 GPM (Ford 1942, 1½-ton, 4×2; Maxim equipment) 8-cyl., 100 bhp, 4F1R, wb 134 in. Fire pump: American Marsh Barton F-300, centrifugal type, front-mounted. Also on Chevrolet and Dodge chassis.

Truck, Fire, Pumper, Class 325, 4×4, 300 GPM (Chevrolet, 1½-ton, 4×4; Maxim equipment) 6-cyl., 83 bhp, 4F1R×2, wb 145 in. Fire pump: AM Barton F-300 or Darley, centrifugal type, front-mounted. Capacity 300 gpm @ 120 lb/sq in or 150 @ 200.

Truck, Fire, Pumper, Class 500, 4×2, 500 GPM (Ford 1941, 1½-ton, 4×2; Maxim equipment) 8-cyl., 100 bhp, 4F1R, wb 158 in. Fire pump: Hale GSS, centrifugal type, midship mounting. Capacity 500 gpm @ 120 lb/sq in, 167 @ 250.

Truck, Fire, Pumper, Class 500, 4×2, 500 GPM (Chevrolet 1942, 1½-ton, 4×2; Maxim equipment) 6-cyl., 81 bhp, 4F1R, wb 160 in. Class 300 brush-fire truck was rather similar in appearance but had front-mounted pump

Truck, Fire, Pumper, Class 750, 4×2, 750 GPM (Maxim, 1941) Special chassis manufactured by Maxim Motors. Maxim or Watrous pump, midship mounting. Capacity 750 gpm @ 120 lb/sq in, 375 @ 200, 250 @ 250. Note 'sawn off' type cab.

Truck, Fire, Pumper, Class 750, 4×2, 750 GPM (Pirsch 15, 1940) Hercules RXDLS engine. Special chassis, manufactured by Peter Pirsch & Sons, Co., in Kenosha, Wis. Capacity 750 gpm @ 150 lb/sq in (net) pump pressure.

Truck, Fire, Crash, Class 155, 6×6, High Pressure Fog Foam (Brockway C-666 and Ward LaFrance B-666 6-ton 6×6) Chassis of M1 wrecker truck (Kenworth) was used for first 100 vehicles. Pumping unit had separate engine, Continental R-602 6-cyl. or American LaFrance V-12.

Truck, Fire, Pumper, Class 530, 6×6, 500 GPM, Overseas Type (GMC CCKW-353) 6-cyl., 91 bhp, 5F1R×2, wb 164 (BC 44) in. Equipment by General Detroit Co. Note combat-type wheels and tyres. USMC had similar units on International M-5-6 chassis.

Truck, Fire, Class 150, Type O-1, 6×6, Low Pressure Carbon Dioxide (Reo 29FF/Cardox 9508) Produced by Cardox Corp. on this and on the standard USAAF 7½-ton 6×6 Reo 27XFS chassis, for Army Air Forces. Mack NO (7½-ton, 6×6) version had early NM type front end and cab.

Truck, Fire, Class 150, 6×6, Low Pressure Carbon Dioxide (Sterling DDS-235/Cardox 9506) Produced for US Navy. Equipment almost identical to Cardox 9508 (on Reo chassis). Waukesha 6 WAL 6-cyl., 200 bhp, 5F1R. This Sterling chassis was also used with Gar Wood wrecker equipment.

Plow, Snow, Rotary, Gasoline Engine Driven, Truck Mounted (Truck, 7½-ton, 4×4, Oshkosh) Hercules RXC 6-cyl., 131 bhp, 5F1R×2. Klauer TU-3 snow-plow, driven by Climax R6I 6-cyl. 175-bhp engine. Various types produced, also on similar FWD chassis (model SU).

Plow, Snow, Speed and Roto Wing, Truck Mounted (Walter FGBS) Two Waukesha 145GKU 6-cyl. 175-bhp engines, one driving the Roto Wing. Walter 8-ton 'Four Point Positive Drive' (4×4) chassis, 1942. Used by US Navy and Army airports and Royal Canadian Air Force.

USA
AMPHIBIANS

Makes and Models: Amphibian Car Corp. 'Aqua-Cheetah', 1942–43. Ford GPA, 1942 (G504). GMC DUKW-353, 1942–45 (G501). Lima T34, 1945. Studebaker M29C 'Water Weasel', 1943–45 (G179).

General Data: Purpose: to transport personnel and cargo on land and water. Development on some of the above amphibious vehicles started in 1941. The ¼-ton 4×4 model, or 'amphibious Jeep', was designed by Marmon-Herrington in conjunction with the boat-designing firm of Sparkman & Stephens in New York. The pilot model was designated QMC-4 and it was based on the 'Jeep' chassis. Series production started in September, 1942, by the Ford Motor Co. 12,778 were ordered but only less than half this number were produced. They were also supplied to Britain and the USSR under Lend-Lease. The Russians later produced their own version, designated GAZ-46 (MAV). The 2½-ton 6×6 model was also designed in conjunction with Sparkman & Stephens and used the standard GMC 6×6 truck chassis. It was standardised in October, 1942, and was in production until the end of the war for the US and Allied Armies. From 4508 in 1943 production increased to 11,316 in 1944 and by mid-1945 a total of 21,147 had been produced. It was an extremely successful vehicle, and remained in use for many years. The Russians started producing a copy of it, after the war, featuring a larger cargo space and a tail gate. It was designated ZIL-485 (BAV). The Studebaker 'Weasel' was a tracked amphibian and was a modification of cargo carrier M29(T24), featuring bow and stern extensions, air tanks, track side panels and cable-actuated rudders. It was also used by the British ('Amphibian, 10-cwt, Tracked, GS'). Total production was approx. 15,000 units. The 'Aqua-Cheetah' was a commercial venture. This vehicle featured independent suspension front and rear and final drive with chains (enclosed in trailing arm casings) to each wheel. 75% of its moving parts were similar to the Dodge ¾-ton 4×4 truck. Not many were produced. Lima Locomotive Works produced the Carrier, Amphibian, Full-Track, T34, in 1945.

Vehicle shown (typical): Truck, ¼-ton, 4×4, Amphibian (Ford GPA)

Technical Data:
Engine: Ford GPW 442 4-cylinder, I-L-W-F, 134·2 cu in, 54 bhp @ 4000 rpm.
Transmission: 3F1R×2.
Brakes: hydraulic.
Tyres: 6.00–16.
Wheelbase: 84 in.
Overall l×w×h: 182×64×69(54) in.
Weight: 3660 lb.
Note: GPA = General Purpose, Amphibious. Popularly known as 'Seep', 'Duckling', 'Waterbug', etc.

111

Truck, 2½-ton, 6×6, Amphibian, COE (GMC—Pilot) This was the pilot model, built on the 2½-ton 6×6 model AFKWX-353 chassis in 1941/42. Production model was on CCKW-353 chassis and designated DUKW-353 (D = 1942 model year, U = amphibian, K = all-wheel drive, W = dual rear axles).

Truck, 2½-ton, 6×6, Amphibian (GMC DUKW-353 'Duck') 6-cyl., 104 bhp, 5F1R×2, wb 164 (BC 44) in, 372×99×106(90) in, 15,080 lb (early models 14,360 and 14,710 lb). 11.00–18 tyres with central pressure control. Engine-driven propeller. Early models had vertical windshield (screen).

Truck, ½-ton, 4×4, Amphibian (Amph. Car Corp. XAC-2 'Aqua-Cheetah') Dodge 6-cyl., 105 bhp, 4F1R×1, wb 96 in, 196×90×64(50½) in, 5500 lb. Crew: 6 (in emergency 12, in water 21) plus driver. Engine at rear. 17-in propeller, vacuum-retractable. Speed in still water 6·5 mph.

Carrier, Cargo, Amphibian, M29C (Studebaker 'Water Weasel') 6-cyl., 65 bhp, 3F1R×2 (2-speed final drive at rear), 192×67¼×71(54) in, 4778 lb. 20-in tracks Ground pressure 1·93 lb/sq in. Water propulsion effected by the tracks. Speed in water approx. 4 mph. Controlled differential.

USA

GUN MOTOR CARRIAGES
WHEELED

Makes and Models (Ordnance No., armament, chassis): *Note:* most of the models listed were only pilot models or projects. T1 (twin 0·50 cal. mgs, Dodge ½-ton 4×4). T2, T2E1 (37-mm gun, Bantam ¼-ton 4×4) (see page 20). T8 (37-mm gun, Ford 4×4 'Swamp Buggy'). T13, T14 (37-mm gun, Willys/Ford 6×6, modified ¼-ton). T15 (3-in gun, Ford 6×6). T21 (37-mm gun, Dodge ¾-ton 4×4; became M6) (G121). T27 (75-mm gun, Studebaker 4×4). T33 (37-mm gun, Ford ¾-ton 4×4). T44 (57-mm gun, Ford ¾-ton 4×4). T45 (105-mm howitzer, Mack). T46 (75-mm howitzer, Studebaker 4×4). T55, T55E1 (3-in gun, Allied Machinery/Cook 8×8). T66 (75-mm gun, Chevrolet 6×6). T69 (four 0·50 cal. mgs, Ford 6×6). T74 (twin 0·50 cal. mgs, Dodge 6×6) (see page 50).

General Data: Vehicles in this class are also referred to as self-propelled guns, or just 'SPs'. The only model of the above listing of wheeled gun motor carriages to become standardised was the M6 (T21), 5380 of which were built. All the others remained in the experimental stage. Most, if not all, of those models where the armament was mounted on an existing truck chassis were engineered at Aberdeen Proving Ground. The T-numbers which are lacking in the above listing were allocated to similar experimental vehicles using half- or full-track chassis. The advantages of gun carriages on wheeled chassis generally were their high strategic mobility and relatively low cost. The main disadvantages were their poor tactical mobility and crew protection. With tracked chassis the reverse was true. The SPs to enter active service were almost invariably of the full- and half-track configuration. The standardised 37-mm model, M6 (Dodge), was declared obsolete with the standardisation of the light armoured car, M8, vehicles in stock being converted to ¾-ton Weapons Carriers.

Vehicle shown (typical): Carriage, Motor, 37-mm Gun, M6 (Dodge, ¾-ton, 4×4, T214-WC55).

Technical Data:
Engine: Dodge T214 6-cylinder, I-L-W-F, 230·2 cu in, 92 bhp @ 3200 rpm.
Transmission: 4F1R×1.
Brakes: hydraulic.
Tyres: 9.00–16.
Wheelbase: 98 in.
Overall l×w×h: 178×88×82 in.
Weight: 5850 lb.
Note: first designated M4(12/41-6/42).

Carriage, Motor, 57-mm Gun, T44 (Ford, ¾-ton, 4×4) 6-cyl., 90 bhp, 4F1R×1, wb 92 in. Basically the chassis of the experimental ¾-ton model GAJ cargo truck. T33 was similar but mounted 37-mm gun. Ford T8 'Swamp Buggy' had engine and driver's seat at rear, 37-mm gun at front.

Carriage, Motor, 75-mm Gun, T27 (Studebaker, 1½-ton, 4×4) Hercules 6-cyl., 93 bhp, 5F1R×2, wb 124 in, 185×95 in, 6890 lb. This model had duplicate driving controls and steered on all four wheels. Single driver's seat with swivelling back rest. Engine at rear.

Carriage, Motor, 37-mm Gun, T14 (Willys MT-TUG) 4-cyl., 54 bhp, 3F1R×2, wb 100 (BC 36) in, 227×60×40½ in, 2845 lb. 6×6 conversion of standard 'Jeep'. T13 was basically similar but chassis used was Ford GP (converted to 6×6). Only six Willys 6×6 units were made, in 1942.

Carriage, Motor, 3-in Gun, T55 (Allied Machinery/Cook 8×8) Twin Cadillac V-8-cyl., 130 bhp each, Hydramatic 4F1R×4, 307½×111½×105½ in, 38,300 lb (gross). Cook Bros tandem axle bogies with chain drive. Turntable steering. T55E1 was lower and lighter. Known as 'Cook's Cozy Cabin'.

USA

SCOUT CARS and ARMOURED CARS

Makes and Models: *Scout Cars:* Dodge T-211, 1941 (pilot, 4×4). Dodge T-223, 1942 (pilot, 6×6). White M3A1, M3A1E1, E2, E3, 1937–43 (G67, 4×4). Willys/Smart T25, T25E1, E2, E3, 1941–42 (pilots, 4×4). Willys T24, 1942 (pilot, 6×6).
Armoured Cars: Chevrolet M6 (T17E1), T17E2, 1942–44 ('Staghound', 4×4, G122). Chevrolet T19, T19E1, 1942 (pilots, 6×6). Chevrolet M38 (T28), 1942–43 (later produced by Ford, 6×6). Dodge T23, 1942 (pilot, 6×6). Dodge T23E1, 1942 (pilot, 4×4). Ford M5 (T17), 1942 (pilot, 6×6). Ford T22, 1942 (pilot, 6×6). Ford T22E1, 1942 (pilot, 4×4). Ford M8 (T22E2), 1942–45 ('Greyhound', 6×6, G136). Ford M20 (T26), 1942–45 (Arm. Utility, 6×6, G136). GMC T18, T18E2, 1942 ('Boarhound', pilots, 8×8). Studebaker T21, 1942 (pilot, 6×6). Studebaker T27, 1943 (pilot, 8×8). Trackless Tank Corp. and Reo T13, 1940–42 (pilots, 8×6). *Note:* T22, T22E1, T23 and T23E1 were originally known as 37-mm gun motor carriages, with the same ordnance numbers. T21 was originally known as 37-mm gun motor carriage, T43.

General Data: Of the above models only a few reached series production for the US forces (Scout Car M3A1, Armoured Cars M8 and M20). The others either remained in the experimental stage or were produced for other nations. The Scout Car M3A1 was also used by the British, Canadian and Russian armies; over 20,000 of them were produced and some were fitted with a diesel engine (Buda-Lanova and Hercules DJXD). In the British and Canadian armies the M3A1 was classed as a Truck, 15-cwt, 4×4, and used for various roles (APC, Command, Ambulance, etc.). The Chevrolet T28 was a very advanced model, and despite US Army prejudice which favoured full-track combat vehicles, it was assigned an ordnance number, M38, and approved for production. Owing to the volume of war work at Chevrolet the contract was given to Ford, but few were made. The names 'Deerhound', 'Staghound', 'Boarhound' and 'Wolfhound' were given by the British to the T17, T17E1/E2, T18E2 and T28 respectively.

Vehicle shown (typical): Car, Scout, 4×4, M3A1 (White)

Technical Data:
Engine: Hercules JXD 6-cylinder, I-L-W-F, 320 cu in, 110 bhp @ 2800 rpm.
Transmission: 4F1R×2.
Brakes: hydraulic (hydrovac).
Tyres: 8.25–20 (later 9.00–20).
Wheelbase: 131 in.
Overall l×w×h: 221¼×77×83 in.
Weight: 8900 lb.
Note: developed from T7, M2(A1) and M3. 20,918 produced until 1944.

Car, Scout, 4×4, T25 pilot (Willys MA/Smart) 4 cyl., 54 bhp, 3F1R×2, wb 80 in, 128½×61×62 in, gross weight 3255 lb. Based on early Willys MA ¼-ton 4×4 'Jeep'. Produced in 1941. Armour plating only applied to front end and doors.

Car, Scout, 4×4, T25E3 (Willys MB/Smart) 4-cyl., 54 bhp, 3F1R×2, wb 80 in, approx. 144×65×59 in. Latest development in the 'Jeep'-based range of experimental scout cars. Excessive weight caused the end of these projects in 1943.

Car, Scout, 6×6, T24 (Willys MT-TUG) 4-cyl., 54 bhp, 3F1R×2, wb 100 (BC 36) in, 167¾×69½×59 in, gross weight 5450 lb. Same chassis as self-propelled gun carriage T14. Produced in 1942 on modified 'Jeep' chassis, fitted with dual-drive tandem rear bogie. Crew three.

Tender, Observation Post, Armored, T2 (Ford) 6-cyl., 90 bhp, 4F1R, wb 92 in, 153½×80×72 in, 5800 lb. Two produced in 1941 on Ford ¾-ton 4×4 truck/SP gun chassis (T33, T44). Other vehicle had completely open top and conventional (folding) windscreen.

Car, Armored, Utility, M20 (T26) (Ford GAK-M20) Hercules JXD 6-cyl., 110 bhp, 4F1R×2, wb 104 (BC 48) in, 197×100×91 in, gross weight 15,650 lb. High-mobility armoured personnel and cargo carrier and field commander's car. Based on Car, Armored, M8. One 0·50 cal. mg. 3791 made.

Car, Armored, Light, M8 (T22E2) (Ford GAK-M8) Hercules JXD 6-cyl., 110 bhp, 4F1R×2, wb 104 (BC 48) in, 197×100×88½ in, gross weight 17,200 lb. Armed with 37-mm gun and 0·30 cal. mg. Rear-engined. Known by British forces as 'Greyhound'. Total of Ford-built armoured cars was 12,564.

Car, Armored, Light, T21 (Studebaker) Hercules JXD 6-cyl., 112 bhp, 5F1R×2, wb 103½ (BC 47) in, 194×90¾×73¼ in, 14,540 lb. Formerly 37-mm Gun Motor Carriage, T43. Rather similar in design to Ford-built M8. Engine at rear. Top speed 50 mph, range of action 290 miles. No series production.

Car, Armored, Light, M38 (T28) (Chevrolet, Ford) Cadillac V-8-cyl., 110 bhp, Hydramatic 4F1R×2, wb 118 in, 219½×96×76¾ in, 12,700 lb. 12.50–20 tyres. Max. speed 59 mph, range of action 300–325 miles. Crew four. Chevrolet design. Standardised but no series production.

Car, Armored, M6 (T17E1) (Chevrolet) Twin GMC 270 6-cyl., 2×104 bhp, twin Hydramatic trans., wb 120 in, 212×106×92½ in, gross weight 29,100 lb. 14.00–20 tyres. Used mainly by British ('Staghound' Mk I). Armament: one 37-mm gun, three 0·30 cal. mgs, one 0·30 cal. sub-mg.

Car, Armored, T17E2 (Chevrolet) Technical data as M6 (T17E1) except gross weight 26,000 lb. Twin 0·50 cal. AA mgs and one 0·45 cal. sub-mg. The British also equipped Staghounds with the 3-in howitzer and later with a Crusader tank type turret with 75-mm gun. Total production T17E1 and E2: 3844.

Car, Armored, T13 (Reo) Guiberson diesel 9-cyl. radial, 250 bhp, 5F1R, approx. 205×100×84 in, 32,000 lb. Chain drive to six rear wheels. Two front wheels steer. Originally (1940/41) a commercial venture by the Trackless Tank Corp. using riveted hull, later (1942) produced by Reo as T13.

Car, Armored, Heavy, T18E2 (GMC) Twin GMC 6-cyl., 2×150 bhp, wb 197 in, 246×121×103¼ in, gross weight 53,000 lb. 14.00–20 tyres. 57-mm 6-pdr gun. Like South African Marmon-Herrington VI designed for use by the British Army in North Africa. Known as 'Boarhound'. Only few produced.

USA

TRACTORS and TRUCKS
HALF-TRACK

Makes and Models: *Tractors:* Allis-Chalmers Snow Tractor M7 (T26E4), 1943–44 (G194, 195). Willys Snow Tractors T28, T29, T29E1, 1943 (pilots, conversions of MB). *Trucks:* Autocar T17, 1943 (pilot). Diamond T T16, 1943 (pilots). Linn (pilot). Mack T19, 1942–43 (pilots). White T17, 1943 (pilot).

General Data: Purpose: tractors—provision of over-snow transport; trucks—provision of improved mobility over rough terrain. The over-snow tractors were produced mainly for use by USAAF ground search and rescue units in the far North and were often painted orange for easy spotting against the snow from the air. Experiments were first carried out with relatively simple half-track conversions of the standard Willys MB 'Jeep'. Standardised, however, was a special Allis-Chalmers tractor, fitted with tracked rear bogies and powered by a Willys 'Jeep' engine. Front wheels could be fitted, with or without skis, or could be carried on special brackets on the sides, in which case only the skis were used under the front axle. Unit ground pressure (with skis, fully loaded at 2935-lb GVW) was 1·48 lb/sq in. The crew of two sat one behind the other. Maximum speed and cruising range were 41·3 mph and 200 miles respectively. The much heavier armoured half-track trucks were produced in 1942–43. They could carry a cargo load of just over 2½ tons or a crew of 14. Their development was discontinued in 1944. The half-track truck T17, pilot models of which were produced by Autocar and White, was powered by a horizontally opposed 12-cylinder 'Pancake' engine which had been introduced by White in 1934 for use in White heavy-duty commercial trucks, following more than two years of use in huge 100-passenger White City Coaches in large city traction fleets. There was also a half-track radio carrier, T17, which consisted of a standard-type half-track chassis, M3, with an SCR299 radio set in an unarmoured truck body.

Vehicle shown (typical): Truck, Armored, Half-Track, T16 (Diamond T)
Technical Data:
Engine: Hercules RXLD 6-cylinder, I-L-W-F, 529·2 cu in, 174 bhp @ 2600 rpm.
Transmission: Spicer auto. torque converter.
Brakes: hydraulic.
Tyres: 9.00–20.
Tracks: endless band, 12-in.
Wheelbase: 120⅜ in.
Bogie length: 112¾ in.
Overall l×w×h: 247⅜×94×94 in.
Weight: 24,510 lb (combat loaded: 30,010 lb).
Note: Controlled diff. steering. Crew 14. Armour thickness ¼ in. Two pilot models built in 1943, tested until mid-1944.

Truck, Armored, Half-Track, T17 (Autocar, White) White 24 12-cyl., 210 bhp, 4F1R×1, 281¼×102½×90¼ in, 25,261 lb. 10.00–20 tyres. Two pilots built in 1943, one by Autocar, one by White. 72-in wb Kégresse-type bogies with six pairs of rollers.

Truck, Armored, Half-Track, T19 (Mack) Continental R6572 6-cyl., 215 bhp, 4F1R×1, wb 114½, 246⅝×101×86 in, 23,292 lb. Bogie wb 79 in. 9.00–20 tyres. Engine at rear. Controlled differential steering. Air brakes. Two pilots built in 1942–43. Medium tank-type bogie components.

Truck, 2½-ton, Half-Track, Cargo (Autocar) This half-track load carrier was developed by the Autocar Co. of Ardmore, Pa, for Lend-Lease shipment to Russia during the latter part of the war. Basically similar to armoured personnel carriers M2 and M3. Radio carrier, T17, was probably similar.

Truck, Chassis, Half-Track (Linn) This chassis was developed for the US Army by the Linn Mfg Co. During the 1930s Linn had built various types of half-tracks. This model differed in having front-wheel drive, dual front tyres, overlapping bogie wheels, etc. It was about 37 ft long.

Snow Tractor, Half-Track, T28 (Willys MB, modified) 4-cyl., 63 bhp, 3F1R, $160 \times 72\frac{3}{4} \times 92$ in, 3000 lb approx. Tyres: four 4.00–12, two 4.75–19, two 7.50–16 (front). Tracks: Bombardier or Chase endless-band type with metal crosspieces. Experimental 'Penguin Jeep'. Found derelict in 1971 and now preserved.

Snow Tractor, Half-Track, T29E1 (Willys MB, modified) 4-cyl., 63 bhp, 3F1R, $158 \times 72\frac{3}{4} \times 72$ (windshield) in, 2940 lb. T29 and T29E1 were much like T28 but with small pickup body. 'Jeep' engine was uprated (63 bhp @ 3900 rpm, torque 108 @ 1800 rpm); transfer case was not used. Brakes were mechanical.

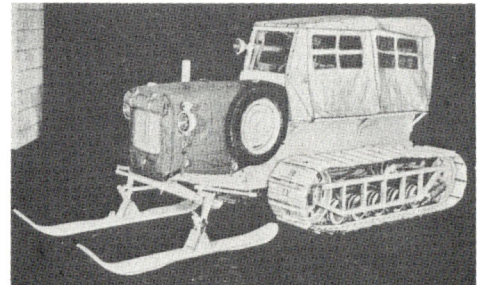

Snow Tractor, Half-Track, M7 (T26E4) (Allis-Chalmers) Willys MB 4-cyl., 63 bhp, $3F1R \times 1$, $136 \times 63 \times 64(43)$ in, 2500 lb approx. (gross 3050 lb). Ground pressure 0·75 lb/sq. in. Unsprung walking-beam type rear bogie suspension. A special Allis-Chalmers 1-ton snow trailer, M19, could be towed.

Snow Tractor, Half-Track, M7 (Allis-Chalmers) Same as shown on left but with cold-weather equipment installed. For use in deep snow the front wheels could be replaced with skis, as shown. When not in use the skis were carried at sides, serving as mudguards. See also General Data.

USA

COMBAT VEHICLES
HALF-TRACK

Makes and Models: (*Note:* space does not permit complete listing; only standardised types are included. Makes are abbreviated as follows: AUC = Autocar, DT = Diamond T, IHC = International Harvester, WI = White; SNL No. is G102, except for IHC which is G147.) Car, Half-Track, M2, M2A1 (AUC, WI), M9, M9A1 (IHC). Carrier, Personnel, Half-Track, M3, M3A1 (AUC, DT, WI), M5, M5A1 (IHC). Carrier, 81-mm Mortar, Half-Track, M4, M4A1 (WI), M21 (WI). Carriage, Motor, 75-mm Gun, M3, M3A1 (AUC). Carriage, Motor, Multiple Gun, M13 (WI), M14 (IHC), M16 (WI), M17 (IHC). Carriage, Motor, Combination Gun, M15 (AUC).

General Data: Purpose: cars and carriers (APCs—Armoured Personnel Carriers)—to transport cargo and personnel in combat zone: carriages (SPs—Self-propelled weapons)—to provide mobility for mounted weapons. Half-track vehicles were widely used by the US forces in most theatres of war. Many, especially types of IHC manufacture, were supplied to America's allies under Lend-Lease (UK, Canada, USSR, etc.). Some countries, notably Israel, are still using them. Altogether 41,170 of all types were produced, mainly during 1941–44. Many were modified or re-manufactured to improved specifications or to fulfil other roles. The half-track car, M2 (10-seater APC), was originally designated T7 and combined the main features of the 4×4 scout car, M3, and the half-track truck, M2 (Marmon-Herrington, 1937). The half-track carrier, M3 (13-seater APC), was originally known as T14 (1939/40). These two standardised models were produced by Autocar, Diamond T and White. International Harvester produced similar basic models (Carrier, M5, and Car, M9) and derivations but using their own engine and running gear. To improve cross-country performance, particularly on wet clay, half-track vehicles could be equipped with special track chains, which were simply longer editions of ordinary single-tyre snow chains, long enough to fit right round the tracks. Special 'overall chains' were also developed to convert 6×6 trucks into 'half-tracks'. Altogether, including prototypes, there were about 70 half-track variations.

Vehicle shown (typical): Carrier, Personnel, Half-Track, M3 (Autocar, Diamond T, White) (later renamed: Carrier, Personnel, Half Tracked: M3)

Technical Data:
Engine: White 160 AX 6-cyl., I-L-W-F, 386 cu in, 147 bhp @ 3000 rpm.
Transmission: 4F1R×2.
Brakes: hydraulic (hydrovac) (with controller for electric trailer brakes).
Tyres: 8.25–20 (front).
Tracks: rubber, endless band, 12¾ in.
Wheelbase: 135½ in.
Overall l×w×h: 242⅝×77½×89 in.
Weight: 14,800 lb.
Note: Crew 13. M3A1 had armoured mg ring mount. If fitted with winch, vehicle is 7 in longer, 500 lb heavier.

Car, Half-Track, M2A1 (Autocar, White) Data as given for APC M3, except: dimensions $235 \times 77\frac{1}{2} \times 100$ in, weight 14,600 lb. Crew 10. Developed from T7 (1939–40). M2 similar but without armoured ring mount (height 89 in, weight 14,400 lb). No rear door.

Carrier, 81-mm Mortar, Half-Track, M4A1 (White) Data as given for APC M3, except: dimensions $244 \times 87\frac{1}{2} \times 89$ (hull) in, weight 15,320 lb. M4 did not have the rear hull extensions. Both were based on car, M2, and originally known as T19. Mg track rail round upper edge of hull. Crew 6.

Carrier, Personnel, Half-Track, M5 (International) IHC 6-cyl, 143 bhp, 4F1R×2, wb $135\frac{1}{2}$ in, $242 \times 87 \times 91$ in, 15,400 lb. 9.00–20 tyres. Equivalent to std APC M3, but IHC engine and axles. Welded hull. 4296 supplied to UK in 1943. Crew 13. M5A1 had ring mount. Originally M3E2.

Car, Half-Track, M9A1 (International) Technically and outwardly similar to M5A1. Equivalent to std APC M2A1 (with crew of 10), but fitted with rear door. Note flat-section mudguards. 1419 supplied to UK, where they were converted to M5(A1) spec., seating 13. M9 similar but w/o ring mount. Originally M2E5.

Carriage, Motor, Multiple Gun, M13 (White) Data as given for APC M3, except: dimensions 250 × 77½ × 92 in, gross weight 19,800 lb. Based on 75-mm gun motor carriage M3. Originally known as T1E4. Two 0·50 cal. AA mgs in turret. Crew five. Note hinged top sections of hull sides.

Carriage, Motor, Multiple Gun, M14 (International) Similar to M13 but based on APC M5. Dimensions 249 × 85¾ × 74½ (hull) in, 16,200 lb. No rear door. 1600 supplied to British, who removed the guns and converted them into trucks, APCs (wooden seats, crew 13), and command vehicles.

Carriage. Motor, 75-mm Gun, T12 (Autocar) Data as given for APC M3 (on which this SP was based), except: dimensions 245 × 77½ × 94 in, weight 17,450 lb. First SP adopted by US Army in World War II. Developed at Aberdeen Proving Ground, 1941. Standardised as M3 and first used in Pacific area.

Carriage, Motor, 75-mm Howitzer, T30 (White) Data as given for APC M3 (on which this SP was based), except: dimensions 236¾ × 77½ × 96¾ in, gross weight 19,500 lb. Pack howitzer, with or without shields. Although not standardised, 500 were built. Note windscreen armour, hinged at bottom.

Carriage, Motor, Multiple Gun, M15A1 (Autocar) Data as given for APC M3 (on which this SP was based), except: dimensions 243½ × 88 × 92¾ in, 17,600 lb. One 37-mm AA gun and two 0·50 cal. mgs, all synchronised, in rotating turret. Developed from T28 (M15) and T28E1.

Car, Half-Track, T16 (Autocar) Basically a half-track car M2 but experimentally fitted with overhead armour, consisting of hinged panels, and longer track bogies with larger rollers and double support rollers. Produced in 1942 but not progressed beyond prototype stage.

Car, Armored, Half-Track (Marmon-Herrington DHT-5) This experimental model was built in 1941/42 by one of the pioneers of the US half-tracks. It featured a turret-mounted 37-mm gun and the hull resembled that of the T11 armoured cars built in 1933–36 by FWD and M.-H.

Carriage, Motor, 40-mm Gun, T1 (Mack) Self-propelled gun mount produced by Mack, based on the chassis of the experimental APC T3. Rear-mounted engine. 40-mm Bofors gun and Kerrison director.

USA

TRACTORS
FULL-TRACK

Makes and Models: (*Note:* mfr abbreviations: AC = Allis-Chalmers, CAT = Caterpillar, CL = Cleveland (Cletrac), IHC = International Harvester). Tractor, Crane, 1-ton, M1 (IHC T6, G108), 2-ton, M3 (CAT D6, G69; IHC TD14, G132), 2-ton, M5 (IHC T9, G99), 6-ton, M4 (CAT D7, G126). Tractor, Light (CAT D4, G151; CL 20C, G36; IHC TD9, G99), Light, M2 (CAT 20, G7; IHC T6, G113). Tractor, Medium, M1 (AC HD7W, G125; CAT D6/RD6, G69; CAT 30, G21; CAT 35, G47; CL 35, G49; CL BD; IHC TD14, G132). Tractor, Heavy, M1 (AC L, G49; AC HD10W, G98; CAT D7, G126; CAT RD7, G89; IHC TD18, G101). Tractor, Amphibian (Roebling, G156). Tractor, High-Speed, 7-ton, M2* (CL MG1, MG2, G96), 13-ton, M5,** M5A1/A2/A3/A4 (IHC, G162), 18-ton, M4(C), M4A1(C) (AC, G150), 38-ton, M6 (AC, G184). Tractor, Snow, T30, T30E1, T37 (Crosley), T36 (Iron Fireman), T38 (Lima Locomotive).

General Data: Except for the high-speed types, all the above tractors were basically commercial types. Most medium M1 and heavy M1 types had a front-mounted winch. The high-speed types were purely military models, developed for tactical use. The 7-ton, M2, was used by the Air Forces, the heavier types, M4, M5, M6, by the army for artillery towing. The prototype for the 13-ton high-speed tractor was built on the chassis of the 7-ton M2 model (Cletrac) but for the production models use was made of light tank-type bogies (T21, M5). A cargo carrier (T13) was also built on this chassis, but only as a pilot model. The experimental prototypes for the heavier M4 and M6 tractors were the T9E1 and T23 respectively. Certain types of full-track tractors were used with bulldozer or angledozer blades and used by the Corps of Engineers for towing, grading and earthmoving. Marmon Herrington produced commercially a series of light (TBS-30) and medium (TBS-45) 'track-laying' tractors for the Netherlands East Indies in 1941. These were shipped to Java.

* Formerly known as: Tractor, Light, M2.
** Formerly known as: Tractor, Medium, M5.

Vehicle shown (typical): Tractor, Medium, M1 (Allis-Chalmers HD7W) (later renamed: Tractor, Crawler Type, Diesel Engine Driven, 46 to 60 DBHP, w/Artillery Towing Attachment, Winch, 1-Drum, Front Mounted).

Technical Data:
Engine: GMC 3-71 RC14 2-stroke diesel, 3-cylinder, I-I-W-F, 213 cu in, 54 drawbar hp.
Transmission: 4F1R.
Brakes: mechanical, controlled differential.
Tracks: steel-link, 18-in.
Wheelbase: 67 in.
Overall l×w×h: 145×81×86(70½) in.
Weight: 17,500 gross.
Note: winch capacity 25,000 lb.

Tractor, Light, M2 (International TD9) 4-cyl. diesel, 38 drawbar hp, 5F1R, 126½×76×79(66½) in, 11,325 lb (gross). Tractor shown was in service with USMC, 1941. Also used by Army, for artillery towing, etc. Later renamed: Tractor, Crawler Type, Diesel Engine Driven, 36 to 45 DBHP.

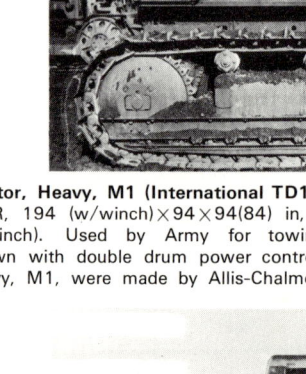

Tractor, Heavy, M1 (International TD18) 4-cyl. diesel, 80 bhp, 6F2R, 194 (w/winch)×94×94(84) in, 30,275 lb (fighting wt, w/winch). Used by Army for towing 155-mm gun, etc. Shown with double drum power control unit. Other Tractors, Heavy, M1, were made by Allis-Chalmers and Caterpillar.

Snow Tractor, Full-Track (Crosley 'Mechanical Dog') One of several very light tractors developed during 1943 and intended to replace the husky at hauling sleds. Model shown had 4-cyl. Crosley engine. The driver could ride it, as shown, or walk near the front and steer with the vertical lever.

Snow Tractor, Full-Track, T36 (Iron Fireman Mfg Co.) Dodge T214 6-cyl., 99 bhp, 4F1R, 172×69×82 in, 7500 lb (gross). Intended for use in Alaska and on Alcan Highway but instead served with Air Forces for rescuing fliers downed in arctic territories. Similar tractor produced with 6-cyl. Ford engine.

127

Tractor, High Speed, 7-ton, M2 (Cletrac MG1, MG2) Hercules 6-cyl., 137 bhp, 4F1R, 170×70×68 (min.) in, 14,700 lb. Introduced in 1941. Crew three. Aux. generator driven off front pto. Rear-mounted compressor driven by main engine. Also used by RAF in Britain for aircraft towing.

Tractor, High Speed, 13-ton, M5 (International) Continental 6-cyl., 207 bhp, 4F1R×2, wb 108½ in, 191×100×104 in, fighting weight 28,572 lb. Crew 9, front entrances. M5A1 had hardtop cab like M4, crew 11. M5A2 had horizontal volute spring suspension. After 1945 developed into M5A3 and A4.

Tractor, High Speed, 18-ton, M4 (Allis-Chalmers) Waukesha 6-cyl., 190 bhp, 3F1R×3 with torque converter, wb 124 in, 209¾×97×107 in, fighting weight 31,400 lb. M4C and M4A1(C) differed mainly in stowage and type of ammunition carried. Fitted with shell handling hoist. Crew 11 or 9.

Tractor, High Speed, 38-ton, M6 (Allis-Chalmers) Two Waukesha 6-cyl., 190 bhp each, 2F1R×2 with torque converter, wb 176⅜ in, 258×120½×104 in, fighting weight 76,000 lb. Crew 10. 30-ton winch. Towed load 50,000 lb gross. Electric and air brake connections for trailers. Various types of track.

USA

COMBAT VEHICLES
FULL-TRACK

Makes and Models: (*Note:* the list below is intended only to provide a brief summary. There were many variants and developments based on these models.)
Cargo Carriers: T14/M30 (Pressed Steel, G158). T15/M28 (Studebaker, G154). T16 (Ford 'Universal', G166). T24/M29 (Studebaker, G179). M39 (Buick, G163).
Light Tanks: M3, M3A1, M3A3 (Am. Car & Fdry, G103). M5, M5A1 (Cadillac, Am. Car & Fdry, Massey-Harris, G103). T9E1/M22 (Marmon-Herrington, G148). M24 (Cadillac, Massey-Harris, G200).
Medium Tanks: M3, M3A1-5 (6 Mfrs, G104). M4, M4A1-6 (10 mfrs, G104). M26 (Fisher, Chrysler, G226) (shown on page 101).
Gun Motor Carriages: M7 (105-mm How. on M4 medium tank chassis, G128). M8 (75-mm Gun on M5, light, G127). M10 and M10A1 (3-in Gun on M4A2 and A3, medium, G130 and G170). M12 (155-mm Gun on M3, medium, G158). M18 (76-mm Gun, Buick 'Hellcat', G163). M36 (90-mm Gun, on SP M10A1, G210).
Landing Vehicles: LVT1, 2, 3, 4 (Food Machinery Corp.; certain models also by other mfrs; G156, 167, 209). LVT(A)1, 2, 3, 4, 5 (armoured versions; G168, 214).

General Data: During World War II the United States produced 88,410 tanks. The best-known American tank of this period was the M4 series medium ('Sherman'). It succeeded the M3 series ('Lee' and 'Grant') early in 1942 and first saw active service with the British Army in North Africa. About half the total US tank production during 1940–45 consisted of the M4 and its many variants. The light tanks, M3 and M5 series, were built in large numbers during the early years of the war, after which production dropped off in favour of the medium types. Several other vehicle types were based on tank chassis, notably gun carriages (SPs), open-top carriers and heavy recovery vehicles. The tracked landing vehicles ('Alligator', 'Water Buffalo') were used mainly by the USMC and the British (Walcheren, Rhine, Northern Italy). 18,620 of various open-top and turreted versions were produced during 1941–45. Some heavy tanks were developed (M6, M26) but relatively few were produced.

Vehicle shown (typical): Vehicle, Armored, Utility, M39 (Buick)

Technical Data:
Engine: Continental R975C4 9-cylinder, R-I-A-R, 973 cu in, 400 bhp @ 2400 rpm.
Transmission: 3F1R $\times$ 1.
Brakes: mechanical, controlled differential.
Track width: 12 in.
Overall l $\times$ w $\times$ h: 214 $\times$ 113 $\times$ 80½ (77½) in.
Fighting weight: 35,500 lb.
Note: based on Buick 'Hellcat' tank destroyer (M18) chassis; used for transport of cargo and personnel and to tow 3-in gun, M6.

129

Carrier, Light Cargo, M28 (Studebaker) 6-cyl., 65 bhp, 3F1R×2, 132×60×67 in, 3390 lb. Predecessor of M29. Used for transport over snow. Originally known as T15. Engine back-to-front at rear. 18-in wide steel/rubber tracks. Some units fitted with cargo boxes at rear (as shown).

Carrier, Cargo, M29 (Studebaker 'Weasel') 6-cyl., 65 bhp, 3F1R×2, 126×66×71(54) in, fighting weight 4541 lb. Early models had 15-in tracks, later 20-in. Studebaker 'Champion' engine, alongside driver. Also used by British (Truck, 10-cwt, Tracked, GS). For amphibious version see page 111.

Carrier, Universal, T16 (Ford) V-8-cyl., 100 bhp, 4F1R, wb 78 in, 152½×83×61 in, 7700 lb. Produced for British. Differed from British version in many respects (welded, longer, hull; extra bogie wheels; lever steering/braking of different design). Ford US produced 13,893 'Universals'.

Landing Vehicle, Tracked, Mk 3 LVT(3) Two Cadillac V8 engines and transmissions, mounted in the sides. All other LVT except the very early models had a Continental 7-cyl. radial engine. Like the LVT(4) this model had a rear-loading ramp. 6000-lb payload. Integral and 'pin-up' armour.

Tank, Light, M3A1 (American Car & Foundry Co.) Continental 7-cyl. radial, 242 bhp, or Guiberson Buda 9-cyl. radial diesel, 5F1R, 178×88×90 in, fighting weight 28,515 lb. First used in Libya, Nov., 1941. Popularly known as 'Honey' by British, 'Stuart' in US Army. Developed into Cadillac-built M5.

Tank, Light, M22 (Marmon-Herrington) Lycoming 6-cyl., 4F1R, 155×87¾×68½ in, fighting weight 17,000 lb. Air-borne tank, used by British (Rhine crossing). Originally T9E1, later known as 'Locust'. Introduced in 1943. 37-mm gun, one mg, one sub-mg. M.-H. built total of 1070 tanks.

Tank, Light, M24 (Cadillac, Massey-Harris) Two Cadillac V8, 110 bhp each, 4F (twin Hydramatic) plus 2F1R transfer case, 198×116×109 in, fighting weight 40,500 lb. Introduced in 1943, later also used in Korea. Specially designed 75-mm tank gun. Superseded M3 and M5 series. Known as 'Chaffee'.

Carriage, Motor, 155-mm Gun, M12 (Pressed Steel Car Co.) Continental 9-cyl. radial, 400 bhp, 5F1R, 265×105×106(96) in, 58,000 lb gross. Modification of medium tank, M3, utilising obsolete 155-mm gun, M1917A1 of World War I vintage. Introduced in 1942 and used in Italy, France and Germany.

Tank, Medium, M3A5 (Baldwin Locomotive) Twin GM 2-stroke diesel, 210 bhp each, 5F1R, 222×107×122 in, fighting weight 64,000 lb. Known by British as 'Lee'. 75-mm gun with counter balance on muzzle and Westinghouse stabiliser. Riveted hull. M3 series medium tanks were also supplied to Russia.

Tank, Medium, M4A1 (various mfrs) Continental 9-cyl. radial, 400 bhp, 5F1R, 230×102×112 in, fighting weight 67,300 lb. One of the series of tanks known as 'Sherman'. Cast hull. 75-mm gun and three mgs. Produced in large numbers during 1942 and 1943. (Early M4 with welded hull shown on page 101.)

Tank, Medium, M4 (105-mm Howitzer) (Chrysler) Continental 9-cyl. radial, 400 bhp, 5F1R, 244×118×133 in, fighting weight 72,335 lb. Shown fitted with horizontal volute spring suspension. British designation Sherman IBY. 2539 produced by Chrysler in Detroit Tank Arsenal, Oct., 1944, to June, 1945.

Carriage, Motor, 90-mm Gun, M36 (Fisher) Ford V-8-cyl., 450 bhp, 5F1R, 235×120×125 in, fighting weight 61,000 lb. Produced by Fisher Tank Div. of GM. Originally known as T71. Like British 'Achilles' (17-pdr gun; see page 134) it was developed from gun motor carriage M10(A1).

GREAT BRITAIN

During the inter-war period army motorisation in Britain—as indeed in most other countries—had been slow. Numerous prototypes had been produced and evaluated, but few reached the production stage and consequently only relatively small numbers reached the troops. One result was that some private hiring firms who had built up impressive fleets of suitable vehicles for use in army manœuvres found business booming right up until World War II. The most common types held by the British Army in 1939 were the 15-cwt 4×2, the 30-cwt 4×2 and 6×4 and the 3-ton 4×2 and 6×4.

At the time war broke out in September, 1939, the War Department held a total of approximately 85,000 motor vehicles, over 26,000 of which were impressed. This figure included some 21,500 motorcycles (almost one-third of which were impressed), and slightly under 7000 trailers. A large proportion of these vehicles was shipped out to France with the BEF and only just over 5000 came back. The remainder was left behind in the evacuation of France (Dunkirk) in May/June, 1940. Meanwhile procurement of additional vehicles had started on a large scale. The British motor industry worked at full capacity (albeit mainly on production of pre-war types, continuation of which was the logical thing to do under the circumstances). In addition, large quantities of vehicles were ordered and obtained from the USA and Canada. The eventual result was an enormous variety of vehicle types. This, however, was unavoidable. The military operations of World War II were so widely dispersed and of such enormous overall size that vehicle production proper had priority over development of new tactical types and standardisation. By VE Day the total number of military vehicles (excluding AFVs) was in the region of one and a quarter million.

The above also applied, generally, to AFVs. Britain had very few tanks and armoured cars at the outbreak of the war. Production was immediately stepped up and several new types were designed and put into production with little or no time for field trials. A typical example was the Churchill tank.

'We had in the hands of the troops in the UK fewer than a hundred tanks. These, and those under production at the time, were of a type which had proved in battle in France to be too weak to stand up to the German tank guns. . . .' These grave words were spoken, in Parliament, by the late Sir Winston Churchill, shortly after the Dunkirk retreat, at a time when Britain was daily expecting invasion.

Vauxhall Motors was asked to design and produce an entirely new tank engine, developing 350 bhp, for use in a new tank, the A20. The biggest engine they were making at the time was the 72 bhp Bedford truck engine, so it was a remarkable achievement that the new tank engine was running on the test bed in 89 days. By the time it was ready for production, the A20 tank had been scrapped (after five prototypes had been built) and an improved type was to be developed. This was also done by Vauxhall, starting in July, 1940. The first tank, designated A22, was to be in quantity production within a year. In fact, the first pilot model was running in December of the same year and the first batch of productio.. models was delivered six months later. This tank became known as the 'Churchill'. The Churchill appeared in many variants and, with the help of some ten sub-contractors, 5640 were produced.

Vehicle Types

In the British forces the various categories of vehicles were divided as follows:

(a) Army 'A', 'B', 'C' and 'RASC' vehicles (WO, War Office).
(b) Royal Air Force (RAF) vehicles (AM, Air Ministry).
(c) Royal Navy (RN) vehicles (Admiralty).

'A' Vehicles This category comprised all armoured fighting vehicles (AFVs), wheeled and tracked, i.e. scout cars, armoured cars, carriers (Universal, Bren, etc.), LVTs, tanks and their derivatives (SPs, APCs, ARVs, BARVs, etc.). Some vehicles which, although armoured, were excepted: light reconnaissance cars, US-built half-tracks, and armoured trucks.

'B' Vehicles Under this heading were grouped: motorcycles, cars (incl. utilities and light recce cars), ambulances,

amphibians (except LVTs), trucks of all types (wheeled and half-tracked), tank transporters, artillery and AA tractors, breakdowns, gun portees, wheeled SP guns, trailers and motorcycles. *Note:* trucks with over 15-cwt payload capacity were termed 'lorries'.

'C' Vehicles Special-purpose vehicles as used mainly by the Royal Engineers, such as mobile cranes (except those mounted on truck chassis), earthmoving equipment (excavators, graders, etc.), dumpers, industrial-type tractors, etc.). This classification was introduced later than the 'A' and 'B' classifications. 'C' vehicles becoming more numerous as time went on.

RASC Vehicles Before the war the Royal Army Service Corps was the largest user of motor transport vehicles in the British Army. The RASC was responsible for all 'second line' MT vehicles, ambulances, fire engines, etc., operated by corps units. These vehicles were termed 'RASC vehicles'. The RAOC (Royal Army Ordnance Corps) was responsible for 'first line' vehicles (including AFVs) for operational units. The enormous growth of army mechanisation during the early war years called for a reorganisation of vehicle supply, maintenance and repairs. From 1942 all 'RASC' vehicles were included in the 'B' vehicle classification. The MoS (Ministry of Supply) was formed as a provision agent for all vehicle types, the REME (Royal Electrical and Mechanical Engineers) for all the army's electrical and mechanical maintenance and repair activities. The RAOC became the exclusive source of vehicle supply, with headquarters and main depot (Central Ordnance Depot) at Chilwell, Notts.

RAF Vehicles The Air Ministry was responsible for the supply and development of vehicles for the Royal Air Force until early in 1941, after which there was close co-operation with the War Office. Before 1941 the RAF had its own distinctive vehicle types, for example the Albion AM463 and certain types of Crossley, Dennis and Ford. In fact, during the late 1930s, a large proportion of the total output of Albion and Crossley was allocated to the Air Ministry. After the outbreak of war the RAF acquired many vehicles which were also used by the Army, including some Canadian and US Lend-Lease types. The British Ford WOT1 range of 6×4 vehicles and the Crossley 4×4 types were among the vehicles used exclusively by the RAF. They were standardised RAF chassis and used for the mounting of a wide variety of load-carrying and special bodies. In RAF parlance a truck was called Tender (Van if under 1-ton capacity). Special-equipment and special-purpose vehicles were referred to as 'technical types'. From 1935 to 1965 the RAF vehicle fleet grew from 2000 to 22,000, although during WWII well over 100,000 vehicles were in RAF service.

RN Vehicles The Royal Navy used relatively few motor vehicles and the majority were standard commercial types with only minor modifications. Some special types were supplied by Scammell, including open-cab 3×2 and 4×4 tractors and special torpedo-carrying semi-trailers.

Basic Army MT Chassis Types
8-cwt 4×2 and 4×4 (staff cars, trucks, light 2-stretcher ambulances).
15-cwt 4×2 and 4×4 (trucks).
30-cwt 4×2, 4×4, 6×4 (trucks, heavy 4-stretcher ambulances).
3-ton 4×2, 4×4, 6×4 (trucks).
6-ton 4×2 (trucks), 4×2-2 (tractors with semi-trailer).
10-ton 6×4 (trucks).
Field Artillery Tractors (4×4, 6×6).
Medium Artillery Tractors (4×4).
Heavy Artillery Tractors and Breakdowns (6×4, 6×6).
Tank Transporters (6×4, tractors with semi-trailer).

Colours
Army: basic colour was dark earth or dark brown with an overlaying camouflage pattern of black or dark green based on varying diameter circles ('Micky Mouse ears'); in the Middle East and North Africa a sand colour was used. *RAF:* 'RAF blue' or special camouflage. *RN:* 'battleship grey' or special camouflage. There were exceptions for special vehicles (fire fighters, airfield vehicles, etc.) and special operating conditions.

For identification by friendly aircraft, British vehicles were originally provided with the RAF roundel, painted on the bonnet. Later all Allied vehicles had five-pointed white stars in prominent positions. The largest, usually on the bonnet or cab roof, was often surrounded by a white circle.

Dunkirk, June 1940

Normandy, June 1944

The Netherlands, Sept. 1944

(IWM photo B10243)

Registration Numbers

British Army vehicles carried a registration number on both sides of the bonnet or cab, and on the rear of the body or in equivalent positions. These numbers were officially known as WD numbers and commonly referred to as census numbers. They normally were $3\frac{1}{2}$-in white figures and were prefixed with a letter denoting the vehicle type, as follows:

A   Ambulances.
C   Motorcycles.
F   Armoured cars, Scout cars.
H   Tractors (incl. Breakdown tractors).
L   Trucks (1-ton and over).
M   Cars (incl. Light utilities).
P   Amphibians.
S   SP mountings.
T   Carriers, Tanks.
V   Vans (RASC).
X   Trailers (all types).
Z   Trucks (under 1-ton).

The RAF and RN had their own registration number systems and these appeared on the vehicles in much the same way as on civilian vehicles, prefixed by the letters RAF or suffixed by the letters RN. These letters also often appeared on the cab doors or in other prominent positions.

Who's Who in the British Motor Industry (1939–45)

The following list gives the major manufacturers of British military vehicles (excluding tanks) *during World War II*. Some of these have now ceased to exist or have merged with others. For example, AEC, Crossley, Maudslay and Thornycroft formed the ACV (Associated Commercial Vehicles), merged with the Leyland group in 1962 and are now, together with Albion, Scammell, Austin, Morris, etc. part of the British Leyland Motor Corporation.

AEC	The Associated Equipment Co. Ltd, Southall, Middlesex.
Albion	Albion Motors Ltd, Scotstoun, Glasgow.
Alvis	Alvis Ltd, Coventry.
Alvis-Straussler	Alvis-Straussler Ltd, Coventry.
Ariel	Ariel Motors Ltd, Selly Oak, Birmingham.
Austin	Austin Motor Co. Ltd, Longbridge, Birmingham.
Bedford (GM)	Vauxhall Motors Ltd, Luton, Bedfordshire.
BSA	BSA Cycles Ltd, Birmingham.
Commer (Rootes)	Commer Cars Ltd, Luton, Bedfordshire.
Crossley	Crossley Motors Ltd, Gorton, Manchester.
Daimler	Daimler Motor Co. Ltd, Coventry.
David Brown	David Brown Tractors Ltd, Meltham, Huddersfield, Yorkshire.
Dennis	Dennis Bros Ltd, Guildford.
Dodge	Dodge Brothers (Britain) Ltd, Kew, Surrey.
ERF	ERF Ltd, Sandbach, Cheshire.
Foden	Fodens Ltd, Sandbach, Cheshire.
Ford(son)	Ford Motor Co. Ltd, Dagenham, Essex.
Guy	Guy Motors Ltd, Wolverhampton.
Hillman (Rootes)	The Hillman Motor Car Co. Ltd, Coventry.
Humber (Rootes)	Humber Ltd, Coventry (Humber armoured cars produced by Karrier Motors Ltd).
Karrier (Rootes)	Karrier Motors Ltd, Luton, Beds. (armoured cars named Humber).
Leyland	Leyland Motors Ltd, Leyland, Lancs. (also Kingston-upon-Thames, Surrey).
Matchless	Matchless Motor Cycles, London SE18.
Maudslay	The Maudslay Motor Co. Ltd, Alcester, Warwickshire.
Morris (Nuffield)	Morris Motors Ltd, Cowley, Oxford.
Morris-Commercial (Nuffield)	Morris-Commercial Cars Ltd, Adderley Park, Birmingham.
Norton	Norton Motors Ltd, Birmingham.
Royal Enfield	The Enfield Cycle Co. Ltd, Redditch, Worcs.
Scammell	Scammell Lorries Ltd, Watford, Herts.
Standard	Standard Motor Co. Ltd, Coventry.
Straussler	Straussler Mechanisation Ltd, London SW1
Thornycroft	John I. Thornycroft & Co. Ltd, Basingstoke, Hants.
Tilling-Stevens	Tilling-Stevens Ltd, Maidstone, Kent.
Triumph	Triumph Engineering Co. Ltd, Coventry.
Vauxhall (GM)	Vauxhall Motors Ltd, Luton, Beds. (trucks named Bedford).
Velocette	Veloce Ltd, Hall Green, Birmingham.
Wolseley (Nuffield)	Wolseley Motors Ltd, Birmingham.

GB

MOTORCYCLES
SOLO and WITH SIDECAR

Makes and Models: Ariel W/NG, 350 cc (solo). BSA B30 L/wt, 350 cc OHV (solo); B30, 350 cc OHV (solo); M20(B), 500 cc (solo and w/SC). Excelsior, 98 cc (solo). James ML L/wt, 125 cc (solo). Matchless G3L L/wt, 350 cc OHV (solo); G3, 350 cc OHV (solo); 38G/3, 350 cc OHV (solo). Norton 16H, 500 cc (solo and w/SC); 633 cc (w/SC; 3×2). Royal Enfield WD/RE L/wt, 125 cc (solo); WD/C, 350 cc (solo); WD/CO and WD/CO/B, 350 cc OHV (solo). Triumph 3SW, 350 cc (solo); 3HW, 350 cc OHV (solo). Velocette MAF 350 cc (solo).

General Data: When war broke out in September, 1939, the British Forces had well over 21,000 motorcycles of all types. This total included some 6345 impressed machines. By June, 1940, the total had risen to just under 50,000 and at VE Day the total number of motorcycles in service was almost 270,000. During the war years the British industry had produced a total of 425,000 motorcycles. In addition to supplying its own forces, Britain also delivered motorcycles to Russia and to British Commonwealth armies. Canada, for example, used the Matchless G3L (featuring 'teledraulic' front fork) and the Norton 16H to provide transportation for despatch riders and for other purposes. In addition to British machines, the British and Canadian Armies and the RAF used the American Indian and Harley-Davidson. Only Norton produced a motorcycle with sidecar-wheel drive, a type which the German Army used in relatively large numbers (BMW, Zündapp). Specially developed for airborne use was a folding motor scooter or 'Parascooter'. It was known as the 'Welbike' because it was designed at the Research Station at Welwyn, Herts. It had collapsible handlebars and with steering column and saddle lowered it would fit in a cylindrical container. It was designed to withstand severe shock and its 98-cc Excelsior engine was good for a top speed of approx. 30 mph. The 6½-pt fuel tank gave it a range of 90 miles.

Vehicle shown (typical): Motorcycle Combination, 3×2 (Norton 633) (IWM photo H23377)

Technical Data:
Engine: Norton single-cylinder, L-C-A, 633 cc, 14·5 bhp @ 4000 rpm.
Transmission: 4F, chain drive (shaft drive to sidecar wheel).
Brakes: mechanical.
Tyres: 4.00–18.
Wheelbase: 4' 6".
Overall l×w×h: 7' 2"×5' 6½"×3' 10½" (3' 3¾").
Weight: 679 lb, gross 1232 lb.
Note: Clips on rh side of sidecar for Bren mg.

Motorcycle, Solo (Ariel W/NG) Single-cyl., 12·5 bhp, 4F, wb 4′ 7″, 7′ 0″×2′ 5½″×3′ 6½″, 351 lb. 348 cc. Overhead valves. Burman gearbox. 3.25–19 tyres. Girder-type front fork with adjustable steering damper. Home Guard used Ariels with special Lewis mg-carrying sidecars.

Motorcycle, Solo (BSA B30 WD) Single-cyl., 15 bhp, 4F, wb 4′ 5″, 6′ 9½″×2′ 3½″×3′ 3″, 350 lb. 348 cc. Overhead valves with hairpin-type springs. 3.25–19 tyres. Girder-type front fork with single coil spring. Pillion seat. BSA supplied 126,334 motorcycles during 1939–45.

Motorcycle, Solo (Triumph 3SW) Single-cyl., 10 bhp, 4F, wb 4′ 4½″, 6′ 9″×2′ 6″×3′ 4″, 317 lb. Here shown with armour plating and Bren mg (not included in dimensions and weight) for airfield defence. 343 cc. Side valves. 3.25–19 tyres.

Motorcycle Combination, 3×1 (BSA M20) Single-cyl., 12 bhp, 4F, wb 4′ 6″, 7′ 0″×5′ 6½″×3′ 3″. 496 cc. Side valves. Sidecar built by SS Cars Ltd, featuring mica windscreen, quick-lift hood and watertight hinged scuttle. Locker at rear with hinged lid.

138

Motorcycle, Solo, Lightweight (Matchless G3/L) Single-cyl., 16 bhp, 4F, wb 4′ 3″, 6′ 10½″×2′ 6½″×3′ 5″, 328 lb. 347 cc. Overhead valves. 3.25–19″ tyres. Differed from Model G3 mainly in having 'teledraulic' front forks, more ground clearance, lower weight and lower fuel consumption.

Motorcycle, Solo (Velocette MAF) Single-cyl., 14·9 bhp, 4F, wb 4′ 4¼″, 7′ 1″×2′ 3½″×3′ 1½″, 340 lb. 349 cc. Overhead valves. 3.25–19″ tyres. Girder-type front forks. Ground clearance 5″. Militarised version of civilian Velocette Model MAC. Used by Army and RAF.

Motorcycle Combination, 3×1 (Norton 16H) Single-cyl., 12 bhp, 4F, wb 4′ 6″, 7′ 1½″×5′ 6½″×3′ 3″. Side valves. Single-seater sidecar body (6′ 1½″×1′ 9″×1′ 10″) with door, hinged watertight scuttle and locker at rear. Used by RAF. Also used as solo machine (Army and RAF).

Motorcycle, Folding (Excelsior 'Welbike') Single-cyl. 98-cc two-stroke engine, wb 3′ 3½″, 4′ 4″(4′ 3″)×1′ 10″(1′ 0″)×2′ 6¾″ (1′ 3″), 70 lb. Dunlop 'Carrier' tyres. Collapsible in order to be carried in small parachutable container. Known also as 'Parascooter'. After the war a two-speed version, the Corgi, appeared.

GB

CARS, 2-SEATER and LIGHT UTILITY

Makes and Models: *Cars, 2-seater, 4×2:* Austin 8 HP Open Tourer (AP). *Cars, 2-seater, 4×2, Light Utility:* Austin 10 HP (G/YG). Hillman 10 HP (Mks I, IA, II, IIA, IIB, IV). Morris 8 HP and 10 HP (Series M). Standard 12 HP.

General Data: The Austin 8 HP two-seater was produced during the early war years for liaison and communication work. It was basically a civilian-type tourer, from which it differed mainly in having two near-vertical louvres in the engine-compartment side panels. Behind the two seats was a loading space. The windscreen could be folded down. The Light Utilities were also two-seaters, based on civilian-type cars, but had a pickup type body with canvas tilt (usually with opening sections at forward end on roof and either side). When used without the tilt, the three hoopsticks could be carried bunched together in sockets at the forward end. The back of the cab was open to the main body. Some had two additional folding seats in the main body. The spare wheel was carried on the cab roof except on the Standard, which had a full-length tilt. At first all models had a civilian-type radiator grille. Later models, except the Austin, had a simple wire mesh grille. There were also other changes in production, including differences in bonnet (hood) side panels. The Hillman and the Standard were also supplied with all-steel van body. The RAF used the following light vans: 5-cwt: Singer 9 HP; 10-cwt: Ford E83W, Morris 12 HP, Standard 12 HP; 12-cwt: Standard 14 HP. The Morris 8 HP Light Utility was based on the 8 HP GPO Van, but only a few of these were produced. The Hillman 10 HP was also produced as a Ladder Van in 1942 (steel-panelled van, carrying extendible ladder on roof) and of the same van there was an estate-car conversion with rear seat and side windows (RAF). The Standard 12 HP continued in production for a while after the war and fitted with civilian-type radiator grille and simple bodywork was used mainly for export to the Netherlands.

Vehicle shown (typical): Car, 2-seater, 4×2 (Austin 8 HP Series AP).

Technical Data:
Engine: Austin Eight 4-cylinder, I-L-W-F, 900 cc, 23·5 bhp @ 4000 rpm.
Transmission: 4F1R.
Brakes: mechanical.
Tyres: 5.00–16.
Wheelbase: 7′ 5″.
Overall l×w×h: 12′ 4″×4′ 7″×5′ 2″ (4′ 2″).
Weight: 13½ cwt.
Note: militarised civilian 1939 Austin Eight Two-seater, with vertical instead of horizontal bonnet louvres.

Car, Light Utility, 4×2 (Austin 10 HP G/YG) 4-cyl., 29.5 bhp, 4F1R, wb 7' 10¾", 13' 1"×5' 3¼"×6' 5½", 19¾ cwt. Tyres 6.00–16. Used by British and several other armies. Vehicle shown has one-piece tilt and was produced in 1943. These light utilities were popularly known as 'tillies'.

Car, Light Utility, 4×2 (Hillman 10 HP) 4-cyl., 30 bhp, 4F1R, wb 7' 8", 12' 7"×5' 3"×6' 3", 1 ton 1 cwt approx. Six marks, all differing in detail. Used by Army and RAF. Mk IIB (shown), introduced in 1943, had removable cab top. Two folding seats in rear. Based on Minx car (see page 142).

Car, Light Utility, 4×2 (Morris 10 HP Series M) 4-cyl., 37.2 bhp, 4F1R, wb 7' 10", 13' 5½"×5' 0"×6' 4", 1 ton 1¼ cwt. Tyres 6.00–16. Hydraulic brakes. Later production had wire mesh radiator grille and one-piece canvas tilt.

Car, Light Utility, 4×2 (Standard 12 HP) 4-cyl., 44 bhp, 4F1R, wb 9' 0", 13' 10"×5' 4½"×6' 8"(4' 6"), 1 ton 3¼ cwt. Tyres 6.00–16. Based on Flying Twelve car but rigid front axle. Early types had civilian-type radiator grille, no louvres in bonnet sides and independent front susp. Bodies also varied in detail.

CARS, 4-SEATER 4×2
SALOONS and TOURERS

Makes and Models: *Saloons:* Austin 10 HP G/RQ (4-str). Ford WOA1, WOA1/A (4-str). Hillman 10 HP Minx and 14 HP (4-str). Humber Snipe (4/5-str). Standard 10 HP (4-str). Vauxhall 14 HP and 25 HP (4-str). *Tourers:* Ford WOA2 (modified) (4-str). Humber Snipe (4-str).

General Data: The cars listed above were standard types used by the British Army and/or RAF. In addition there were military cars of pre-war manufacture (Austin, Ford, Hillman, Humber, Packard (USA), Riley, etc.) and many impressed civilian-type cars, especially during the early part of the war. In September, 1939, the army had more impressed cars than WD types (4285 as against 3803), but by June, 1940, the figures were 2315 and 12,660 respectively. At VE Day, 134,573 cars were in WD service. Some 8000 cars were lost during the evacuation of France. In addition to the British types, the American Chevrolet (1941) and Ford (1940, 1942) were used. A number of Fords were assembled at Dagenham and modified for desert use (oversize tyres, etc.; see page 88). The Ford WOA2 Tourer was a modification of the WOA2 Heavy Utility (see next section), the roof of which had been removed and replaced by a convertible top during the North African campaigns. A similar field modification was carried out on the Humber Snipe Heavy Utility, on the similar Canadian Fords (Canadian Army) and even on the armoured Humber Light Reconnaissance car. The Ford WOA1 and WOA1/A staff cars (which differed in wheel and tyre equipment) were basically civilian-type saloons with military-pattern front end and other modifications. After the war, in 1948, it was de-militarised and launched as Ford of Britain's 'first post-war model': the $3\frac{1}{2}$-litre V8 Pilot. The bodywork of the Humber Snipe Tourer was produced by Thrupp & Maberly.

Vehicle shown (typical): Car, 4-seater, 4×2, Saloon (Ford WOA1/A)

Technical Data:
Engine: Ford 8-cylinder, V-L-W-F, 3·6 litres, 85 bhp @ 3600 rpm.
Transmission: 3F1R.
Brakes: mechanical.
Tyres: 6.50–16 (WOA1: 9.00–13).
Wheelbase: 9' $0\frac{1}{4}$".
Overall l×w×h: 14' 0"×6' 4"×6' 4".
Weight: 1 ton $7\frac{3}{4}$.
Note: developed from 1938/39 Model 62 '22 HP' saloon, but powered by '30 HP' engine.

Car, 4-seater, 4×2, Saloon (Austin 10 HP Series GRQ) 4-cyl., 29·5 bhp, 4F1R, wb 7' 9¾", 13' 2"×4' 10½"×5' 4½". 1·125-litre engine. Tyres 5.25–16. Few were supplied directly for military service. Some were impressed, but not in such relatively large numbers as Ford 10 HP saloons.

Car, 4-seater, 4×2, Saloon (Hillman 10 HP) 4-cyl., 30 bhp, 4F1R, wb 7' 8", 12' 9"×5' 1"×5' 3", 18 cwt. 1·18-litre engine. Mech. brakes. Tyres 5.00–16. Supplied mainly to RN (shown) and RAF. The larger Hillman 14 HP was also used. Both were slightly modified civilian cars.

Car, 4-seater, 4×2, Saloon (Standard 10 HP) 4-cyl., 36 bhp, 4F1R, wb 7' 6", 12' 8"×4' 10"×5' 2", 18 cwt. 1·26-litre engine. Mech. brakes. Tyres 5.00–16. Sliding roof. Independent front suspension with transverse leaf spring. 150 were supplied to the RAF.

Car, 4-seater, 4×2, Saloon (Vauxhall 14 HP Series JI) 6-cyl., 48 bhp, 3F1R, wb 8' 9", 14' 1"×5' 4"×5' 4", 21 cwt. 1·78-litre OHV engine. Hyd. brakes. Tyres 5.75–16. Sliding roof. Independent front suspension with torsion bars. The larger Vauxhall 25 HP Series G was also used.

143

Car, 4-seater, 4×2, Saloon (Humber Snipe Mk 2) 6-cyl., 85 bhp, 4F1R, wb 9′ 6″, 15′ 0″×5′ 10″×5′ 11″, 1 ton 13 cwt. 9.00–13 tyres (first production had 7.00–16 tyres). Military version of Super Snipe Saloon. Entered production in Dec., 1939. Roof luggage rail with cover.

Car, 4-seater, 4×2, Saloon (Humber Snipe, modified) 6-cyl., 85 bhp, 4F1R, wb 9′ 6″, 15′ 0″×5′ 10″×6′ 0″. Modified 8-cwt FFW truck, fitted with German Karmann Saloon bodywork. Produced in small series for British forces in Germany, shortly after the war.

Car, 4-seater, 4×2, Open Tourer (Humber Snipe) 6-cyl., 85 bhp, 4F1R, wb 9′ 6″, 15′ 0″×5′ 10″×5′ 10½″(5′ 7½″), 1 ton 13 cwt. Basically as Snipe Saloon. Vehicle shown is 1941 'Old Faithful', used by FM Lord Montgomery in 8th Army, from Alamein to the Sangro, and still exists.

Car, 4/5-seater, Armoured, 4×4 (Humber) 6-cyl., 85 bhp, 4F1R×2, wb 9′ 3¾″, 14′ 4″×6′ 2″. One of several staff-car conversions of the Humber light recce car. Another specimen had an ex-'Jeep' folding windscreen. Note RAF roundel on bonnet. North Africa, 1942.

GB 144

CARS, HEAVY UTILITY
4×2 and 4×4
and 6-SEATER, 4×2

Makes and Models: *Heavy Utility, 4×2:* Ford WOA2. Humber Snipe.
Heavy Utility, 4×4: Humber F.W.D.
Car, 6-str, 4×2: Humber Pullman Limousine.

General Data: The Heavy Utility was the military type 'estate car', featuring four side doors and full-width rear doors, split horizontally. The spare wheel was carried on the back. A folding map table was fitted behind the bucket-type front seats. They were used chiefly as staff cars. The chassis were classed as 8-cwt and also used as the basis for other bodies. Some Fords and Humbers were converted into convertible-type tourers for use in the North African campaigns. The Humber F.W.D. Utility was the only British-built 4×4 type in this class. In addition to the above British types, the British forces used several imported types, namely the Canadian Ford C11ADF (1941) and C29ADF (1942) militarised station wagons and the C011DF (1941) and C291D (1942) military-pattern F8 models (fitted with British bodywork), the Canadian Chevrolet military-pattern C8A (4×4) and the American Dodge ½-ton and ¾-ton 4×4 Carryall (WC26 and WC53). A civilian-type Dodge 4×2 station wagon was used in small numbers. The Humber Pullman limousine, which had bodywork by Thrupp & Maberly, was used by the Army and RAF for transportation of high-ranking officers. The Humber Snipe and Pullman models were based on the 4-litre Super Snipe saloon which had been introduced by the Rootes Group late in 1938. A modification of the Humber F.W.D. Utility, used by General Staff officers, featured a forward-mounted sliding roof, strengthened windscreen pillars, improved door window-winding mechanisms, map reading lamp, armrests, etc. In 1943 Thrupp & Maberly designed a 'Cross-Country Saloon' on this chassis, featuring car-type boot and fabric-covered roof. In 1945 an experimental model was produced with coil-spring suspension.

Vehicle shown (typical): Car, 4×4, Heavy Utility (Humber F.W.D.).

Technical Data:
Engine: Humber 6-cylinder, I-L-W-F, 4.08 litres, 85 bhp @ 3400 rpm.
Transmission: 4F1R×2.
Brakes: hydraulic.
Tyres: 9.25–16.
Wheelbase: 9' 3¼".
Overall l×w×h: 14' 1"×6' 2"×6' 5".
Weight: 2 tons 7½ cwt.
Note: same chassis also used for various other types, including Light Reconnaissance Car Humber Mk III(A). Independent front suspension.

145

Car, 4×2, Heavy Utility (Ford WOA2) V-8-cyl., 85 bhp, 3F1R, wb 9' 0$\frac{1}{4}$", 14' 5"×6' 3$\frac{1}{2}$"×5' 10", 1 ton 11$\frac{3}{4}$ cwt. 9.00–13 tyres. All-steel body. Folding-type middle seats, two tip-up seats in rear corners. Sliding panel in centre of roof. Produced from May, 1941, to July, 1944. Some had 6.50–16 tyres.

Car, 4×2, Tourer (Ford WOA2, modified) As WOA2 but with convertible-type canvas top. Similar conversion carried out on Humber 4×2 and Canadian Ford C11ADF and C29ADF. After 1945 Ford produced a pickup truck using this chassis. IWM photo NA3675.

Car, 4×2, Heavy Utility (Humber Snipe) 6-cyl., 85 bhp, 4F1R, wb 9' 6", 15' 4"×6' 1$\frac{1}{2}$"×6' 1", 1 ton 14$\frac{1}{3}$ cwt. 9.00–13 tyres. Apart from bodywork, similar to Snipe saloon and tourer, with ifs and extra wide mudguards. IWM photo H23371.

Car, 6-str, 4×2, Limousine (Humber Pullman) Chassis generally as Snipe saloon, but wb 10' 7$\frac{1}{2}$", 16' 5"×6' 1" ×6' 2$\frac{1}{2}$", 1 ton 14$\frac{1}{3}$ cwt, 7.00–16 tyres. Roof luggage rail with waterproof cover. Division with sliding windows separating front seats from rear compartment.

GB

CARS, 5-CWT and LIGHTWEIGHT 4×2 and 4×4

Makes and Models: SS Cars VA and VB, 4×2. Standard 4×2; 4×4 JAB; 5-cwt, 4×2.

General Data: The above vehicles, with the exception of the 5-cwt, were designed, developed and tested for possible airborne use in the Far East. They were very compact in dimensions and low in weight, and intended to be capable of being manhandled when in difficulty. The Standard 'Jungle Bugs', produced in 1944, had motorcycle-type saddles and pillions. They were powered by a special lightweight version of the firm's 8 HP car engine. A special amphibious trailer was produced, into which the vehicle could be placed for ferrying across rivers. The 4×4 model was developed into a peacetime 'Farmers' General Purpose Vehicle' in early 1945, but it did not go into series production. The SS VA, first of the two prototypes designed and developed by SS Cars Ltd (now Jaguar), had a rear-mounted 10·9 HP JAP engine with motorcycle gearbox and chain final drive. The differential was lockable by means of a set screw. It had left-hand drive, originally with handle-bar steering, later with a steering-wheel. It appeared early in 1943 as a private venture and was given limited tests to compare it with the heavy motorcycle with sidecar combination. The VB was more conventional in layout and had a Ford 10 HP engine-cum-gearbox with a two-speed auxiliary gearbox. The Standard 5-cwt 4×2 car was clearly intended to be a British version of the American 'Jeep'. It looked a promising vehicle but lacked front wheel drive. Sufficient supplies of the American model probably ruled out the need to put this car into production. Large numbers of Ford and Willys 'Jeeps' were supplied to Great Britain under Lend-Lease. One was experimentally converted by Nuffield Mechanizations Ltd into a lightweight (see page 23). Many 'Jeeps' were modified extensively for special purposes such as cable-laying, desert patrolling and SAS (Special Air Services). After the war the 'Jeep' in Britain was gradually replaced by the Austin 'Champ' and the Land-Rover.

Vehicle shown (typical): Car, Ultra Lightweight, 4×4 (Standard JAB)

Technical Data:
Engine: Standard 8 HP Special, 4-cylinder, I-L-W-F, 1·02 litres, 22 bhp @ 3500 rpm.
Transmission: 3F1R×1.
Brakes: mechanical.
Tyres: 7.50–12 (modified aero tyres, cross-country tread).
Wheelbase: 3′ 11″.
Overall l×w×h: 6′ 6″×3′ 7″×3′ 6″.
Weight: 742 lb.
Note: Shown with amphibious trailer.

Car, 5-cwt, 4×2 (Standard 12 HP) 4-cyl., 44 bhp, 4F1R, wb 6' 3", 10' 4"×5' 3"×5' 4", 17 cwt. 6.00–16 tyres. Produced experimentally in 1943 as British equivalent of the 'Jeep'. Independent front suspension. Lacked front-wheel drive.

Car, Ultra Lightweight, 4×2 (SS VA) JAP 2-cyl., 10·9 HP, 3F1R, wb 4' 7", 7' 6" (approx.)×4' 10½"×3' 6", 7 cwt 3¼ qtrs. Rear-mounted air-cooled 1096-cc V-twin engine. Chain final drive to lockable differential. Track, front 4' 5", rear 4' 4". Left-hand drive. Machine gunner sat on right-hand side.

Car, Ultra Lightweight, 4×2 (Standard) 4-cyl., 22 bhp, 4F1R, wb 3' 11", 6' 0"×3' 4"×3' 4", 6 cwt. 4.00/4.25–15 tyres (later 7.50–12 rear). Carrying capacity 6 cwt or four men plus equipment. Magneto ignition; no other electrical equipment.

Car, Ultra Lightweight, 4×2 (SS VB) Ford 10 HP 4-cyl, 30 bhp, 3F1R×2, wb 5' 1¼", 8' 0"×4' 6"×6' 0"(3' 0"), 8¼ cwt. 5.00/5.75–16 tyres. Track, front and rear 4' 0". Independent suspension front and rear. Removable steering-wheel. Built in 1943, scrapped in 1947.

GB
148

COACHES and TROOP CARRIERS

This category can be divided into three groups: 'crew coaches' (RAF), 'troop carriers' (Army, RAF) and 'coaches' (all services). Crew coaches were used by the RAF mainly to transport crews to and from aircraft on airfields. They had special bodywork, mounted on 4×2 and 6×4 truck chassis. Troop carriers were based on a special version of the Bedford QL 3-ton 4×4 chassis. Coaches were mainly impressed 32-seater civilian types of various makes. New Bedford OWB 32-seaters were supplied also. During 1943–45 some 55 Green Line coaches were modified and used as 'Clubmobiles' by the American Red Cross.

Truck, 3-ton, 6×4, Crew Coach (Fordson WOT1) V-8-cyl., 85 bhp, 4F1R, wb 13' 8½" (BC 3' 6¼"), 24' 0"×7' 4"×9' 3", 6 tons. Seating for 2+23. Double rear doors. Speaking tube to cab. Similar body on Austin K6 (3-ton 6×4) and US Dodge VK62B (see page 87). Some used as recruiting offices.

Truck, 3-ton, 4×4, Troop Carrying (Bedford QLT) Modified lengthened QL chassis. Main body seats 29 troops and their kit. AA hatch at forward end of body with mg mounting. Side seats folding, centre seats assembly removable. Popularly known as 'Drooper'. 22' 0"×7' 7"×9' 10" (7' 0"), 3 tons 18¼ cwt.

Coach, 32-seater, 4×2 (Leyland 'Tiger' TS6) 6-cyl. diesel, 98 bhp (or 104-bhp 6-cyl. petrol). Slightly modified civilian coach, impressed for Army service, suitably camouflaged and fitted with roll-up canvas side curtains. Several makes were used. 1941. IWM photo H14389.

AMBULANCES
LIGHT and HEAVY

Makes and Models: Albion AM463 (30-cwt, 4×2). Austin K2/Y (2-ton, 4×2). Bedford ML (2-ton, 4×2). Fordson E88W (25-cwt, 4×2). Fordson WOT1 (3-ton, 6×4). Humber F.W.D. (8-cwt, 4×4). Morris TWV Series III (10-cwt. 4×2). Morris Y (10-cwt, 4×2). Morris-Commercial CS8 (15-cwt, 4×2, conversion). Morris-Commercial CS 11/30F (30-cwt, 4×2). Morris-Commercial CD (30-cwt, 6×4). Standard 14 HP (12-cwt, 4×2).

General Data: All the above models, except the Humber, were four-stretcher types. The Humber was a 4×4 light field ambulance with two stretchers. The Austin K2 was the most widely used ambulance and was also supplied to other nations, including France, Norway and Russia, and to the US forces (reversed Lend-Lease). From 1940 until VE day, 13,102 Austin K2 ambulances were produced and similar bodywork was also fitted on other chassis, viz. Bedford ML, Morris-Commercial CS 11/30F and CD. The Albion AM463 and the Ford WOT1 were chassis standardised by the RAF and were also used with various other types of bodywork. The Standard 14 HP was another RAF exclusive and superseded the Morris Y type which had a similar body. In addition, the British forces used several imported types, viz. the Canadian Chevrolet C8A-HUA (4×4), Ford F602L (4×2) and F60L (4×4), and the American Chevrolet (4×2), Dodge WD21 (4×2), WC27 and 54 (4×4), and Ford RO1T (4×2, assembled at Dagenham). Modifications of the Canadian GM C15TA and the American White scout car M3A1 and IHC half-track were used as armoured ambulances. At the outbreak of war the British Army had some 1700 ambulances, about one-fifth of which were impressed vehicles. At VE Day the total number (excl. RAF) was 15,309. Some of the Humber F.W.D. ambulances were converted into recording vans and used by BBC war correspondents for recording eyewitness accounts of fighting and other activities in the actual battle zones.

Vehicle shown (typical): Ambulance, Heavy, 4×2 (Austin K2/Y)

Technical Data:
Engine: Austin 6-cylinder, I-I-W-F, 3·46 litres, 60 bhp @ 3000 rpm.
Transmission: 4F1R.
Brakes: hydraulic.
Tyres: 10.50–16.
Wheelbase: 11′ 2″.
Overall l×w×h: 18′ 0″ × 7′ 3″ × 9′ 2″.
Weight: 3 tons 1½ cwt.
Note: occasionally used for other purposes like mobile office, loudspeaker van, etc.

Ambulance, Heavy, 4×2 (Bedford ML) 6-cyl., 72 bhp, 4F1R, wb 11′ 11″, 19′ 0″×7′ 3″×9′ 3″, 3 tons 1 cwt. 10.50–16 tyres. Mann Egerton Body No. 2 Mk 1/L, also fitted on Morris 4×2 and 6×4. Four stretchers or 10 sitting patients.

Ambulance, Heavy, 4×2 (Morris-Commercial CS11/30F) 6-cyl., 60 bhp, 4F1R, wb 11′ 2″, 17′ 9″×7′ 5″×9′ 2″, 3 tons 1 cwt. 10.50–16 tyres. Mann Egerton body. Many of these were presented by private individuals, institutions, etc., this particular one by the Naiyasha District, Kenya Colony.

Ambulance, Light, 4×4 (Humber F.W.D.) 6-cyl., 85 bhp, 4F1R×2, wb 9′ 3¾″, 13′ 9″×6′ 0″×7′ 5″, 2 tons 4¼ cwt. 9.25–16 Runflat tyres (if fitted with conventional tyres, spare could be carried in N/S locker). Bodywork by Thrupp & Maberly.

Ambulance, Heavy, 6×4 (Fordson WOT1) V-8-cyl., 85 bhp, 4F1R, wb 13′ 10½″ (BC 3′ 6¼″), 22′ 0″×7′ 5″×9′ 6″, 4 tons 4½ cwt. 9.00–16 tyres. Speaking-tube from cab to main body. Fitted with two-tier stretcher elevating gears, heater, washbasin, electric gong, searchlight, etc.

Ambulance, Light, 4×2 (Morris TWV, 11·9 HP) 4-cyl., 30 bhp approx., 3F1R, wb 7' 6", 13' 8"×5' 6"×7' 10", 1 ton 3 cwt. Tyres 5.00–18. Bodywork by Composite Vehicles Ltd (Brittain 'Patent'). Used by Home Guard, RASC, etc. Accommodation for eight sitting or four stretcher cases.

Ambulance, Light, 4×2 (Standard 14 HP) 4-cyl., 48 bhp, 4F1R, wb 9' 0". Early production had civilian Flying Standard car-type radiator grille. Similar body on Morris Y 14 HP chassis. Double-hinged tailboard (forming step) and canvas curtain at rear. Four stretchers. RAF.

Ambulance, Heavy, 4×2 (Commer Q3) 6-cyl., 66 bhp, 4F1R, wb 10' 0". Four-stretcher ambulance bodywork by Strachans. Vehicle shown was presented to the Royal Navy by the British Red Cross Society in Uruguay in commemoration of the Battle of the River Plate, 13th December, 1939.

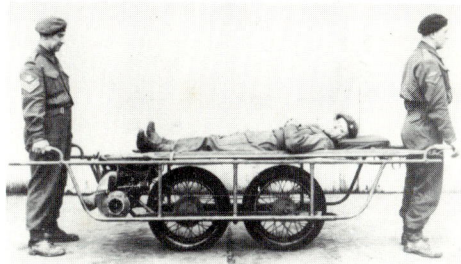

Powered Stretcher, 2×2, SP 'Chinese Wheelbarrow' (TT) 1-cyl. 147-cc 2-stroke engine, chain drive to both wheels (67:1 gear reduction), wb 2' 6", 10' 0"×2' 5"×2' 6½", 2¾ cwt. Payload 350 lb. Tyres 4.00–18. Could be converted for use as cargo carrier. Designed for jungle warfare, 1944.

GB

TRUCKS, 8-CWT
4×2 and 4×4

Makes and Models: Ford WOC1 (4×2). Humber Snipe (4×2) and F.W.D. (4×4). Morris-Commercial PU (4×2) and PU8/4 (4×4).

General Data: The 8-cwt range was produced during the early years of the war and later eliminated from the Army's requirements with a view to cutting down the large overall number of types. The chassis in the 8-cwt range, produced by Ford, Rootes (Humber) and Nuffield (Morris-Commercial), were used for a variety of bodies, including Staff cars (Heavy Utility) and Ambulances, Personnel or GS (General Service) trucks and FFW trucks (fitted for wireless). The latter two types were similar in external appearance. The Personnel/GS version had a well-type body, providing seating for three men, with waterproof cover on tubular superstructure. This superstructure had folding legs and could be used off the vehicle as a shelter. The FFW version was used for the carriage of a No. 11 wireless set. The main body had seating for two men, a table for wireless transmitting and fittings for batteries and equipment. An auxiliary generator, driven from the power take-off on the gearbox, was fitted for charging wireless batteries. The above British types were supplemented by the Canadian Chevrolet C8 (4×2) and C8A-HUW (4×4 with Heavy Utility body), Dodge T212 (4×4) and Ford F8 (4×2) and F8A (4×4) types. Apart from the C8A all these 8-cwt trucks had a similar body to the British types. The American Dodge ½-ton 4×4 (1941 pattern) was also used, as 'Truck, 10-cwt, 4×4', with various types of bodywork, incl. Panel FFW. The Rootes Group and Vauxhall Motors supplied quantities of 8-cwt and 12-cwt GS vans (Commer, Bedford) during the very early part of the war, but these vehicles were pure commercial types as in production at the time. The Canadian FFW type trucks did not have the PTO drive for the auxiliary generator but were equipped with a separate 'chorehorse' installed between the cab and the body (carried inside on later models with longer body).

Vehicle shown (typical): Truck, 8-cwt, 4×2, FFW (Morris-Commercial PU/Mk 2)

Technical Data:
Engine: Morris OH 6-cylinder, I-L-W-F, 3·48 litres, 60 bhp @ 2800 rpm.
Transmission: 4F1R.
Brakes: hydraulic.
Tyres: 9.00–13.
Wheelbase: 9' 0".
Overall l×w×h: 13' 10"×6' 7"×6' 5" (5' 8").
Weight: 1 ton 16¼ cwt.
Note: produced from 1936 until c. 1941.

Truck, 8-cwt, 4×2, Personnel/GS (Ford WOC1) V-8-cyl., 85 bhp, 4F1R, wb 9' 4", 14' 10½"×6' 2½"×6' 1⅕"(5' 8"), 1 ton 14 cwt. 9.00-13 tyres. British-built but based on US 1940 style Ford V8 chassis, fitted with truck gearbox. Limited production. Bottom gear was often blanked off.

Truck, 8-cwt, 4×2, FFW (Humber Snipe) 6-cyl., 85 bhp, 4F1R, wb 9' 6", 14' 7"×6' 5"×6' 2½"(5' 4"), 1 ton 14¾ cwt. 9.00-13 tyres. Ifs. Based on Super Snipe saloon chassis. Early production had louvres in bonnet sides.

Truck, 8-cwt, 4×4, Personnel/GS (Humber F.W.D.) 6-cyl., 85 bhp, 4F1R×2, wb 9' 3¾", 14' 0"×6' 5½"×6' 9"(5' 11"), 2 tons 6¼ cwt. 9.25-16 tyres. Independent front suspension. Produced on same chassis: Heavy Utility Staff car and Ambulance.

Truck, 8-cwt, 4×4, Personnel/GS (Morris-Commercial PU 8/4) 6-cyl., 74 bhp, 4F1R×1, wb 8' 0¼", 13' 4"×6' 7"×6' 8" (6' 3"), 2 tons 2¼ cwt. 9.25-16 tyres. Like all British-built models in this class, fitted with soft-top cab and half-doors with Perspex sidescreens. Limited production 1940/41.

TRUCKS, 15-CWT
4×2, 4×4 and 6×6

Makes and Models: Austin BYD (4×2) and K7 (4×4). Bedford MW Series (MWC/D/G/R/T/V; 4×2). Commer 'Beetle', Q15, and Q2 (4×2). Ford EO1T, EO1Y, and RO1T Van (4×2). Fordson WOT2 Series (WOT2A/B/C/D/E/F/H; 4×2). Guy Ant (4×2) and Quad-Ant (4×4). Humber 'Hexonaut' (6×6). Morris-Commercial CS8 Mk I, II, III, C4 Mk I and II (4×2), C8/GS (4×4).

General Data: All basic models were supplied with WD-pattern GS bodies, but many were also used for fitment of special bodywork (Water Tank, Office, FFW, Wireless (House type), Air Compressor, AT Portee and Tractor, etc.). The 15-cwt truck was one of the types most widely used by the British forces. The 15-cwt WD-type infantry truck had been evolved during the early 1930s and in 1934 Morris started series production. Two years later Guy Motors joined in with the 'Ant', followed by Bedford, Commer and Ford, all large-volume producers of trucks. In 1939 the WD held approx. 15,000 15-cwt trucks. This figure had risen to over 230,000 when the war ended. The majority of the 1939 holdings were lost during the evacuation of France (Dunkirk) in 1940 and the Canadian industry (Ford, GM and Chrysler) was called upon to assist in making up for these losses. The following Canadian types were supplied, starting from 1940: Ford F15 and F15A, Chevrolet C15 and C15A, Dodge D15. In addition, the Ford Motor Co. in Dagenham produced the partly American EO1T and EO1Y from October, 1940, to July, 1941. The former of these two had dual rear tyres, an unusual feature in this class. Under Lend-Lease, the US Government supplied Dodge ¾-ton 4×4 trucks (Weapons Carriers and several other types). The Dodge Weapons Carrier was also produced in Canada, as a 15-cwt airportable truck. Austin produced a British equivalent, the K7, which looked much like the Weapons Carrier. Its lightweight body could be removed and manhandled. The front bumper was specially shaped to facilitate loading into aircraft. A 2-ton winch was fitted at front. 15-cwt trucks which remained in use after the war were up-rated to 1-tonners.

Vehicle shown (typical): Truck, 15 cwt, 4×2, GS (Bedford MWD)

Technical Data:
Engine: Bedford 6-cylinder, I-I-W-F, 3·5 litres, 72 bhp @ 3000 rpm.
Transmission: 4F1R.
Brakes: hydraulic.
Tyres: 9.00–16.
Wheelbase: 8′ 3″.
Overall l×w×h: 14′ 4½″×6′ 6½″×7′ 6′ (5′ 3″).
Weight: 2 tons 5 cwt.
Note: late-type cab shown. Also with other body types, including 200-gallon water tank (Model MWC).

155

Truck, 15-cwt, 4×2, AA, Mk IA (Bedford MWC) 6-cyl., 72 bhp, 4F1R, wb 8' 3", 14' 4"×6' 10"×7' 0"(6' 2"), 2 tons 3½ cwt. Fitted with 20-mm AA gun. Same body on Ford WOT2; Mk II and III on CMP Chevrolet C15 and Ford F15. Early type cab with side curtains and two folding windscreens.

Truck, 15-cwt, 4×2, GS (Commer 'Beetle') 6-cyl., 60 bhp, 4F1R, wb 8' 0", 1 ton 13¼ cwt. 9.00–16 tyres. Like the Bedford MW and Guy Ant basically a militarised and shortened version of a commercial-type truck. Limited production. 1939.

Truck, 15-cwt, 4×2, Van (Commer Q2) 6-cyl., 60 (later 80) bhp, 4F1R, wb 10' 0", 15' 9"×6' 6"×7' 6", 2 tons 8 cwt. 10.50–16 tyres. Cab integral with body. Also produced with closed cab for GS and special bodies incl. aircraft pre-heater van. Q15 had 4-cyl. engine, 7.50–20 tyres.

Truck, 15-cwt, 6×6, GS (Humber 'Hexonaut') Two Hillman 14 HP engine/transmission units, each driving the wheels on one side. Skid steering by levers operating throttles and brakes. Truck could float and was designed to be carried in Dakota aircraft. Believed only three built. Prototype No. 1 shown.

Truck, 15-cwt, 4×2, GS (Morris-Commercial CS8) 6-cyl., 60 bhp, 4F1R, wb 8' 2", 13' 10½"×6' 6"×6' 6", 1 ton 18¼ cwt. Early-type infantry truck, many of which were lost in France and subsequently used by the Germans (see illustrations on pages 134 and 301). Also with various other bodies.

Truck, 15-cwt, 4×2, Wireless (House Type) (Morris-Commercial C4 Mk I) 4-cyl., 60 bhp, 4F1R, wb 8' 2", 14' 9"×6' 9"×8' 9". Similar to CS8, except 3·5-litre, 4-cyl. engine, soft-top cab. Mk II similar except wb 8' 11". This type carried No. 9 wireless set and equipment.

Truck, 15-cwt, 4×2, GS (Fordson WOT2A) V-8-cyl., 85 bhp, 4F1R, wb 8' 10", 14' 9"×6' 6½"×6' 6", 2 tons. Earliest-type Ford infantry truck. WOT2C was similar but had horizontally mounted windscreens and other detail mods. WOT2B and 2D had tilt-covered van body.

Truck, 15 cwt, 4×2, GS (Fordson WOT2E) V-8-cyl., 85 bhp, 4F1R, wb 8' 10", 14' 9"×6' 6½"×7' 6", 2 tons 0¾ cwt. Late-type cab. WOT2H similar. WOT2F similar but steel body and no tilt. Also used by other UN forces. Engine speed was governed to 2840 rpm at which output was 60 bhp.

Truck, 15-cwt, 4×2, GS (Guy Ant) Meadows 4-cyl., 55 bhp, 4F1R, wb 8′ 5″, 14′ 0″×6′ 6″×6′ 3″, 2 tons 1 cwt. Introduced 1935/36, using approx. 90% Guy commercial units ('Otter' engine, 'Vixen' axles, etc.). 'Vixant' was wartime de-militarised long-wheelbase Ant for essential civilian use.

Truck, 15-cwt, 4×4, GS (Guy Quad-Ant) Meadows 4-cyl., 58 bhp, 4F1R×1, wb 8′ 5″, 14′ 9″×7′ 5″×7′ 11″(6′ 6″), 3 tons. First British 15-cwt 4×4. Superseded Quad-Ant FAT in production, *circa* Jan., 1944. Also produced as 6-pdr AT tractor, w/winch.

Truck, 15-cwt, 4×4, GS, APT, w/Winch (Austin K7) 6-cyl., 80 bhp, 4F1R×1, wb 7′ 7″, 2 tons 2 cwt. Special 9.00–16 lightweight tyres. Easily removable body. Designed for quick loading into aircraft. 2-ton winch with fairleads for front and rear pulls. 3·46-litre engine. Pilots only.

Truck, 15-cwt, 4×4, GS (Morris-Commercial C8/GS) 4-cyl., 70 bhp, 5F1R×1, wb 8′ 3″, 15′ 2″×6′ 10″×7′ 0″ (6′ 0″), 2 tons 11 cwt. Introduced in late 1944 to supersede the CS8 4×2 model. Chassis generally similar to MCC C8 FAT except 9.00–16 tyres and no winch. Steel or wooden body.

UCKS, 30-CWT and 2-TON 4×2, 4×4, 6×4

Makes and Models: Albion AM463 (2-ton, 4×2). Austin K30 (30-cwt, 4×2), K2/YF (2-ton, 4×2). Bedford MSC, OXA, OXC, OXD (30-cwt, 4×2), ML (2-ton, 4×2). Commer Q2 (30-cwt, 4×2). Dennis AM (30/40-cwt, 4×2). Ford EO18T (30-cwt, 4×2). Fordson WOT3 (30-cwt, 4×2), WOT8 (30-cwt, 4×4). Karrier Bantam Mk 1 and 2 (2-ton, 4×2). Leyland (30-cwt, 4×2). Morris-Commercial CS11/30(F) (30-cwt, 4×2), C8/P, C9/B (30-cwt, 4×4), CD, CDF, CDFW, CDSW (30-cwt, 6×4). Thornycroft GF/TC4, HF/TC4 (30-cwt, 4×2).

General Data: At the beginning of the war the 30-cwt truck (or 'light lorry') was a widely used type in the British Army. There were well over 10,000 of them, but over half of these were lost in the Dunkirk retreat. Production of this type slackened off considerably as the war drew on. Like the 8-cwt truck, the 30-cwt was largely replaced by a heavier type, in this case the 3-tonner, in the interest of standardisation. The 30-cwt, both in 4×2 and 6×4 configuration, was very much an inter-war type of vehicle and many of them were in fact either built during that period or continued in production. Many were fitted with oversize tyres to get maximum cross-country performance from a type of vehicle that was already in production and continuation of which posed no great problems (this also applied to many 3-ton 4×2 types). The Albion (first introduced in 1934 as an ambulance chassis), Dennis and Karrier were purely RAF types. The Bedford OX and Commer Q were also used as tractors for semi-trailers, chiefly by the RAF and RN. War-time production of the Morris-Commercial 6×4 was largely carried out by Austin (6686 units, 1940–44) and Wolseley (almost 6000, 1939–42). Large numbers of vehicles in this class were also supplied by Canada (Chevrolet and Ford, 4×4) and the USA (1½-ton Chevrolet, Dodge and GMC 4×4, Dodge 6×6). Some of the US supplies were diverted shipments (ex-French contracts). A number of Bedfords (Model MSC) were supplied to the Greek Army in 1940/41.

Vehicle shown (typical): Truck, 30-cwt, 4×2, GS (Austin K30)

Technical Data:
Engine: Austin 6-cylinder, I-I-W-F, 3·46 litres, 60 bhp @ 3000 rpm.
Transmission: 4F1R.
Brakes: hydraulic.
Tyres: 10.50–16.
Wheelbase: 11′ 2″.
Overall l×w×h: 17′ 11″×7′ 1½″×9′ 11″.
Weight: 2 tons 14¾ cwt.
Note: 4625 produced 1939–41. K2 was 2-tonner with closed cab, 14,685 produced 1939–45. Early K30 trucks had 32×6 tyres, dual rear.

Truck, 30-cwt, 4×2, GS (Bedford OXD) 6-cyl., 72 bhp, 4F1R, wb 9′ 3″, 16′ 3″×7′ 1½″×9′ 9″, 2 tons 13½ cwt. 10.50–16 tyres (early types 34×7). OXA had armoured body (airfield defence); OXC was chassis for semi-trailer tractor and mobile canteen.

Truck, 30-cwt, 4×2, GS (Commer Q2) 6-cyl., 66 bhp, 4F1R, wb 10′ 0″, 17′ 6″×7′ 0″×9′ 4½″, 2 tons 4 cwt. 32×6 (2DT) tyres. Also produced with other bodies, incl. explosives van (with diesel engine) and as 15-cwt van (RAF) with 10.50–16 tyres. Basically commercial type. Some had longer wheelbase.

Truck, 30-cwt, 4×2, GS (Thornycroft GF/TC4) 4-cyl., 60 bhp, 4F1R, wb 10′ 1½″, 17′ 0″×7′ 0″×9′ 6″(7′ 7″), 2 tons 12 cwt. 8.25–20 tyres, singles all round. Militarised version of Thornycroft 'Nippy'. RAF had 'Nippy' chassis with dropside and tipper bodies at airfields in Malta.

Truck, 30-cwt, 4×2, GS (Thornycroft HF/TC4) 4-cyl., 60 bhp, 4F1R, wb 12′ 6″, 19′ 9″×7′ 0″×9′ 0″(7′ 5″), 2 tons 12½ cwt. 8.25–20 tyres. FC lwb version of 'Nippy', having 3 ft longer body. Another example of a militarised commercial 3-ton load carrier. 150 built in 1939.

Truck, 30-cwt, 4×2, GS (Bedford MSC) 6-cyl., 72 bhp, 4F1R, wb 10′ 0″, 17′ 6″×6′ 9″×9′ 8″, 2 tons 6 cwt. Termed 'GS Tender' by RAF. Spurling-built D/S body. Note supplemental engine-cooling air louvres. Also built as tipper (2DT). MSC GS trucks were also supplied to Greece.

Truck, 2-ton, 4×2, Refueller, 350-gal. (Albion AM463) 4-cyl., 65·5 bhp, 4F1R, wb 12′ 0″. Special AM chassis, introduced in 1934 for ambulance role, but later also used for several other types incl. GS, crane, tractor, van, etc. 1900 chassis supplied.

Truck, 30-cwt, 4×2, GS (Dennis AM 30/40-cwt) 4-cyl., 75 bhp, 4F1R×2, wb 12′ 4″, 19′ 10″×7′ 6″×10′ 3″, 3 tons. 10.50-20 tyres. Special Air Ministry chassis, used to tow trailers (home use only). Vacuum pipeline to rear for trailer-brake operation.

Truck, 30-cwt, 4×2, Workshop (RAF) (Fordson WOT3) V-8-cyl., 85 bhp, 4F1R, wb 11′ 11½″, 18′ 11½″×7′ 2½″×8′ 11″, 2 tons 18 cwt. One of several body types. GS type most common. Early GS trucks had 32×6 (2DT) tyres. Supplied to Indian Army with open cab.

161

Truck, 30-cwt, 4×2, GS (Morris-Commercial CS11/30) 6-cyl., 60 bhp, 4F1R, wb 11' 2", 17' 0"×7' 0"×9' 6". 10.50–16 sand tyres. Produced 1935–39. Also supplied with closed cab and with 4-cyl. engine (C11/30) and with semi-FC (ambulance).

Truck, 30-cwt, 4×4, GS (modified) (Fordson WOT8) V-8-cyl., 85 bhp, 4F1R×2, wb 9' 10", 16' 8½"×7' 6"×9' 1¾" (GS), 3 tons 16 cwt. Only British-built 30-cwt 4×4. Derived from 3-ton WOT6. Non-detachable cab roof. 10.50–20 tyres. This unit converted into caravan. IWM photo E16978.

Truck, 30-cwt, 6×4, GS (Morris-Commercial CDF) 4-cyl., 55 bhp, 5F1R, wb 10' 7½" (BC 3' 4"), 17' 4"×6' 2"×8' 8", 2 tons 14 cwt. 7.50–20 tyres. CDFW had 4-ton winch. Produced 1933–40. Also produced as mobile office. Normal control version was CD (1933–39).

Truck, 30-cwt, 6×4, Light Breakdown (Morris-Commercial CDSW) 6-cyl., 60 bhp, 5F1R, wb 9' 7½" (BC 3' 4"), 17' 4½"×6' 10½"×7' 9" (6' 3"), 3 tons 8¾ cwt. Jib with 1-ton hoist (w/pulley blocks) at rear, removable counterbalance weights at front. 4-ton winch. Worm-drive rear axles. 9.00–16 tyres.

GB

TRUCKS, 3-TON 4×2

Makes and Models: Albion CL125, KL127. Austin K3. Bedford OYC, OYD. Commer Q4. Dennis WD Tipper. Dodge 80B, 82, 82A. Guy PE. Leyland Lynx WDZ. Morris-Commercial CS 10/80 (WD 10/40), CVS 11/40, CVS 13/50. Thornycroft WZ/TC4, ZS/TC4. Tilling Stevens TS20/2, 3, 4.

General Data: The 3-tonner in its various configurations was the backbone of British military motor transport in World War II. From just under 10,000 in 1939 the total number of vehicles in this class rose to almost 390,000 at VE day, which was more than two and a half times the total number of motor vehicles of all types in British service at the end of World War I. 3-ton 4×2 types were produced by practically all the major British truck manufacturers and large numbers were also supplied by Canada and the USA, both directly and via Lend-Lease (Canadian and US Chevrolets, Dodges and Fords, as well as US Brockways and GMCs). Several early British types were commercial trucks, militarised by fitting open-type cabs and WD GS bodies. Later only closed cabs were used. It can be said that the 3-tonner was for the British what the 2½-tonner was for the Americans; in both cases it was the heaviest type that could be produced in large quantities by the biggest truck producers. A great variety of body types was mounted on this type of chassis, including GS (general service), Battery Slave, Light Breakdown, Machinery, Petrol Tanker, Searchlight, Store, Tipping, Water Tanker, Workshop, etc. A large number of British three-tonners was supplied to the USSR during 1941–44 (Albion, Austin, Bedford and Ford). Some types remained in use in the British Army after the war, notably the Bedford OY.

Vehicle shown (typical): Truck, 3-ton, 4×2, GS (Commer Q4)

Technical Data:
Engine: Commer 6-cylinder, I-L-W-F, 4·08 litres, 81 bhp @ 3200 rpm.
Transmission: 4F1R.
Brakes: hydraulic.
Tyres: 10.50–16.
Wheelbase: 13′ 9″.
Overall l×w×h: 21′ 1″×7′ 1½″×9′ 11″.
Weight: 2 tons 19½ cwt.
Note: early production (from 1939) had 34×7 tyres, dual rear.

Truck, 3-ton, 4×2, GS (Austin K3) 6-cyl., 60 bhp, 4F1R, wb. 13' 1¼", 20' 11"×7' 6"×9' 11", 2 tons 17 cwt. Early type with open cab and 34×7 (2DT) tyres. Later production had split-top closed cab, 10.50–16 tyres. 17,097 produced from 1939 until mid-1945 (K3 Series YB, ZC and ZR).

Truck, 3-ton, 4×2, GS (Bedford OYD) 6 cyl., 72 bhp, 4F1R, wb 13' 1", 20' 5"×7' 1½"×10' 2", 2 tons 13 cwt. 10.50–16 tyres. Early models had 32×6 (2DT) tyres. Chassis/cab also used for mounting of several types of special bodies.

Truck, 3-ton, 4×2, Water, 350-gal. (Bedford OYC) As OYD but dimensions 20' 1"×6' 7"×7' 7", weight 3 tons 2¼ cwt. Camouflage cover (tilt) carried on detachable superstructure. Superseded in 1943 by 500-gal. model. Also fitted with 800-gal. petrol tank. Commonly known as 'bowsers'.

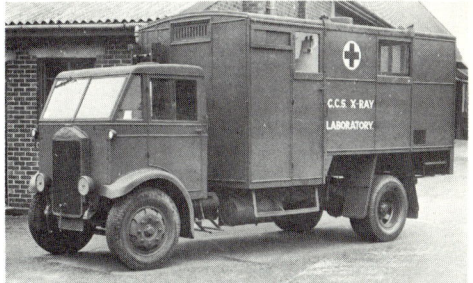

Truck, 3-ton, 4×2, X-Ray Laboratory (Albion KL127) 4-cyl., 59·3 bhp, 4F1R, wb 14' 0", 21' 6"×7' 6"×10' 3", 4 tons 12 cwt. 10.50–20 tyres. Mobile X-Ray facility for CCS (Casualty Clearing Station) on commercial-type Albion chassis. Bodywork by Lagonda. Superseded in 1943 by Bedford OY.

164

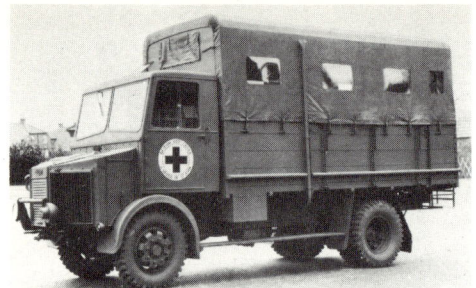

Truck, 3-ton, 4×2, Workshop (Tilling-Stevens TS20) 4-cyl., 70 bhp, 4F1R×2, wb 12′ 0″. 24-kW generator ahead of engine. Used by British Red Cross and Order of St John. Army used similar chassis for Searchlight truck. Tyres 9.00–20. Servo-assisted mech. brakes. 4·57-litre engine.

Truck, 3-ton, 4×2, Tipping (Dennis) 4-cyl., 75 bhp, 4F1R, wb 9′ 8″, 17′ 9″×7′ 3″×7′ 4″, 2 tons 15 cwt. 32×6 (2DT) tyres. Semi-forward control end-tipper with dropside body and Edbro hydraulic gear. Developed from commercial type. Known popularly as 'Dennis Pig'.

Truck, 3-ton, 4×2, Workshop (RAF) (Dodge 82, 82A) 6-cyl., 92 bhp, 4F1R, wb 13′ 6″, 20′ 8″×6′ 10″×12′ 0″ approx. 32×6 (2DT) tyres. 'A'-frame with lifting tackle at rear. Also short wheelbase (9′ 8″) chassis for tipper body, model 80. Built by Dodge at Kew, Surrey.

Truck, 3-ton, 4×2, GS (Leyland Lynx WDZ1) 6-cyl., 76·6 bhp, 5F1R, wb 12′ 0″, 20′ 3″×7′ 6″×10′ 3″, 3 tons 5¼ cwt. 8.25–20 (2DT) tyres. Also produced with civilian-type closed cab. Some 1500 were built, at Leyland's Kingston (Surrey) factory, during the early part of the war.

Truck, 3-ton, 4×2, Tipping (Thornycroft WZ/TC4) 4-cyl., 60 bhp, 4F1R, wb 9' 0", 15' 4"×7' 1"×7' 9", 3 tons 4¾ cwt. 10.50–20 tyres. Three-way tipper, hydraulically operated from PTO on transmission. Cab top was detachable. Also lwb (GS). Note two 'gas detector trays' below windscreen.

Truck, 3-ton, 4×2, Generator Water Tank (Thornycroft ZS/TC4) 4-cyl., 60 bhp, 4F1R, wb 13' 4", 21' 8"×6' 7"×8' 6", 7 tons 9¼ cwt. Generator in front of engine and directly coupled to it. Used by RAF in conjunction with semi-trailer-mounted Tender, Photographic, Mks I and II. 1800 produced.

Truck, 3-ton, 4×2, Searchlight (Thornycroft ZS/TC4) 4-cyl., 60 bhp, 4F1R, wb 13' 4", 21' 7½"×7' 4"×10' 8", 4 tons 3½ cwt. Generator mounted in front of engine. 90-cm projector carried in main body (loading ramps under floor). 1576 produced. Similar vehicles supplied by Guy (PE), Tilling-Stevens (TS20).

Truck, 3-ton, 4×2, GS (Thornycroft WZ/TC4) 4-cyl., 60 bhp, 4F1R, wb 13' 4", 21' 0"×7' 4"×9' 8", 3 tons 6¼ cwt. 10.50–20 tyres. Militarised version of civilian 'Sturdy' forward-control 5/6 tonner. Overhead worm-drive rear axle. Cab top non-detachable. Also swb version (tipper). 1800 produced.

TRUCK, 3-TON 4×4

Makes and Models: AEC Matador. Albion FT11N. Austin K5 Series (K5/YN, ZD, ZK, ZT). Bedford QL Series (QLB/C/D/T/R/W). Crossley Q (RAF only). Fordson WOT6. Guy Lizard. Karrier K6. Thornycroft Nubian TF/AC4/1.

General Data: The 4×4 3-ton load carrier, development of which was sadly neglected during the 1930s, was intended to eventually replace the 4×2 and 6×4 types. Owing to production difficulties of this for the British motor industry a new type of vehicle, the 4×2 type, had to be kept in production until the end of the war. It was also found that the 4×4 could not accommodate certain specialist bodies for which the 6×4 type chassis was being used and it was not until after the war that the 3-ton 4×4 became what it is today—a universal chassis, used by all services for a multitude of roles. Nevertheless, large numbers of 3-ton 4×4s were produced during the war, notably by Vauxhall Motors (Bedford), Ford and Austin. Vauxhall produced 52,245 QLs, from February, 1941, and after the war supplied many thousands of its successor, the RL. Austin, also starting in 1941, turned out 12,280 K5s (and in addition 3373 troop-carriers on the Bedford QLT chassis). The RAF, who until early 1941 had been responsible for their own vehicle developments, had their first 3-ton 4×4 (the Crossley Q) in full-scale production in April, 1940. Albion and Thornycroft were also early and had their first 4×4s on the road in February and June, 1940, respectively. The Nubian was the only type to remain in production after the war in substantially the same form (and as 6×6). Large numbers of Canadian Chevrolets (C60L & S) and Ford (F60L & S) were also used. These complied with the same British WD specifications as the British-built types. British bodies on the 3-ton 4×4 chassis included: GS, AT Portee, Cipher Office, Command, Fire Tender, Kitchen, Machinery, Mobile Terminal Carrier, Recorder AA Mk I (Westex Recorder), TEV, Troop Carrier, Wireless, etc.

Vehicle shown (typical): Truck, 3-ton, 4×4, GS (Albion FT11N).

Technical Data:
Engine: Albion EN280A 6-cylinder, I-L-W-F, 4·57 litres, 90 bhp @ 2800 rpm.
Transmission: 4F1R×2 (w/third diff.).
Brakes: mechanical (servo-assisted).
Tyres: 12.00–20.
Wheelbase: 12′ 0″.
Overall l×w×h: 20′ 3″ × 7′ 6″ × 9′ 10″ (7′ 3″).
Weight: 4 tons 10 cwt.
Note: produced Feb., 1940, to Aug., 1944.

167

Truck, 3-ton, 4×4, GS (Austin K5/ZD, ZK, ZT) 6-cyl., 85 bhp, 4F1R×2, wb 12′ 0″, 19′ 8″×7′ 3″×9′ 11″(6′ 8″), 3 tons 19 cwt. Also with half-cab and with other types of bodywork. K5/YN had non-detachable cab top. Servo-assisted hyd. brakes. 10.50–20 tyres. 3·99-litre engine.

Truck, 3-ton, 4×4, GS (Bedford QLD) 6-cyl., 72 bhp, 4F1R×2, wb 11′ 11″, 19′ 8″×7′ 5″×9′ 10″, 3 tons 5 cwt. Largest-production British 3-ton 4×4. Also used with other types of bodywork, incl. Bofors Tractor, Petrol Tanker, Troop-carrier, FFW, Fire Tender, Field Kitchen, Airportable Tipper w/Winch, etc.

Truck, 3-ton, 4×4, GS (Fordson WOT6) V-8-cyl., 85 bhp, 4F1R×2, wb 11′ 11½″, 19′ 10½″×7′ 6″×10′ 6″(6′ 9″), 3 tons 18 cwt. 10.50–20 tyres. Like late-production QL, had detachable top to cab to reduce shipping height. Also with various types of Machinery bodies. Production 1/42–9/45.

Truck, 3-ton, 4×4, GS, w/Winch (Karrier K6) 6-cyl., 80 bhp, 4F1R×2, wb 13′ 0″, 18′ 11½″×7′ 2½″×10′ 4″, 3 tons 19 cwt. 4¼-ton power winch fitted under flat-floor body. 120-ft winch cable. Models without winch had wheel-arch body. Split-top type cab. Produced from December, 1940.

Truck, 3-ton, 4×4, GS (Crossley Q Type 2, 3) 4-cyl., 96 bhp, 4F1R×2, wb 11′ 6″, 20′ 3″×7′ 6″×11′ 0″, 4 tons 14½ cwt. 12.00—20 tyres. Standardised RAF chassis, also used for various other bodies, incl. Fire Tender, MT Breakdown and Workshop. Tractor version had 8′ 5¾″ wb.

Truck, 3-ton, 4×4, GS (Guy Lizard) Up-rated version of the 4×4 Quad-Ant and intended as a GS load carrier. A few prototypes were produced in 1938/39, but no series production followed. The chassis, suitably modified, was used for a series of armoured command vehicles (ACV).

Truck, 3-ton, 4×4, GS (Thornycroft Nubian TF/AC4/1) 4-cyl., 85 bhp, 4F1R×2, wb 12′ 0″, 20′ 3½″×7′ 6″×10′ 3″, 4 tons 15 cwt. Continued after the war as 4×4 and 6×6 with Thornycroft, AEC or RR engine (incl. diesel). Featured epicyclic hub reduction gears. About 5000 produced, from 8th June, 1940.

Truck, 3-ton, 4×4, GS (Bedford QL) 6-cyl., 72 bhp, 4F1R×2, wb 11′ 11″, 19′ 3″×7′ 6″×10′ 0″ approx. Originally built as AT 6-pdr Portee and Fire truck, but later during the war re-bodied. Soft-top cab with folding windscreen. Same applied to similar Austin K5. See also illustrations on pages 134 and 187.

Truck, 3-ton, 4×4, Machinery (Albion FT11) 6-cyl., 90 bhp, 4F1R×2, wb 12' 0", 20' 8"×7' 9½"×10' 2"(6' 10"), 4 tons 16¼ cwt. No. 4 Mk V body for various workshop roles. Split-top cab. Hinged body sides for horizontal position or double fold. Also on Ford WOT6.

Truck, 3-ton, 4×4, Wireless, House Type (Bedford QLR) Standard QL chassis with PTO-driven aux. generator. Three basic body shell types; original type (shown) used for Cipher Office, Command, Mobile Terminal Carrier, TEV (shown), Wireless. 20' 6"×7' 7"×9' 7", 3 tons 19¼ cwt (less equipment).

Truck, 3-ton, 4×4, Armoured, Ammunition (AEC Matador 0853) Also listed as 4-ton. Basically similar to the 'Deacon' AT 6-pdr gun carrier (SP) (see page 187). At least one AEC Matador appeared with soft-top cab and CMP 3-ton GS body (probably modified from amn truck or 'Deacon').

Truck, 3-ton, Half-track, GS (Bedford QL, modified) QLD with hind quarters of Bren carrier ('Bedford-Bren') suggested by MoS as method of saving rubber tyres. Pilot only. Other QL mods. included the 'Giraffe' (elevated engine and cab) and a twin-engined GS truck with armoured cab.

GB

TRUCKS, 3-TON 6×4

Makes and Models: AEC Marshal. Albion WD 131, BY1, BY3, BY5. Austin K3/YF/YX/YY/ZB, K6, K6/A, K6/ZB. Crossley IGL (NC, RAF), IGL8 (FC, WD). Ford E917T (RAF), Fordson WOT1, 1A, 1A/1 (RAF). Guy FBAX. Karrier CK6, FM6A. Leyland Retriever WLW1, 1C, 2A, 3, 3A, 4, 4A, 4B, Terrier TSE4. Thornycroft Tartar WO/AC4, WOF/AC4, WOF/DC4.

General Data: Apart from GS bodies a great variety of specialist WD and RAF bodies were fitted on the above chassis, incl. Bridging, Breakdown, Crane, Crewbus (RAF), Derrick, Machinery, Printing, Searchlight, Signals, Wireless, etc. The 6×4, or 'rigid six-wheeler', was first introduced during the mid-1920s and was highly favoured by the British War Office. The industry was encouraged to produce these chassis by means of special subsidies for customers and the free use of the WD-patented 'articulating rear bogie'. The light (30-cwt) 6×4 was produced during the late 1920s and the 1930s by Crossley, Garner, Morris-Commercial and Vulcan. The medium type (3-ton) by Albion, Crossley, Guy, Karrier, Leyland, Thornycroft and Vulcan. Heavier 6×4 and 6×6 types were produced by FWD, Guy and Scammell. At the outbreak of war the production of the 3-ton type by some of the above manufacturers was increased, since the 4×4 type, which was intended to supersede it, was not ripe for production yet. As it was, several 6×4s remained in production throughout the war, mainly as basis for specialist bodies. Austin and Ford produced the largest numbers, chiefly for the RAF. These two chassis were fitted with a multitude of body types and remained in use for many years. The only types to be produced in this class after the war were the Ford Thames ET6 Sussex for the RAF and the Albion FT103N for the Army. During the war the following American types were imported: Chevrolet/Thornton, Dodge/Thornton WK60 Special, and GMC AFWX-354, as well as the 2½–5-ton 6×4 GMC and Studebaker. Canada used the Leyland Retriever for certain specialist roles. Britain also imported quantities of 2½-ton 6×6 trucks from the USA (GMC and Studebaker).

Vehicle shown (typical): Truck, 3-ton, 6×4, Breakdown Gantry (Austin K6/A)

Technical Data:
Engine: Austin 6-cylinder, I-I-W-F, 3·99 litres, 72 bhp @ 2800 rpm.
Transmission: 4F1R×2.
Brakes: hydraulic, servo-assisted.
Tyres: 9.00–20.
Wheelbase: 12′ 9″ (BC 4′ 0″).
Overall l×w×h: 20′ 3″×7′ 6″×11′ 0¼″.
Weight: 5 tons 18 cwt (incl. ballast).
Note: introduced Feb. 1944. Similar body on Crossley, Dodge (US), Guy, Leyland. 5-ton winch under body. 10½ cwt of ballast at front (half of which detachable). Vacuum servo brake connection (for 7½-ton 6-wh. Light Recovery Trailer).

Truck, 3-ton, 6×4, GS (Austin K3/YF) 6-cyl., 72 bhp, 4F1R×2, wb 12′ 9″ (BC 4′), 21′ 11″×7′ 6″×10′ 4½″, 4 tons 8 cwt. 9.00–20 tyres. Early type with open cab, civilian-type front end. Superseded by K6. Austin produced 13,279 6×4s of various types, 1939–45.

Truck, 3-ton, 6×4, Signals (Austin K6/ZB) 6-cyl., 83 bhp, 4F1R×2, wb 12′ 9″ (BC 4′). One of several types of signals shell bodies fitted on this standard RAF chassis. This is a Type F Mk 3, also mounted on the Fordson WOT1A/1.

Truck, 3-ton, 6×4, MT Breakdown (Crossley IGL) 4-cyl., 90 bhp, 4F1R×2, wb 12′ 6″ (BC 4′ 0″), 22′ 3″×7′ 6″×10′ 10″, 5 tons 1 cwt. Gnuss air springs at front. Body contained workbench and equipment. Harvey Frost 5-ton hand-operated 'Ambulance Crane'. Also other body types. RAF.

Truck, 3-ton, 6×4, Balloon Winch (Dodge Kew) Most RAF balloon winches were mounted on Fordson Sussex and WOT1 6×4 chassis. Few were based on other chassis, incl. Austin K6 (with ballast weights at front, as K6/A Breakdown) and Kew-built Dodge (shown).

172

Truck, 3-ton, 6×4, Searchlight (Fordson (E)917T) V-8-cyl., 85 bhp, 4F1R, wb 13′ 1″. Militarised British-built Fordson Sussex chassis. 1939 (shown) and 1938 (817T) models were used, mostly by the RAF. Superseded by the military WOT1. On both there were several body types.

Truck, 3-ton, 6×4, GS (Fordson WOT1) V-8-cyl., 85 bhp, 4F1R, wb 13′ 8½″ (BC 3′ 6¼″), 23′ 3″×7′ 0″×10′ 2½″(6′ 10″), 3 tons 6 cwt. Tyres 9.00 or 10.50–16. Middle axle had two crown wheels and two pinions, one for driving the back axle. Some had two-speed aux. gearbox.

Truck, 3-ton, 6×4, Floodlight (Fordson WOT1A) V-8-cyl., 85 bhp, 4F1R(×2), wb 14′ 7⅝″ (BC 3′ 6½″), 24′ 6″×6′ 10″ ×11′ 7″, 5 tons 4½ cwt. Same equipment on Brockhouse 4-wh. trailer and on WOT1 chassis. Some had floodlight mounted centrally, generating set at rear. Late type cab shown.

Truck, 3-ton, 6×4, Dental Surgery (Fordson WOT1) WOT1, 1A and 1A/1 chassis were used for many special bodies incl. Office, Parachute Drying, Power and Charging Equipment, etc. WOT1 and 1A (LWB) models had mech. brakes, WOT1A/1 had vacuum servo assisted type. Dim. 24′ 4″×7′ 6″×10′ 8″.

173

Truck, 3-ton, 6×4, GS (AEC Marshal) 4-cyl., 70 bhp, 4F1R×2, wb 12' 8½", (BC 4'), 22' 1"×7' 6"×10' 4½"(8' 2"), 5 tons 8 cwt. 600 built as GS and Bridging (Small Box Girder and Trestle or Sliding Bay) units. 5·1-litre engine.

Truck, 3-ton, 6×4, Bridging (Albion BY3) 6-cyl., 80 bhp, 4F1R×2, wb 13' 0" (BC 4'), 22' 4"×7' 6"×8' 3", 4 tons 10¼ cwt. BY1 similar but 4-cyl. 63·5-bhp engine. Trestle and Sliding Bay body (also on AEC and Leyland). BY1 produced 1938–40, BY3 6/1940–11/1941.

Truck, 3-ton, 6×4, Bridging (Albion BY5) 6-cyl., 96 bhp, 4F1R×2, wb 13' 0" (BC 4'), 21' 9"×7' 6"×10' 3", 4 tons 10 cwt. Folding Boat Body No. 6 Mk I (Weymanns), also on BY1 and BY3. BY5 produced 5/41–8/45. Similar: Pontoon Body No. 5 Mk I (Raft Unit) on Albion and Leyland.

Truck, 3-ton, 6×4, Searchlight (Crossley IGL8) 4-cyl., 75 bhp, 4F1R×2, wb 12' 10" (BC 4'), 22' 0"×7' 6"×11' 7", 6 tons 9 cwt. Early type. Superseded by Guy and Leyland. Other bodies on FC Crossley: GS, Breakdown, Workshop, Derrick, Coles Crane.

174

Truck, 3-ton, 6×4, Searchlight (Guy FBAX) 4-cyl., 76 bhp, 4F1R×2, wb 12' 6" (BC 4'), 21' 3"×7' 6"×11' 8", 6 tons 9¾ cwt (all incl. body). 24-kw PTO-driven generator visible behind cab. Other bodies on Guy chassis: GS, Breakdown, Workshop, Wireless.

Truck, 3-ton, 6×4, Bridging (Karrier CK6) 6-cyl., 80 bhp, 5F1R, wb 12' 6" (BC 4'), 21' 10"×7' 7½"×8' 5", 4 tons. Similar body on AEC and Albion. Carried pontoons and folding boat equipment. Also on Karrier CK6 chassis: GS, Workshop.

Truck, 3-ton, 6×4, Searchlight (Leyland Retriever WLW3) 4-cyl., 73 bhp, 4F1R×2, wb 13' 0" (BC 4'), 22' 3"×7' 6" ×11' 10", approx. 6 tons. Other bodies: GS, Machinery, Breakdown, Workshop, Derrick, Bridging, Wireless, Coles Crane, Gun Mounts, etc. Total produced: 6542.

Truck, 3-ton, 6×4, GS (Thornycroft Tartar WOF/AC4/1) 4-cyl., 77 bhp, 4F1R×2, wb 13' 0" (BC 4'), 22' 5"×7' 6" ×10' 8½", 4 tons 15¼ cwt. Superseded WO model in Oct., 1941. Some fitted with closed cab after 1945. WOF/DC4/2 had 4-cyl. 60-bhp diesel engine. Four other body types.

GB

TRUCKS, 5- and 6-TON 4×2

Makes and Models: AEC (5-ton). Bedford OWL OWST (5-ton). Commer Q6 (6-ton). Dennis (5-ton), Max Mk I and II (6-ton). Dodge 100, 101 (5-ton), 120, 121 (6-ton). ERF 2CI4 (6-ton). Foden DG4/6 (6-ton). Fordson Thames 7V (4–6-ton). Leyland Badger (6-ton). Maudslay Militant (6-ton).

General Data: The above 5- and 6-ton trucks were basically commercial types. The 6-ton class was a standardised WD payload class, but little attempt was made to convert the commercial type lorry into a militarised vehicle. Only the 6-ton Dennis Max, ERF, Foden and Maudslay had been modified to suit military requirements (WD GS bodies, radiator guards, etc.), and the last three of these had much in common. Some Dennis 5-tonner semi-forward control types were used in small numbers for specialist bodies like Explosives Van (WD) and 750-gal. Sullage Tender (RAF). AEC and Leyland (Badger) commercial types were used by the RAF and RN. From the US the following commercial-type heavy trucks were obtained: Autocar 5-ton, 4×2 (1250-gal. petrol tanker), Brockway 6-ton, 4×2 (1600-gal. petrol tanker), and 10-ton, 4×2 (tipper), International 6- and 7-ton, 4×2 (GS), Mack 5-ton, 4×2 (GS), and 10-ton, 4×2 (tipper), and White 10-ton, 4×2 (GS and Aircraft Carrier). At the beginning of the war a number of heavy commercial trucks (four- and six-wheelers) was used for the mounting of concrete pillboxes. These vehicles, like several other types of improvised armoured vehicles, were intended for aerodrome defence. The concrete armour was proof against armour-piercing ammunition and vehicles thus equipped were known as 'Bisons' after the trademark of their producers, the firm of Concrete Ltd. The front ends of these mobile pillboxes were also protected with concrete armour.

Vehicle shown (typical): Truck, 6-ton, 4×2, GS (Foden DG4/6)

Technical Data:
Engine: Gardner 4LW 4-cylinder diesel, I-I-W-F, 5·5 litres, 68 bhp @ 1700 rpm.
Transmission: 5F2R.
Brakes: hydraulic.
Tyres: 36×8 (2DT).
Wheelbase: 12′ 7″.
Overall l×w×h: 21′ 9″×7′ 6″×10′ 1″ (7′ 9″).
Weight: 5 tons 1½ cwt.

176

Truck, 6-ton, 4×2, GS (ERF 2CI4) Gardner 4LW diesel, 4-cyl., 68 bhp, 5F1R, wb 12' 5½", 21' 7"×7' 8½"×10' 5½"(8' 0"), 4 tons 18 cwt. 36×8 (2DT) tyres. Worm-driven rear axle. Overdrive gearbox. Militarised model CI4 commercial lorry.

Truck, 6-ton, 4×2, GS (Maudslay Militant) Gardner 4LW diesel, 4-cyl., 68 bhp, 4F1R, wb 14' 0", 21' 7"×7' 7" ×10' 9"(7' 4"), 4 tons 19¾ cwt. Standard flat-floor GS body. After the war this vehicle developed into the commercial Militant Mk II. Note short rear overhang.

Truck, 6-ton, 4×2, GS (Dennis Max Mk I) 4-cyl. diesel, 77 bhp, 4F1R, wb 14' 0", 21' 10½"×7' 6"×10' 6"(8' 5"), 4 tons 16 cwt. Dennis O4 6·5-litre engine. Tyres 36×8(2DT). Servo-assisted mechanical brakes. Overhead worm-drive rear axle. Flat-floor body.

Truck, 6-ton, 4×2, GS (Dennis Max Mk II) 4-cyl. diesel, 77 bhp, 5F1R (OD top), wb 14' 0", 21' 10"×7' 6"×9' 5"(7' 0"), 5 tons 4 cwt. Introduced in June, 1944. Split-type cab with hip ring in roof. Wheel-arch body. Detail changes to instruments and controls to facilitate waterproofing.

Truck, 6-ton, 4×2, GS (Commer Q6 Superpoise) Rootes' Commer Superpoise range was introduced in April, 1939, replacing the 'N' series. Many were supplied for the Services. The Q6 was one of the heaviest. Truck shown had double-D/S body with tilt, for the MoS.

Truck, 5-ton, 4×2, 750-Gallon Cesspool Emptier (Dennis) 4-cyl., 75 bhp, 4F1R, wb 13' 6", 22' 4"×7' 4"×8' 2", 3 tons 10 cwt. Tyres 34×7. Hyd. brakes. PTO-driven rotary-type air pump. 10' lengths of 3" suction hose carried in side lockers. Used by RAF (shown) and RASC.

Truck, 4–6-ton, 4×2, Tipping (Fordson Thames 7V) V-8-cyl., 85 bhp, 4F1R, wb 9' 10", 15' 0"×7' 1½"×6' 11½", 2 tons 9 cwt. 4-cu yd all-steel body, hydraulically operated. 7V was used for various roles with short and long wb, incl. Allen Taylor tructor (RAF).

Truck, 6-ton, 4×2, Tipper (Leyland Badger) 4-cyl., 68 bhp, 4F1R. Basically commercial truck, with double-D/S tipper body for Royal Navy Torpedo Depot in Plymouth. Swb version with double-D/S wooden GS body was also used by RN. Unladen weight 5 tons 17 cwt and 3 tons 17 cwt resp.

GB

TRACTOR TRUCKS and SEMI-TRAILERS
3×2–2, 4×2–2, 4×4–2*

Makes and Models: *Tractors:* Bedford OXC (4×2), QLC (4×4). Commer Q2 and Q4 (4×2). Crossley Q (4×4). Fordson Thames 7V (4×2). Scammell MH6, MH3, MH6-3 (3×2, 'Mechanical Horse').
Semi-trailers: Carrimore; Glover, Webb & Liversidge; Scammell; Scottish Motor Traction; Tasker, etc.

General Data: Truck tractors were not widely used in the British Army, at least not for the GS load-carrying role. Many more were employed by the RAF and RN, e.g. for the conveyance of aircraft and aircraft components on a special 40-ft long semi-trailer, known as the 'Queen Mary'. These semi-trailers, as well as several other RAF types, were coupled to the Bedford OXC and Commer Q tractive units, which had a special Tasker coupling. In October, 1943, the Crossley Q-type heavy tractor was introduced and with it came a new up-rated 5-ton 'Queen Mary'. Scammell, pioneer of British 'articulated six-wheelers', fitted the Scammell automatic coupling device on a specially modified version of the Bedford OXC, the Bedford/Scammell, and supplied about 6000 semi-trailers, of many different types, including torpedo carriers, petrol tankers, etc. The Army's only British 4×4 tractor unit was the Bedford QLC. Its semi-trailer was produced by Glover, Webb & Liversidge and Scottish Motor Traction to a common design, and the outfit was generally similar to the Canadian Ford F60T 4×4–2, which was also used. Imported from Canada were two types of FWD truck tractors with GS semi-trailers: the HAR (6-ton, 4×4–2) and the SU-COE (10-ton, 4×4–2). These had been originally ordered for Canadian Army requirements, but a number were taken over by the WD. Amongst US tractor types, supplied under Lend-Lease, were the Autocar (4×4), Brockway (4×2), International (KR8 and 8R, 4×2), Mack (4×2) and Studebaker (6×4), as well as some heavier types for tank transporters.

Vehicle shown (typical): Truck, Tractor Chassis, 4×2 (Commer Q2) (used with Tasker Articulated Trailers)

Technical Data:
Engine: Commer 6-cylinder, I-L-W-F, 4·085 litres, 80 bhp @ 3000 rpm.
Transmission: 4F1R.
Brakes: hydraulic (servo-assisted).
Tyres: 10.50–16 (earlier 10.50–20).
Wheelbase: 8' 0".
Overall l×w×h and weight (complete with 3-ton long low-loading Tasker semi-trailer 'Queen Mary'): 50'×8' 6"×7' 3", 5 tons. Produced from Sept. 1939. Complete unit is shown on page 105.

*4×4–2 indicates four-wheel drive tractor with two-wheel semi-trailer.

Truck, 6-ton, 4×2–2, Petrol, 1750-Gal. (Bedford/Scammell OXC) 6-cyl., 72 bhp, 4F1R, wb 9′ 3″, overall dim. 26′ 5″ × 7′ 4″ × 8′ 8″(7′ 11″), 4 tons 5½ cwt. Double-compartment elliptical tank. Fitted with superstructure for camouflage tilt. One of 409 supplied by Scammell during 1939–45.

Truck, 6-ton, 4×2–2, GS (Bedford/Scammell OXC) 6-cyl., 72 bhp, 4F1R, wb 9′ 3″, overall dim. 25′ 0″ × 7′ 6″ × 10′ 3″, 3 tons 8¼ cwt. 10.50–16 tyres all round. Early models had 32 × 6 (4DT) tyres. Also used with various other semi-trailers incl. torpedo carrier, platform, 'Queen Mary', etc.

Truck, 6-ton, 4×4–2, GS (Bedford QLC) 6-cyl., 72 bhp, 4F1R×2, wb 11′ 11″, overall dim. 30′ 6″ × 7′ 6″ × 10′ 8″, 4 tons 18 cwt. Permanently coupled by Tasker ball-type coupling. Small D/S cargo body (1 ton) behind cab. Vacuum servo-operated trailer brake. 10.50–20 tyres all round.

Truck, 5-ton, 4×4–2, Long Low Loading (Crossley/Tasker) 4-cyl., 91 bhp, 4F1R×2, wb 8′ 5¾″, overall dim. 50′ 0″ × 9′ 0″ × 12′ 6¾″(9′ 0″), 8 tons 5 cwt. 2-ton collapsible ballast body behind cab for use when towing full-trailers. Tractor tyres 10.50–20, semi-trailer 13.50–20. Known as 'Queen Mary'.

TRUCKS, 10-TON 6×4

Makes and Models: Albion CX6N (NC), CX23N (FC). Foden DG/6/10, DG/6/12. Leyland Hippo Mk I, Mk II, IIA, IIB.

General Data: The British-built trucks in the 10-ton payload class were based on pre-war commercial diesel-engined goods chassis. The Albion CX6N and the Leyland Hippo Mk I had a rear bogie with four conventionally mounted semi-elliptic springs, the four inner ends of which were attached to 'rocking bars'. All the other models, including the Hippo Mk II range, had a 'fully articulated' rear bogie with inverted semi-elliptic springs pivoting at their centres, which gave them a certain degree of cross-country performance. All models in this class had a diesel engine and the latest types had large single tyres all round (except Hippo Mk IIA). In its GS form the 10-tonner was used mainly for long-distance haulage of supplies, in which role they were supplemented by American GMCs, Internationals, Macks and Whites, all normal-control 6×4 types. The Foden chassis was also fitted with the following special bodies: Auto Processing, Dark Room, Enlarging and Rectifying, Photo-Mechanical, Platform, Printing, Railway Breakdown (LPTB), and for the mounting of a heavy coastal gun. The latter type had an armoured half-cab. Several of these Foden types were also used by the Canadian Army. The Leyland Hippo Mk II was designed in 1943 to meet the expected needs of the invasion forces. It entered production late in 1944 and by 1945 a weekly output of forty vehicles was being obtained. Over 1000 had been produced by VE Day and it remained in use for many years by the Army and the RAF. Leyland Motors also designed a 12-ton model, but only one such vehicle was produced. The 10-ton 6×4 became a widely used military vehicle in Britain after the war. In addition to the Leyland Hippo (which developed into the 19H Series) large numbers were supplied by AEC, Albion, Thornycroft and others, supplemented by various 6×6 types produced by AEC, Leyland and Scammell. The first post-war Foden military trucks were not acquired until 1967 and were also in the 10-ton 6×4 class.

Vehicle shown (typical): Truck, 10-ton, 6×4, GS (Albion CX6N)

Technical Data:
Engine: Albion EN 244 diesel, 6-cylinder, I-I-W-F, 9·08 litres, 100 bhp @ 1750 rpm.
Transmission: 4F1R×2.
Brakes: hydraulic (servo-assisted).
Tyres: 40×8 (2DT).
Wheelbase: 16' 0", BC 4' 6".
Overall l×w×h: 28' 3"×7' 7"×11' 7½" (9' 6").
Weight: 7 tons 10 cwt.

Truck, 10-ton, 6×4, GS (Albion CX23N) 6-cyl. diesel, 100 bhp, 4F1R×2, wb 16′ 0″ (BC 4′ 6″), 26′ 4½″×8′ 3½″×11′ 0″(9′ 1″), 8 tons 2¼ cwt. Produced in 1941 and used by RASC for heavy convoy work. 13.50–20 tyres, single rear. GS body with well-floor.

Truck, 10-ton, 6×4, GS (Foden DG/6/12) Gardner 6-cyl. diesel, 102 bhp, 5F2R, wb 15′ 8″ (BC 4′ 4″), 26′ 8″×7′ 7″×10′ 6″ (7′ 6″), 8 tons 6 cwt. 13.50–20 tyres. Split-top cab. Well-type body. DG/6/10 had 36×8 (4DT) tyres and flat-floor body.

Truck, 10-ton, 6×4, GS (Leyland Hippo Mk I, WSW17) 6-cyl. diesel, 97·2 bhp, 5F1R, wb 16′ 0″ (BC 4′ 7″), 28′ 3″×7′ 6″×11′ 2″(7′ 9″), 7 tons 4¼ cwt. 36×8 (4DT) tyres. Pre-war commercial type fitted with WD-pattern open cab and GS body. Operated by RASC.

Truck, 10-ton, 6×4, GS (Leyland Hippo Mk II, IIB) 6-cyl. diesel, 101 bhp, 5F1R×2, wb 15′ 6″ (BC 4′ 7″), 27′ 3″×8′ 1″×10′ 11″(7′ 6″). 13.50 or 14.00–20 tyres (Mk IIA: 10.50–22 front, 36×8 twin rear). Introduced in 1944. After 1945 also used with van-type bodies. Mk IIB similar. Over 1000 built.

TANK TRANSPORTERS

Makes and Models: Albion CX24S: Transporter, 20-ton, 6×4–4 (later de-rated to 15-ton). Scammell 'Pioneer' TRMU: Transporter, 20-ton, 6×4–8, Recovery (Scammell semi-trailer); ditto, 30-ton (Shelvoke & Drewry semi-trailer); ditto, 30-ton (Scammell TRCU/30 semi-trailer).

General Data: Tank transporters were used for the haulage of tanks and other full-track AFVs to and from battle zones and most types also for their recovery in cases of breakdown or damage. The most widely used British types were the Scammell 20- and 30-ton tractor/semi-trailer combinations. Large numbers of US types were also used, mainly the Diamond T with either its original 45-ton trailer, M9, or British full- or semi-trailer. The US Federal 604 and Reo 28XS were also used, as well as the Pacific/Fruehauf, M25, and, during the early war years, Mack and White 6×4 tank carriers (see pages 92/93). One Pacific M26 tractor was modified in Britain to have single 21.00–24 tyres and a ballast body. The Albion CX24, some 800 of which were produced, proved inadequate as a tank transporter, especially in North Africa, and was subsequently de-rated to 15-ton payload and used for carriage of cable drums, telegraph poles, etc. With tank weights increasing, heavier transporters were needed. Numbers of Scammell 20-tonners were modified to 30-tonners by replacing the semi-trailer with the 30-ton Shelvoke & Drewry unit, which had originally been designed for coupling to the Diamond T 980 tractor (modified prime mover). It was generally similar to the standard Scammell 30-tonner. British trailer manufacturers turned out some very good types of transporter and recovery trailers (BTC, Crane, Dyson, S & D, etc.) including a tracked type (Boulton & Paul) but, especially for suitable tractors, the MoS was very much dependent on supplies from the US. Requirements ran into thousands, whilst the total production of British 30-tonners (9/39-8/45) was only just under 500 units. An armoured recovery vehicle (ARV) was also used; it was built on a modified US medium tank chassis. Numbers of the Scammell tractors were after the war converted to tractors for full-trailers and carried concrete ballast weights on the chassis. For Albion 8-wheeled tank transporter tractor see page 194.

Vehicle shown (typical): Transporter, 30-ton, 6×4–8, Recovery (Scammell Tractor TRMU/30, Semi-trailer TRCU/30)

Technical Data:
Engine: Gardner 6LW diesel, 6-cylinder, I-I-W-F, 8·369 litres, 102 bhp @ 1700 rpm.
Transmission: 6F1R.
Brakes: air.
Tyres: front and semi-trailer 13.50–20, tractor rear 15.00–20.
Wheelbase: 15' 0", BC 4' 6¼" (semi-trailer, from king pin: 23' 1", BC 5' 1").
Overall l×w×h, tractor: 22' 0"×8' 7"× 9' 5"; *tractor plus semi-trailer:* 49' 8"×9' 5½"×10' 11".
Weight, tractor plus trailer: 19 tons 18 cwt.
Note: Scammell 8-ton vertical winch. 548 produced (all types) during 1939–45.

Transporter, 20-ton, 6×4–8, Recovery (Scammell) Tractor similar to 30-ton model but front tyres 10.50–20. Scammell semi-trailer with 10.50–20 tyres and horizontal tank carrier runways. Overall dimensions: 49' 3" × 9' 2" × 9' 5", 15½ tons. Tractors later modified and coupled to 30-ton S & D S-T.

Transporter, 20-ton, 6×4–8 (Albion CX24S) 6-cyl., 140 bhp, 4F1R×2, wb 14' 9" (BC 4' 6") plus 16' 0" (BC 4' 6"), overall 43' 9¼" × 9' 5" × 8' 6", 15 tons 2¼ cwt. Tyres 36×8 (8DT). Developed from CX22S HAT. Was found too light for its task and de-rated to 15-ton transporter.

Transporter, 30-ton, 6×4–8 (Diamond T/S&D) Diamond T 980 tractor, modified with fifth wheel. Semi-trailer had eight 13.50–20 tyres and measured 35' 5" × 9' 8" × 8' 0". Overall 49' 10" × 9' 8" × 8' 3", 22 tons 9 cwt. Rear bogie similar to that of British 40-ton trailer.

Trailer, 40-ton, 24-Wheeled, Transporter, Mk I (Crane) 31' 11" × 10' 0" × 5' 10", 13 tons 12 cwt. Tyres 36×8 (12DT). Designed for Churchill tanks, by Crane, but also built by other mfrs. Superseded by Mk II (Dyson design; various mfrs) which had welded frame and could accommodate wider tanks.

GB

TRACTORS, PORTEES, SP GUNS, WHEELED
LIGHT AA (BOFORS) and AT

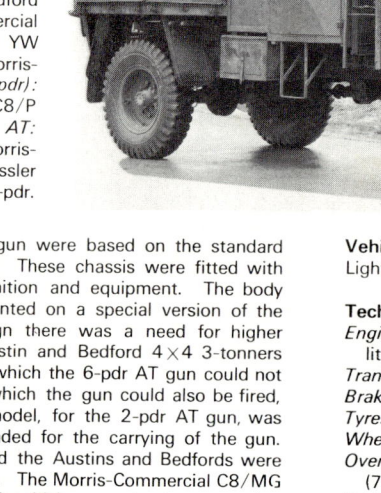

Makes and Models: *Tractors, Lt AA (Bofors):* Bedford QLB (4×4). Guy Quad-Ant (4×4). Morris-Commercial CD/SW (6×4). *Portees, AT:* Austin K5/YU and YW (6-pdr, 4×4). Bedford QL (6-pdr, 4×4). Morris-Commercial C8/MG (2-pdr, 4×4). *Tractors, AT (6-pdr):* Guy Quad-Ant (4×4). Morris-Commercial C8/AT, C8/P (4×4). Standard (WVEE 588) (4×2). *SP Guns, AT:* AEC Matador 0853, 'Deacon' (6-pdr, 4×4). Morris-Commercial C9/B (40-mm AA Bofors, 4×4). Straussler 'Monitor' (17-pdr, 4×3). Thornycroft Amazon (17-pdr, 6×4).

General Data: Tractors for the 40-mm AA Bofors gun were based on the standard Bedford 4×4 and Morris-Commercial 6×4 chassis. These chassis were fitted with special bodywork for the carrying of crew, ammunition and equipment. The body type used on the Morris-Commercial was also mounted on a special version of the Guy Quad-Ant. During the North African campaign there was a need for higher mobility of AT guns. Consequently, numbers of Austin and Bedford 4×4 3-tonners were built with a half-cab and special bodywork on which the 6-pdr AT gun could not only be carried at relatively high speeds but from which the gun could also be fired, either in forward or rearward direction. A lighter model, for the 2-pdr AT gun, was built by Morris-Commercial, but this was only intended for the carrying of the gun. Later these vehicle types were declared obsolete and the Austins and Bedfords were modified by the fitting of a GS body (see 3-ton 4×4). The Morris-Commercial C8/MG became an AT gun tractor, rather similar to the C8/P, which was a tractor conversion of the Predictor vehicle that used to be a companion vehicle to the C9/B SP Bofors. Both the C8/P and C9/B were developed from the MCC C8 FAT. Lt AA Bofors tractors, Portees and SP gun carriages were also supplied by Canada on CMP chassis. The Standard, Straussler and Thornycroft units were only produced experimentally. This also applied to some SPs based on existing British and Canadian armoured chassis.

Vehicle shown (typical): Tractor, 4×4, Light AA (Bofors) (Bedford QLB)

Technical Data:
Engine: Bedford 6-cylinder, I-I-W-F, 3·5 litres, 72 bhp @ 3000 rpm.
Transmission: 4F1R×2.
Brakes: hydraulic (servo-assisted).
Tyres: 10.50–20.
Wheelbase: 11′ 11″.
Overall l×w×h: 18′ 10″×7′ 7″×9′ 3″ (7′ 4″).
Weight: 4 tons 4¼ cwt.
Note: 5-ton winch. Tops of cab and rear compartments detachable.
  Crew: eight plus driver.

185

Tractor, 6×4, Light AA (Bofors) (Morris-Commercial CD/SW)
6-cyl., 60 bhp, 5F1R, wb 9′ 7½″ (BC 3′ 4″), 17′ 2½″ × 7′ 4″ × 7′ 6″ (6′ 7″), 3 tons 12¾ cwt. 9.00–16 tyres. 30-cwt chassis, also used with Breakdown body (see page 161). 4-ton winch. Crew: six plus driver. Earlier version used as tractor for 25-pdr.

Tractor, 4×4, Light AA (Bofors) (Guy Quad-Ant) Meadows 4-cyl., 58 bhp, 4F1R×1. Prototype, no quantity production. Body as MCC CD/SW. Spare gun barrel carried in lower portion of body well. Lockers above rear wheels and at rear for eight cases of ammunition.

Tractor, 4×4, AT (17-pdr) (Morris-Commercial C8/AT) 4-cyl., 70 bhp, 5F1R×1, wb 8′3″, 14′ 3″ × 7′ 3″ × 9′ 0″, gross wt 4 tons 19 cwt. 10.50–16 tyres. Conversion of C8/MG portee. Crew: seven plus driver. Canadian Chevrolet CGT portee also thus converted. No winch.

Tractor, 4×4, AT (17-pdr) (Morris-Commercial C8/P) 4-cyl., 70 bhp, 5F1R×1, wb 8′ 3″, dim. and wt as model shown on left. Conversion of SP Predictor vehicle which used to be operated in conjunction with MCC C9/B SP Bofors vehicle. Some had Warner electric brake controller.

Truck, 4×4, AT Portee (2-pdr) (Morris-Commercial C8/MG) 4-cyl., 70 bhp, 5F1R×1, wb 8′ 3″, 15′ 3″×7′ 0″×8′ 8″, 3 tons 5¾ cwt. Gun is hoisted into vehicle by manual winch, then rests in curved channels. Also 4-ton power winch. Crew: four plus driver. Later converted into AT tractor. IWM photo H19547.

Carrier, SP, 4×4, 40-mm AA (Bofors) (Morris-Commercial C9/B) 4-cyl., 70 bhp, 5F1R×1, wb 9′ 9″, 20′ 2″×7′ 4″×7′ 6″, 3 tons 8 cwt. Developed from C8 FAT Mk III. Tyres 10.50–16. Manual and predictor control. Also electrically operated hydraulic control. Similar body on CMP Ford chassis (F60B).

Tractor, 4×2, AT Gun (6-pdr) (Standard) 4-cyl., 36 bhp, 4F1R. 6.00–16 tyres (gun 8.25–10). Tractor-cum-gun prototype, 1943. Tractor rear-axle sprockets drove gun wheels via detachable chain final drives. When operated solo, tractor wheels were refitted onto axle hubs. IWM photo MH 4726.

Carrier, SP, 4×3, 17-pdr AT Gun (Straussler 'Monitor') Bedford QL running gear. Both rear wheels could be power-turned at 90° to front wheels and gun could thus be rotated 360°, in either direction, by engine-driven rh rear wheel. Produced experimentally in 1943. Good test results. 26′ 5″×9′ 5″×5′ 4″.

187

Truck, 4×4, AT Portee and Fire (6-pdr) (Austin K5/YW)
6-cyl., 85 bhp, 4F1R×2, wb 12′ 0″, 19′ 8″×8′ 5″×11′ 1″ (8′ 5″ plus 14″), gross wt 7 tons 18¼ cwt. K5/YU similar but with timber body frame. Gun could be fired whilst on vehicle. Note gun blast shield over radiator.

Truck, 4×4, AT Portee and Fire (6-pdr) (Bedford QL)
6-cyl., 72 bhp, 4F1R×2, wb 11′ 11″, 19′ 3″×8′ 4½″×11′ 1″, gross wt 7 tons 11¼ cwt. Similar to Austin. Also built on CMP (Ford and Chevrolet) chassis. Later all three types were converted into GS trucks (see '3-ton 4×4' section).

Carrier, SP, 4×4, 6-pdr, Mk I (AEC Matador 0853 'Deacon') 6-cyl. diesel, 95 bhp, 4F1R×2, wb 12′ 7½″, 12 tons. 13.50–20 tyres. 175 produced in 1942. After active service in North Africa they were acquired by Turkey. Could be camouflaged to look like 'soft skin' GS truck.

Carrier, SP, 6×4, 17-pdr (Thornycroft Amazon) 6-cyl., 100 bhp, 4F1R×2, wb 11′ 9″ (BC 4′ 6″), 13 tons 15 cwt (laden). Experimental SP 17-pdr gun designed in 1942. Fully-armoured (50-mm) hull. Poor cross-country performance owing to excessive weight. No series production.

TRACTORS
FIELD ARTILLERY

Makes and Models: Albion FT15N, FT15NW (6×6). Garner-Straussler G3 (4×4). Guy Quad-Ant (4×4). Karrier KT4 (4×4). Morris-Commercial C8/FWD Mk I, II, III (4×4).

General Data: The first of these field artillery tractors (FAT), or 'Quads', were introduced by Guy (Quad-Ant, 1938) and Morris-Commercial (C8 Mk I, 1938/39). The Quad-Ant FAT was discontinued in late 1943, but the chassis (which had also been used, in modified form, for the Guy armoured car) remained in production and was fitted with a GS body. Canada supplied FATs on CMP Chevrolet CGT and Ford FGT chassis; at first these were generally of the same pattern as the British type, but from 1942 they had a standard CMP cab. Later, in 1944/45 both the Morris-Commercial and the CMP types were produced with a new square-contour body with canvas top. The Karrier KT was generally similar to the other early British types and was built for the Indian Army. The Garner-Straussler was an ingenious design, powered by two Ford V8 engines. It had been designed by Mr Nicholas Straussler and the prototype had been produced by Manfred Weiss in Hungary in 1938. Garner produced a series of these, as well as prototypes of the G1 and G2 which were Ford V-8-engined 4×4 GS trucks of 1- and 2-ton capacity. On the G3, the two engines drove separate gear trains in a common transfer case. The left-hand engine drove the front axle, the other the rear axle. Either engine could be used on its own, or the two together, to drive the vehicle. Another Straussler design was the experimental LAC 4×4 produced by Alvis-Straussler Ltd on a projected armoured-car chassis. This pilot model had two engines, each driving the wheels on one side, four-wheel steering and a soft-top body for a crew of seven, incl. driver. Towards the end of the war a new, more powerful 6×6 low-silhouette type was evolved and produced in limited numbers by Albion during 1945.

Vehicle shown (typical): Tractor, 4×4, Field Artillery, w/Tilt (Morris-Commercial C8 Mk III)

Technical Data:
Engine: Morris 4-cylinder, I-L-W-F, 3·52 litres, 70 bhp @ 3000 rpm.
Transmission: 5F1R×1.
Brakes: hydraulic.
Tyres: 10.50–16.
Wheelbase: 8′ 3″.
Overall l×w×h: 14′ 9½″×6′ 11″×7′ 11″.
Weight: 3 tons 5 cwt.
Note: 4-ton winch. Drop tailboard. Earlier models had sloping rear body contour, two side doors and measured 14′ 9″×7′ 6″×7′ 9″. Mks I and II had 10.50–20 tyres and permanent all-wheel drive.

Tractor, 4×4, Field Artillery (Guy Quad-Ant) Meadows 4-cyl., 58 bhp, 4F1R×1, wb 8' 5", 15' 0"×7' 6"×7' 7¼", 3 tons 16¼ cwt. 10.50–16 (early models 10.50–20) tyres. Crew: six. Lockers for 96 rds of amn, 8 rds of AP shot. Also supplied to Egypt. Similar bodywork on Morris-Commercial C8 chassis.

Tractor, 4×4, Field Artillery (Karrier KT4) 6-cyl., 80 bhp, 4F1R×2, wb 9' 2", 15' 6"×7' 3" approx. Tyres 10.50–20. Supplied to Indian Army for towing 18- and 25-pdr field guns. 5-ton vertical winch. Body by BLSP of Slough. Chassis also supplied without winch.

Tractor, 4×4, Field Artillery (Garner-Straussler G3) Twin Ford V-8-cyl., 2×89 bhp, 2×4F1R×1, wb 12' 6⅝", 5 tons. 12.00–20 tyres. Designed by Mr Nicholas Straussler, 53 produced by Garner, originally for Turkish Government. Used in North Africa. Payload 3 tons

Tractor, 6×6, Field Artillery (Albion FT15N, FT15NW) 6-cyl., 95 bhp, 4F1R×2, wb 12' 6", 20' 11"×7' 7½"×7' 8", 5 tons 18 cwt. 10.50–20 tyres. Low-silhouette type, 1945. Turner 8-ton worm-drive winch. Prototype had FT11 cab. FT15NW was fully waterproofed. About 150 built.

TRACTORS
MEDIUM and HEAVY ARTILLERY and HEAVY AA, WHEELED

Makes and Models: AEC Matador 0853 (MAT, HAA, 4×4). AEC Prototypes (HAT, 6×6). Albion CX22S (HAT, 6×4). Scammell Pioneer R100 (HAT, 6×4).

General Data: Of the models listed above, the Scammell Pioneer was the oldest design. In its basic form it had been in production since 1927 and of the heavy artillery tractor shown here 786 were produced during 1939–45. It was also tested with driven front axle (6×6). The AEC Matador was introduced as a medium artillery tractor in 1939. Early Matador MATs, and the later HAA tractors for twin 6-pdr equipment, had a steel panelled roof. A total of 8612 Matador tractors were produced and in addition over 400 similar chassis for the RAF. The latter were mostly fitted with a platform body and used to tow full-trailers. Others had a house-type body. The Albion CX22S was introduced in February, 1944. The prototype, developed from the Albion CX23N 10-tonner, had a 'greenhouse'-type cab, like most of the AEC Matadors. In 1944 the WVEE started tests with an AEC-built 6×6 heavy artillery tractor. Two prototypes were built and one was fitted with an armoured cab. They had been developed as the prime mover for the 7·2 howitzer and the US 155-mm gun. Experiments were also carried out with semi-track types, patterned on German examples. An AEC Matador was also experimentally fitted with tracks in lieu of rear wheels. American types in this category, supplied to Britain under Lend-Lease, were the FWD SU-COE (rather similar to AEC Matador and fitted with British MAT body), the Mack NM5 and Corbitt/White 666 6-ton 6×6, and the Mack NO 7½-ton, 6×6. The latter remained in use with the British Army for many years. A full-track artillery tractor under test in 1945 was the 8-ton Alecto. It was produced by Vickers-Armstrongs, powered by a Meadows 158-bhp engine and could also be used as an SP gun mounting.

Vehicle shown (typical): Tractor, 4×4, Medium Artillery (AEC Matador 0853)

Technical Data:
Engine: AEC 0853 diesel, 6-cylinder, I-I-W-F, 7·58 litres, 95 bhp @ 1780 rpm.
Transmission: 4F1R×2.
Brakes: hydraulic (servo-assisted) or air.
Tyres: 13.50–20.
Wheelbase: 12' 7½".
Overall l×w×h: 20' 9"×7' 10½"×10' 2".
Weight: 7 tons 3 cwt.
Note: also supplied with petrol engine and with 'streamlined' cab roof. Developed from FWD/Hardy design of the early 1930s.

Tractor, 6×4, Heavy Artillery (Albion CX22S) 6-cyl. diesel, 100 bhp, 4F1R×2, wb 14′ 8″ (BC 4′ 6″), 25′ 6″×8′ 9″ ×10′ 4½″, 10 tons 8¾ cwt. 14.00–20 tyres. Well-floor body with crew compartment at front and doors either side. AA ring in roof. Commander's hip ring in cab roof.

Tractor, 6×4, Heavy Artillery (Scammell Pioneer R100) Gardner 6LW 6-cyl. diesel, 6F1R, wb 12′ 2″ (BC 4′ 3¼″), 20′ 7″×8′ 6″×9′ 10″, 8 tons 9½ cwt. 13.50-20 tyres. Steel-panelled body for nine men, ammunition, equipment. Overhead runway with 10-cwt hoist for loading and unloading.

Tractor, 6×6, Heavy Artillery (AEC, prototype) 6-cyl. diesel, 150 bhp, 4F1R×2, wb 14′ 6″ (BC 4′ 2″), 24′ 6″×7′ 8″×8′ 1½″, 9 tons 5 cwt. 14.00–20 tyres. Low-silhouette body. Cab shown was later replaced by one with a smaller area of glass. Total crew 10, eight of which in special compartment.

Tractor, 6×6, Heavy Artillery (AEC, prototype, armoured) 6-cyl. diesel, 150 bhp, 4F1R×2, wb 14′ 6″ (BC 4′ 2″). Similar to 'soft-skin' type shown on left but armoured cab and armour plate protection for crew; intended for a special role. Used same AEC 9·5-litre engine as Valentine tank.

GB

TRACTORS
VARIOUS TYPES

Note: *see preceding sections for artillery tractors, semi-trailers and tank transporters.*

Makes and Models: AEC Matador O853 (4×4 and semi-track prototype), (O)854 (6×6). Albion CX33 (prototypes, 8×8, 8×6). Alvis-Straussler Hefty (4×4), LAC (4×4). Bedford BT 'Traclat' (prototypes, semi-track). Crossley Q (4×4). David Brown AW100, AW500, VIG.1/100, VIG.1/462 (4×2). Dennis 'Octolat' (prototype, 6×6). Fordson Major (4×2). Fordson/Allen Taylor (Thames and WOT3 conv., 4×2). Fordson/Reynolds (WOT3 conv., 4×2). Fordson/Roadless (half-track). 'Loyd' Carrier (full-track; Wolseley and others). Scammell MH6 (3×2), (4×4, RN). Scammell SV/1T, SV/1S, SV/2S (B/D, 6×4). Straussler LT1 Sturdy (4×4).

General Data: The tractors shown and/or listed in this section were used to tow full-trailers or other equipment. Several experimental artillery tractors are also included. The Scammell Breakdowns were officially classified as tractors. This also applied to the AEC 854 and 0854 (diesel), which were basically Matadors with the tandem rear axles of the Marshal and mostly fitted with a 2500-gal. petrol tank (1514 produced). Other RAF types using this 6×6 chassis were mobile oxygen plants (185) and revolving cranes (192). The Albion CX33 was an eight-wheeler intended for tank recovery and transport, powered by two Albion EN248E petrol engines. The Straussler-designed 'Stury' and 'Hefty' tractors were used by the RAF as Straussler tractors Type B (Light) and Type C (Heavy) respectively. The latter was produced by Alvis-Straussler Ltd in Coventry, as was the LAC 4×4 four-wheel steer twin-engined gun tractor. The LAC was based on the Alvis-Straussler armoured-car chassis, but was not produced in quantity. American models in this class: Case D-EX (like numbers of Fordson tractors fitted with Roadless half-track attachments), Clark 'Clarktor-6' (4×2), and Cletrac MG1 (full-track), all used by the RAF.

Vehicle shown (typical): Tractor, Light, 4×2 (David Brown VIG.1/462)

Technical Data:
Engine: David Brown 4-cylinder, I-I-W-F, 2·523 litres, 37 bhp @ 2200 rpm.
Transmission: 4F1R with Turbo Transmitter.
Brakes: mechanical.
Tyres: front 31×9, rear 12.75–24.
Wheelbase: 5' 10½".
Overall l×w×h: 10' 11"×6' 4"×5' 0".
Weight: 3 tons 16½ cwt.
Note: AW100, AW500 and VIG.1/100 similar but without torque convertor. 5-ton chain-driven winch with 100 ft of $\frac{5}{8}$-in dia. rope at rear.

Tractor, 3×2 (Scammell, 'Mechanical Horse' MH6) 4-cyl., 36 bhp, 4F1R, wb 9' 2¾". Tyres, front 8.25–10, rear 10.50–13. Open cab and ballast body, used by RN for towing full-trailers. Note front frame extension with drawbar. Also with closed cab and S-T. 676 produced (all types).

Tructor, 4×2 (Fordson Thames/Allan Taylor) V-8-cyl., 85 bhp, 4F1R×2, wb 8' 3". Tyres 34×7 (2DT). Tructors were conversions of std Fordson chassis by Allan Taylor and Reynolds. Main mods: shortened frame, Warford two-speed aux. gearbox, vacuum trailer brake, RAF drawbar gear.

Tractor, Heavy, 4×4 (Alvis-Straussler Hefty) Ford V-8-cyl., 85 bhp, 4F1R×1, wb 7' 6", 12' 2"×6' 10", 2 tons 3 cwt. Tyres 10.50–16. Used by RAF as: Tractor, Straussler, Type C (Heavy). Tubular frame; rear axle and ballast body articulating. Smaller type used Singer components.

Tractor, Heavy, 4×4 (Scammell, prototype) 4-cyl. 7-litre (5"×5½") ohv petrol engine, 6F1R×1. Tyres 13.50–20. Vertical winch under rear body. Delivered 20th March, 1939, intended for haulage of heavy coastal artillery. Later became showman tractor. Axles with epicyclic gearing.

Tractor, Heavy, 4×4 (AEC Matador 0853) 6-cyl. diesel, 95 bhp, 4F1R×2, wb 12' 7½", 20' 1"×8' 0"×9' 5", 7 tons. Trailer is 20-ton, 16-wheeled, Low Loading, SMT, also known as Multiwheeler (made by SMT to Multiwheeler design). Combined length just over 47'.

Tractor, Heavy Breakdown, 6×4 (Scammell SV/1T, 1S) Gardner 6-cyl. diesel, 102 bhp, 6F1R, wb 12' 2" (BC 4' 3¼"), 20' 3"×8' 8"×9' 5", 9 tons 13 cwt. 13.50–20 tyres. Relatively rare model with collapsible crane jib. 8-ton vertical winch. WD drawbar gear front and rear. 43 produced.

Tractor, Light Artillery, 6×6 (Dennis 'Octolat') Twin Bedford 6-cyl., 2×72 bhp, (2×4F1R)×1, wb 9' 9" (BC 4' 6"), 19' (approx.) ×7' 6"×7' 4¼". Skid steering. Relied on its (10.00–20) tyres for springing. Originally conceived as 8×8. Also built with 9·8-litre Leyland engine and armoured cab. Exp., 1943.

Tractor, Heavy, 8×6, Tank Transporter (Albion CX33) Twin Albion 6-cyl., 2×140 bhp, 13.50–20 tyres. Tank tractor with ballast body and rear cab for winch control. 2 front axles steered. One engine drove outer axles, other No. 3 (originally 8×8, with outer axles steering). Towed 75 tons.

Tractor, Heavy Breakdown, 6×4 (Scammell SV/2S). Basically similar to SV/1T and 1S shown on left but sliding crane jib, extended by means of small hand winch, with three positions: long lift (2 tons), short lift (3 tons) and travelling. Worm-driven manual hoist. Shown fitted with overall chains.

Tractor, Field Artillery, $\frac{3}{4}$-Track (Bedford BT 'Traclat') Twin Bedford 6-cyl., 136 bhp, 5F1R, 21' 0"×7' 6"×8' 11" (7' 6"), 6 tons 16¼ cwt. Six prototypes produced by Vauxhall in 1944, patterned on German type. Intended as FAT for 25-pdr, but also for 17-pdr and Bofors. Initial work by MCC.

Tractor, Light Artillery, $\frac{1}{4}$-Track (AEC Matador) 6-cyl. diesel, 95 bhp, 4F1R×2, 20' 6"×8' 0"×9' 6", 10 tons approx. Experimental semi-track conversion for towing 17-pdr and 6-pdr, to carry stores, ammunition and crew of nine. Components of Valentine tank were used for the bogie.

Tractor, Half-Track, w/Winch (Fordson/Roadless) 4-cyl., 45 bhp, 3F1R, wb 6' 0", 10' 8"×5' 9"×4' 7", 3 tons 5 cwt. Conversion by Roadless Traction Ltd. Hesford winch at front (shown) or rear. RAF. Basic (wheeled) model used also (RASC, RAF). IWM photo H17992.

CRANES, TURNTABLE

Makes and Models (Chassis): AEC 0854 (5-ton, 6×6). Albion AM 463 (2-ton, 4×2). Austin K6 (3-ton, 6×4). Coles PE (2-ton SP crane). Crossley IGL8 (3-ton, 6×4). Leyland Retriever (3-ton, 6×4). Thornycroft Amazon WF/AC6/1 and 2 (swb), WF8/NR6 and WF8/AC6/2 (lwb) (6 ton, 6×4). *Note:* crane equipment by H. J. Coles Ltd, Derby.

General Data: All the above British-built 6-wheeled types were fitted with a Coles electric or petrol-electric crane revolving on a so-called turntable. These cranes were fitted with a DC main generator from which the hoist, the derrick and the slewing motors obtained their power. Automatic electro-magnetic brakes were fitted to the hoisting, derricking and slewing motions. The main generator was driven by a separate gasoline engine, except on the RAF Leyland Retriever and the Thornycroft Amazon, where it was driven by the vehicle's gearbox PTO. The hoist and derrick motors developed 6 hp @ 900 rpm, the slewing motor 1 hp @ 500 rpm. The Albion AM 463 2-ton 4×2 Air Ministry chassis was also fitted with an electric revolving crane. The other 4-wheeler was produced entirely by Coles but was not mounted on a truck chassis (an electric motor was used for travelling but as an alternative a towing bar was fitted for towing by motor vehicles). Most models had special rigid-beam rear suspension, replacing the normal laminated leaf springs. Two American types were also used, chiefly by the RAF, namely the Bay City 18-T50 and the Thew Lorain MC3, both 6×4 models of the usual US pattern. Like the British types, they were used for hoisting aircraft and components, and for salvage and general hoisting duties. The Thornycroft Amazon 6×4 chassis was also used for an experimental SP gun mount (17-pounder AT), in 1942. A modification of the US Mack NO 7½-ton 6×6 chassis was used after the war by the Royal Navy with a revolving crane (Mobile Crane, 12-ton, 6×6, Northwest). In 1940 three Thornycroft Amazons were supplied to the Air Ministry and Admiralty with Neals crane.

Vehicle shown (typical): Truck, 3-ton, 6×4, Crane (Austin K6/Coles EMA Mk VI Series 2)

Technical Data:
Engine: Austin 6-cylinder, I-I-W-F, 3.9 litres, 72 bhp @ 2800 rpm (Crane engine: Ford 10 HP).
Transmission: 4F1R×2.
Brakes: hydraulic (servo-assisted).
Tyres: 9.00–20.
Wheelbase: 12' 9", BC 4' 0".
Overall l×w×h: 23' 4"×7' 2"×12' 0".
Weight: 9 tons 14 cwt.
Note: max. crane capacity 3 tons @ 7' 9" radius.

Truck, 5-ton, 6×6, Crane (AEC 0854/Coles Mk VII, Series 7) 6-cyl. diesel, 95 bhp, 4F1R×2, wb 14' 9⅛" (BC 4' 3¼"), 30' 0"×7' 9"×14' 3", 15 tons 10½ cwt. Crane engine: Ford V8. 192 built. Max. capacity 5 tons @ 7' 6" radius. On same chassis: 2500-gal. Refueller and Mobile Oxygen plant.

Truck, 6-ton, 6×4, Crane (Thornycroft Amazon WF-8/NR6/Coles Mk VII Series 7) 6-cyl. diesel, 105 bhp, 4F1R×2, wb 16' 6" (BC 4' 6"), 30' 0"×7' 0"×14' 0", 13 tons 5 cwt. Crane as on AEC 6×6 chassis. Thornycroft produced 1432 swb and 388 lwb (shown) crane chassis, mainly for RAF. WF8/AC6/2 had petrol engine. 1944.

Truck, 6-ton, 6×4, Crane (Thornycroft Amazon WF/AC6/1/Coles Mk VII, Series 2) 6-cyl., 100 bhp, 4F1R×2, wb 11' 9" (BC 4' 6"), 26' 6"×7' 9"×13' 6", 12 tons 12¼ cwt. Tyres. front 12.75-20, rear 13.50-20. Max. crane capacity 5 tons @ 7' 0" radius. Similar swb chassis with Mk VII, Series 7 crane.

Truck, 3–ton, 6×4, Crane (Crossley IGL8/Coles Mk VI) 4-cyl., 75 bhp, 4F1R×2, wb 12' 10" (BC 4' 0"), 21' 9"×7' 2"×11' 0", 9 tons 4 cwt. Generator driven by Ford 10 HP engine. Fitted with winch. Similar crane on Austin K6 (w/o winch) and Leyland. Used by WD. Crane capacity 2 tons @ 7' 9" radius.

GB

FIRE FIGHTERS and CRASH TENDERS

Makes and Models: Austin K2 (4×2), K6 (6×4). Bedford QL (4×4). Crossley Q 30/100 HP (4×4). Crossley IGL (6×4). Fordson WOT1, 1A, 1A/1 (6×4). Karrier Bantam (4×2). Leyland FKT (4×2).

General Data: Only a few of the above models were used by the Army, the majority being RAF types. The Austin K2 and the Karrier were RAF chassis, later fitted with 'domestic fire tender' bodies. The Austin K6 was also a war time chassis, later fitted with fire crash-tender body. During the early part of the war a large number of Austin K2 trucks was fitted out as auxiliary towing vehicles (ATV) for the National Fire Service (NFS), towing two-wheeled fire pump trailers. These vehicles had a box-shaped open-end body and a bell on the sloping front part of the roof. Similar bodywork was also fitted on the military-type Fordson WOT2 15-cwt chassis, but these vehicles were not used by the military. The Leyland FKT was basically a commercial-type fire appliance and numbers of them were operated by the RASC. The Bedford QL Fire Tenders were used by the Army Fire Service and looked rather like conventional GS trucks although usually painted red. They were introduced towards the end of 1943 and saw active service in the campaigns following D Day. This vehicle had two functions. Firstly it had a water tank, a PTO-driven pump and a first-aid hose reel which could be brought into action instantly on arriving at a fire. Secondly it towed a trailer pump and had lockers for stowage of hoses, nozzles and other equipment required for the trailer pump. Ladders were carried under the tilt. The RAF also used some American types, namely the US Army Chevrolet 1½-ton 4×4 Fire Crash Tender and a commercial-type Federal 10-ton 4×2 (model 89K-145-167) which had an extended six-man forward control cab and a 400-gpm Barton pump located at the front and driven off the 96-bhp Hercules JXD main engine.

Vehicle shown (typical): Truck, 3-ton, 4×2, Fire Engine (Leyland FKT)

Technical Data:
Engine: Leyland E128/3 6-cylinder, I-I-W-F, 7·7 litres, 96 bhp @ 2000 rpm.
Transmission: 4F1R.
Brakes: hydraulic.
Tyres: 32×6 (2DT).
Wheelbase: 12′ 3″.
Overall l×w×h: 25′ 8″×7′ 0″×10′ 4″.
Weight: 4 tons.
Note: 500-gpm pump, two-stage turbine.

Truck, 3-ton, 4×4, Fire Tender (Crossley Q 30/100 HP) 4-cyl., 96 bhp, 4F1R×2, wb 11' 6", 20' 0"×7' 6"×7' 9", laden wt 7 tons 13 cwt. 1000-gpm pump at rear. 300-gal. water and 25-gal. foam compound tank. Four 60-lb cap. CO_2 gas cylinders carried on tank. IWM photo CH5768.

Truck, 3-ton, 4×4, Fire Tender (Bedford QL) 6-cyl., 72 bhp, 4F1R×2, wb 11' 11", 20' 2"×7' 6"×10' 1"(7' 0"), laden wt 6 tons 14¾ cwt. 200-gal. water tank with PTO-driven pump. Tows trailer pump (stowed in main body for shipping purposes). Fully equipped. Introduced in late 1943.

Truck, 3-ton, 6×4, Fire Crash Tender, Type T (Fordson WOT1) V-8-cyl., 85 bhp, 4F1R, wb 13' 8½" (BC 3' 6¼"), 21' 6"×7' 4"×7' 9", 6 tons 7 cwt (gross). Combined 400-gal. main water and saponine tank. Four foam outlets with 2½" bore canvas hose. 1000-gpm PTO-driven foam pump at rear. Power absorption 75 bhp.

Truck, 3-ton, 6×4, Fire Crash Tender, 1944 Monitor (Foam) (Fordson WOT1) V-8-cyl., 85 bhp, 4F1R, wb 13' 8½" (BC 3' 6¼"), 21' 6"×7' 4"×10' 0". Compartmented tank (300 gals water, 100 gals foam compound). Foam pump driven by separate Ford V-8 engine. 1945 model had closed cab, no ladder.

GB

AMPHIBIANS
WHEELED and TRACKED

Makes and Models: Morris 'Gosling' (4×2). Morris-Commercial/Nuffield Mechanizations 'Argosy' and 'Neptune' (full-track). Morris-Commercial 'Terrapin' Mk I (8×8). Opperman 'Scorpion' (8×8). Thornycroft 'Terrapin' Mk I and II (prototypes, 8×8).

General Data: Of the above models, only the Terrapin Mk I reached series production. 500 were built. All others were produced only experimentally. The Terrapins, like the American GMC DUKW ('Duck'), were intended mainly for ship-to-shore transport of supplies. The Neptune was not unlike the American LVT—a fully tracked amphibious vehicle. In both cases efforts were made to get into production British vehicles of the same types as their American counterpart so as not to be entirely dependent on US production and allocation of amphibians. As it was, Lend-Lease supplies did not entirely fulfil British requirements and the British-type amphibians were inadequate in many respects. The Gosling was an amphibious motorised 'Landcart', developed for signals requirements. The prototype was produced by Morris Motors Ltd and trials were in progress when the war ended. It had a good cross-country performance but as an amphibian it was limited to smooth water. The Opperman Scorpion was a private venture, designed to be either a small armoured vehicle or 30-cwt load carrier. It had two large wheels in the centre, overlapping smaller ones fore and aft, and a single wheel mid-front and mid-rear. Water propulsion of this most unusual vehicle was by outboard motor. No series production took place. The US Ford amphibious 'Jeep' (GPA) was used in limited numbers. The 'Duck' and the LVT (Landing Vehicle, Tracked, or 'Water Buffalo') remained for many years. In 1972 the former were still in use by the Amphibious Squadrons of the RCT's Maritime Transport Units. The latter were extensively used by the British in the operations at Walcheren, Netherlands, and the crossing of the Rhine, during the closing stages of the war.

Vehicle shown (typical): Amphibian, 4-ton, 8×8, Terrapin Mk I (Morris-Commercial)

Technical Data:
Engine: Twin Ford V8, V-L-W-C, 2×3·62 litres, 2×85 bhp.
Transmission: 2×3F1R×1 (two lines of worm shafts, each driving the four wheels on one side).
Brakes: mechanical.
Tyres: 12.75–24 (agricultural tractor type).
Wheelbase: 14′ 3″ (axle spacings 4′ 9″).
Overall l×w×h: 23′ 0″×8′ 9″×9′ 7″ (8′ 3″).
Weight: 6 tons 16 cwt.
Note: prototype by Thornycroft; on the level the front pair of wheels was off the ground.

Amphibian, 5-ton, 8×8, Terrapin Mk II (Thornycroft) Twin Ford V-8-cyl., 2×85 bhp, 2×4F1R×2, wb 15′ 3″, 30′ 8″ ×8′ 10″, 11 tons 8 cwt. Tyres 14.25–20. Improved version of Mk I. Forward cab. Larger load area (Mk I had small holds fore and aft, and central driver's position).

Trolley, Motorised, Four-Wheeled, Gosling (Morris) Villiers, 1-cyl., 2 bhp, wb 4′ 2″, 8′ 0″×3′ 0″×3′ 9″, 4 cwt. Payload 400 lb. Tyres 5.50–8 low pressure. Outboard final drive drove rear wheels and propeller, the latter removable. Tiller steering for walking behind or sitting inside. Design A. Issigonis.

Amphibian, Full-Track, Argosy (MCC/Nuffield) Nuffield Liberty V-12-cyl., 340 bhp, 4F1R. Designed in 1942 by W. J. Daniels, later BMC project engineer. 9-ton payload in rough sea. Twin 2-ft dia. propellers. Prototypes only; no production owing to excessive weight and lack of loading ramp.

Amphibian, Full-Track, Neptune (MCC/Nuffield) Meadows 12-cyl., 280 bhp, 4F1R, approx. 30′×11′ 6″×9′ 6″, 13½ tons. Payload 4 to 5 tons. Produced by MCC, sponsored by Nuffield Mechanizations, as equivalent to US LVT. Rear loading ramp. Water propulsion by track shoes. Lightly armoured.

GB

ARMOURED VEHICLES
WHEELED

Makes and Models: *Scout Cars:* Alvis 'Dingo'. Daimler Mk I, IA, IB, II, III. Humber Mk I, II.
Light Reconnaissance Cars: Dodge (4×2). Hillman Gnat (4×2). Humber Humberette (4×2). Humber Mk I, II (4×2), Mk III, IIIA (4×4). Morris Mk I (4×2), Mk II (4×4). Morris Salamander (4×2). Standard Beaverette Mk I, II, III, IV (4×2).
Armoured Cars: AEC Mk I, II, III (4×4). Alvis-Straussler, ACIIID (4×4). Bedford OXA (4×2 armoured truck). Coventry Mk I, II (4×4). Daimler Mk I, II (4×4). Guy Mk I, IA (4×4). Humber Mk I, II, III, IV (4×4). Leyland 'Beaver-Eel' (6×4 armoured truck). Morris-Commercial CS9/LAC (4×2).
Armoured Command Vehicles: AEC HP, LP Mk I, Mk II (4×4); HP, LP (6×6). Guy Lizard (4×4).
Armoured Flamethrowers: AEC 'Basilisk' (4×4). AEC 'Heavy Cockatrice' and Heavy Pump Unit (6×6). Bedford QL 'Cockatrice' (4×4). Commer/Lagonda (4×2).

General Data: Unlike the US Army in World War II, the British used a large variety of armoured wheeled vehicles. First many were improvised on existing chassis (even including concrete pillboxes on commercial vehicle chassis) or designed on existing chassis (e.g. the early Standard Beaverettes), but development of special types started early, and when sufficient numbers became available, many of the early makeshift types were replaced. As the war drew on developments started on universal types in the interest of standardisation. For example, the 'Coventry' armoured car was designed jointly by the Rootes Group and the Daimler Co. to supersede all their existing models and incorporating the best features of both. Much work was done both in the UK and Canada on development of a universal scout car and a universal vehicle intended to fill the roles of Command, APC, LAD, Ambulance, and Demolition, known as 'Caplad'. Several Canadian and US types were also used.

Vehicle shown (typical): Car, Scout, Daimler, Mk II

Technical Data:
Engine: Daimler 18 HP, 6-cylinder, I-I-W-R, 2·52 litres, 55 bhp @ 4200 rpm.
Transmission: 5F5R, pre-selective, self-changing, with 'fluid flywheel', 4×4.
Brakes: hydraulic.
Tyres: 7.00–18.
Wheelbase: 6' 6".
Overall l×w×h: 10' 5"×5' 7¼"×4' 11".
Weight: 3 tons (laden).
Note: Mk III had canvas roof. 6626 produced (all Mks). Also known as 'Dingo'. Originally BSA.

Car, 4×4, Scout, Humber Mk I and II 6-cyl., 87 bhp, 4F1R×2, wb 7′ 7″, 12′ 7″×6′ 2½″×7′ 0″, 3½ tons (laden). 9.25–16 tyres. Differences between Mk I and II were in minor engine and gearbox details. Vehicle shown fitted with Bren mg on ring mount (Dutch army). 4300 built (all Mks).

Car, 4×4, Light Reconnaissance, Humber Mk IIIA 6-cyl., 87 bhp, 4F1R×2, wb 9′ 3¾″, 14′ 4″×6′ 2″×7′ 1″, 3 tons 12 cwt (laden). 9.25–16 tyres. Chassis as Humber 8-cwt 4×4; Mk I (Ironside I) and II as 8-cwt 4×2. Crew three. Developed from open-top Humberette. 3600 built (all Mks).

Car, 4×2, Light Reconnaissance, Morris Mk I 4-cyl., 71 bhp, 4F1R×2, wb 8′ 2″, 12′ 10″×6′ 8″×6′ 0″, 3 tons 14 cwt (laden). 9.25–16 tyres. Rear engine. Mk II was 4×4 with 9.00–16 tyres. Mk I had ifs. 2200 built (both Mks). Crew three. Driver in centre front.

Car, 4×2, Light Reconnaissance, Beaverette Mk II 4-cyl., 45 bhp, 4F1R×2, wb 9′ 0″, 13′ 6″×5′ 3″×5′ 0″, 2 tons. Standard 14 HP car chassis. Named after Lord Beaverbrook. Also used by RAF (airfield defence), who later converted some into pickup trucks. Mk I similar.

Car, 4×2, Light Reconnaissance, Beaverette Mk IV (Standard) 4-cyl., 45 bhp, 4F1R, wb 6' 2", 10' 2½"×5' 10"×7' 1", 2 tons 5 cwt. Tyres 9.00–13. Double-reduction rear axle (8·9 : 1). Hull of 10-mm armour plate. Mk III similar except flat front to hull top (rather than 'stepped'). Shoulder-operated turret.

Car, 4×4, Armoured, Alvis-Straussler IIID Alvis 6-cyl., 120 bhp, 4F4R×2, wb 9' 3". Tyres 9.00–22. Supplied mainly to Netherlands East Indies and Portugal. Twelve of slightly different model were supplied to the RAF (vertical hull sides, tyres 10.50–20). First prototype built in Hungary in 1933. Independent suspension.

Armoured Car, 4×4, Daimler Mk I 6-cyl., 95 bhp, 5F5R, wb 8' 6", 13' 0"×8' 0"×7' 4", 7 tons approx. Tyres 11.00–20. Crew three. Pre-selector gearbox. H-drive transmission layout, engine at rear. Ind. susp. Mk II differed in many details. 2694 produced (both Mks). 2-pdr gun.

Armoured Car, 4×4, Humber Mk III (Karrier) 6-cyl., 90 bhp, 4F1R×2, wb 8' 6¼", 15' 0"×7' 2"×7' 10", 7 tons approx. Tyres 10.50–20. Named Humber to avoid confusion with 'carrier'. Used many std Karrier 4×4 truck/tractor components. Mks II and IV similar. Mk I had same hull as Guy Mk IA.

Armoured Car, 4×4, AEC Mk III 6-cyl. diesel, 158 bhp, 4F1R×2, 18′ 5″×8′ 10″×8′ 10″, 12½ tons approx. Tyres 13.50–20. 75-mm gun. Mk II similar but 6-pdr gun. Same engine as Valentine tank (Mk I: 105 bhp). Originated as private venture of AEC in 1941. 629 produced (all Mks).

Armoured Car, 4×4, Coventry Mk I Hercules RXLD 6-cyl., 175 bhp (150 net), 5F5R, wb 9′ 10″, 15′ 6½″×8′ 9″×7′ 9″, 11½ tons approx. Tyres 13.50–20. 2-pdr gun (Mk II: 75-mm). Designed jointly by Daimler and Rootes Group (see General Data). Relatively few produced, 1944–45.

Carrier, Wheeled, 4×4, Guy Based on similar rear-engined chassis as Guy armoured car (known at first as Wheeled Tank). Hull was similar to that of the full-track Bren Carrier. It was tested in 1940 but remained in the experimental stage.

Car, Armoured, Personnel, LAD, Ambulance, Demolition ('Caplad') Universal vehicle, using CMP Ford 4×4 truck components on self-supporting hull. Production type was to be built by Ford (UK) as Model W0T9 but did not materialise. Roles: command, compressor, LAD, demolition and medical. 1944.

206

Car, Armoured, 4×2 (Morris/Briggs) One of many improvised types used by Home Guard units during 1940 when invasion of Britain was expected. Both car and truck chassis were used. This well-designed specimen was produced by Briggs Motor Bodies Ltd.

Car, Armoured, 4×2 (Humber Snipe Special Ironside) In 1940 Humber produced a few 'Special Ironside Saloons' for conveyance of the Royal Family and Cabinet Ministers. Based on the Ironside Lt Recce Car Mk I, the interior was furnished luxuriously by Thrupp & Maberly.

Truck, Armoured, 4×2, Armadillo (Bedford/LMSR) Armadillos, Mks I to III, consisted of double-skin wooden pillbox on commercial trucks, mainly Bedfords and Fordsons. Space between inner and outer skins was filled with small pebbles. About 660 were made in 1940. Specification varied.

Truck, Armoured, 4×2, Dodge (Dodge/Briggs) Seventy produced in 1940, following Fordson-based prototype designed by Mr Leo Villa, OBE, Sir Malcolm Campbell's chief racing mechanic. Open-top conversion with 6-pdr gun was used by Home Guard. Mr Villa also designed an armoured Fordson tractor.

Vehicle, Armoured Command, LP, AEC (Matador 4×4)
6-cyl. diesel, 95 bhp, 4F1R×2, wb 12' 7½", 20' 0"×8' 6" ×9' 5", 10 tons 4¾ cwt. 13.50–20 tyres. Similar: HP version (usually with dummy front extension), Armoured Personnel and Armoured Demolition. Total produced 416.

Flamethrower, Heavy Pump Unit (AEC 6×6) 6-cyl. diesel, 95 bhp. One of several experimental models devised by AEC and the Petroleum Warfare Dept. This was a 'trailer type'. Pump was driven by a Napier Lion engine. Intended for airfield and coastal defence. 'Basilisk' was modified AEC A/C Mk I.

Tender, Armoured, Leyland Type C (Leyland 'Beaver-Eel') 4-cyl., 73 bhp, 4F1R×2, wb 13' 0" (BC 4' 0"), approx. 23' ×7' 6"×10'. Retriever 6×4 chassis. Over 300 produced for protection of aircraft factories. Some had armour-plate skirting covering the wheels. Armament varied.

Vehicle, Armoured Command, HP and LP (AEC 0856 6×6) 6-cyl. diesel, 150 bhp, 4F1R×2, wb 13' 8½" (BC 4' 0"), 26' 1"×7' 11"×8' 10", 18 ton (laden). 151 produced in 1944/45. Rear body divided in staff room and wireless compartment.

GB

ARMOURED VEHICLES
FULL-TRACK

Principal Types and Models: *Carriers:* Early types: Machine Gun Mk I; Bren Mk I and II; Cavalry Mk I; Scout Mk I; Universal Mk I; AOP Mk I and II. Universal types: Universal Mk II and III; AOP Mk III; 3-in Mortar Mk I and II; MMG Mk I and II; 'Wasp' Flamethrower. Loyd TPC Mk I and II; TS and C Mk I; TT Mk I and II.
Light Tanks: Mk VI; VI A-C. Mk VII Tetrarch; Mk VIII Harry Hopkins.
Cruiser Tanks: Mk I (A9); Mk II, IIA (A10); Mk III (A13); Mk IV, IVA (A13 Mk II). Mk V (A13 Mk III) Covenanter I–IV. Mk VI Crusader I–III. Mk VIII Cavalier I. Mk VIII Centaur I–IV; Cromwell I–VIII; Challenger. Comet I.
Infantry Tanks: Mk I (A11). Mk II Matilda I–V. Mk III Valentine I–XI. Mk IV Churchill I–VIII (exp. models included Valiant A38, Black Prince A43).
Heavy Tanks and Heavy Assault Tanks: TOG I, II; A33; A39 Tortoise (experimental only).
Special-Purpose AFVs: Modifications of several of the above types were used as SP guns, bridge layers, AA tanks, mine-clearing devices, armoured recovery vehicles, armoured searchlights, etc.
American AFVs used by British Forces (in std or modified form): Light Tanks—M2A4; M3 Series (Stuart Mk I–VI); M22 (Locust); M24 (Chaffee). Medium Tanks—M3 Series (Lee Mk I–VI, Grant); M4 Series (Sherman Mk I–V). SP Guns—Priest M7 (105-mm); Achilles M10 (17-pdr). *Others*—Adder Flamethrower; Crab, Bull's Horn, Centipede and other mine-clearing devices; ARVs; BARVs; troop carriers.
Note: all the above SPs and other variants were based on medium tank M3 and M4 series chassis. *Amphibians:* LVT Buffalo.

General Data: Cruiser tanks were designed for a mobile role with stress on high mobility at the expense of weight (relatively thin armour). Infantry tanks, on the other hand, operated with the infantry, where speed was less important than heavy armour. The British devised a wide range of modifications to employ tanks for specialist roles.

Vehicle shown (typical): Carrier, Universal, No. 2, Mk II (Ford).

Technical Data:
Engine: Ford 8-cylinder, V-L-W-C, 3·62 litres, 85 bhp @ 3800 rpm.
Transmission: 4F1R.
Brakes: mechanical, on final drive axle.
Track width: 5' 2¼".
Overall l×w×h: 12' 0" × 6' 9" × 5' 3".
Fighting weight: 3¼–4 tons.
Note: No. 1 had 65-bhp, No. 3 95-bhp engine. Produced by Aveling-Barford, Ford, Sentinel, Thornycroft, Wolseley. The Universal Carrier (or 'Bren Carrier') was used for a variety of roles. Several derivations existed, incl. Armoured Observation Post, Mortar, Flamethrower (Wasp), etc.

Carrier, Full-Track, Loyd (Wolseley, Dennis, etc.) Ford V-8-cyl. (at rear), 85 or 90 bhp, 4F1R, 13′ 7″ × 6′ 9½″ × 7′ 0″ (4′ 8¼″), 3½–4 tons. Introduced in 1940, the Loyd was used for several roles including tractor for 2-pdr AT (Carrier, TT, Nos 1–3, Mk I), 6-pdr AT and 4·2-in Mortar (Carrier TT, Nos. 1(Z)–3(Z), Mk II).

Tank, Light, Mk VIC (various mfrs) Meadows 6-cyl., 88 bhp, 5F1R, 12′ 11½″ × 6′ 9″ × 7′ 0″, 5 tons 4 cwt (laden). Originated from Carden-Loyd series of tracked vehicles of late 1920s. Horstmann suspension. Front drive. Designed by Vickers-Armstrong. Vehicle shown captured by Germans.

Tank, Light, Mk VII, Tetrarch I (Metropolitan-Cammell) Meadows MAT 12-cyl., 165 bhp, 5F1R, 14′ 1½″ × 7′ 7″ × 6′ 11½″, 7 tons 10 cwt (laden). Vickers design. 171 produced, 20 supplied to USSR, 1940–42. Vehicle shown was first tank to be fitted with Straussler DD floatation/swimming equipment.

Tank, Cruiser, Mk IV(A) (A13 Mk II) (Nuffield Mechanizations) Nuffield Liberty V-12-cyl., 340 bhp, 4F1R, 19′ 9″ × 8′ 4″ × 8′ 6″, 14 tons 15 cwt (laden). Improved version of Mk III, the chassis of which, in turn, was patterned on the American Christie tank. Used in France, 1940, Libya, 1941. 2-pdr gun plus co-axial mg.

Tank, Cruiser, Mk VI, Crusader I (Nuffield Mech., nine other mfrs) Nuffield Liberty V-12-cyl., 340 bhp, 4F1R, 19′ 8″×8′ 8″ ×7′ 4″, 18 tons (laden). 2-pdr gun. Crew five. Adaptations later used as AA tank and tractor for 17-pdr. Power unit was modernised WWI US Liberty aero engine.

Tank, Cruiser, Mk VIII, Centaur I (Leyland and other mfrs) Nuffield Liberty V-12-cyl., 395 bhp, 5F1R, 20′ 9″×9′ 6″ ×7′ 8¼″, 24–28 tons (laden). 6-pdr gun. Crew five. Similar in appearance to the Cruiser Mk VII Cavalier. Both were soon superseded by the Meteor-engined Cromwell.

Tank, Cruiser, Mk VIII, Cromwell IV (Leyland and other mfrs) RR Meteor V-12-cyl., 600 bhp, 5F1R, 20′ 10″×9′ 6½″ ×8′ 2″, 27½ tons (laden). 75-mm gun. Crew five. Several variants. One of the fastest tanks of WW II (up to 40 mph). RR Merlin aero-engine, modified for tank use.

Tank, Cruiser, Comet I (Leyland and other mfrs) RR Meteor V-12-cyl., 600 bhp, 5F1R, 21′ 6″×10′ 0″×8′ 9½″, 32½ tons (laden). Heavy cruiser tank with 77-mm gun (modified 17-pdr). Similar in layout to Cromwell but welded hull and turret. Produced from late 1944. Speed 29 mph. Crew five.

Tank, Infantry, Mk II, Matilda IV CS (Vulcan Foundry and other mfrs) Twin Leyland diesel, 95 bhp each, 6F1R (Wilson preselective), 18' 5"×8' 6"×8' 0", 26·5 tons. 3-in howitzer (CS types; std models had 2-pdr gun). Matilda I and II had AEC diesel engines. Almost 3000 produced (Matilda I–V), 1939–43.

Tank, Infantry, Mk III, Valentine (Vickers-Armstrong) AEC or GM diesel engine, 131–165 bhp, 17' 9"×8' 7$\frac{1}{2}$"×7' 5$\frac{1}{2}$", 16–17 tons (laden). 8275 Valentine tanks were produced in the UK and Canada. Many went to Russia. Tank shown is equipped with Straussler DD floatation/swimming equipment.

Tank, Infantry, Mk IV, Churchill I (Vauxhall) Bedford 12-cyl. (flat twin-six), 350 bhp, 4F1R, 24' 5"×10' 8"×8' 2", 38$\frac{1}{2}$ tons (laden). Succeeded Matilda in 1941. 5640 were produced in many variants. Named after Sir Winston Churchill, who is here seen inspecting one. Also supplied to Russia.

Flamethrower, Churchill Crocodile (Vauxhall) Churchill tanks were frequently used with flamethrowing equipment. Fuel supply for the flamethrowers was carried in the armoured 400-gal. trailer. Many other variants appeared incl. bridge-layers (SBG, ARK, etc.), engineer assault vehicles (AVRE), etc.

BRITISH COMMONWEALTH OF NATIONS

The main supplier of military vehicles in the British Commonwealth of Nations during World War II (outside Great Britain itself) was Canada. Vehicles produced in Australia, India, South Africa, New Zealand, etc., were almost invariably of Canadian origin, although the bodywork and certain other components were often of local manufacture.

The Canadian motor industry, in turn, consisted largely of subsidiaries of the US 'giants' (General Motors, Ford and Chrysler). There being only a few manufacturers involved, with large production potential, standardisation was relatively easy.

During the 1930s British War Office policy had been to standardise military equipment as far as possible throughout the Empire. Products of the two biggest Canadian manufacturers, Ford and GM, were already in use in several Commonwealth countries and in 1936 the two firms had agreed to merge their resources for military production. A family of so-called Canadian Military Pattern (CMP) vehicles was subsequently devised and developed, guided to a large extent by developments in Britain. The same range of chassis classification was used: staff cars and utilities, 8-cwt, 15-cwt, 30-cwt and 3-ton trucks, light and heavy ambulances, field and medium artillery tractors, etc. All vehicles destined for overseas use had right-hand drive and with few exceptions large-section tyres were used with single rear tyres on all types.

Canadian forces at home used mainly US type commercial trucks. When war broke out and large numbers of vehicles were needed without delay, it was relatively simple to step up the production of standard types with some detail modification to meet military requirements. Such vehicles were known as Modified Conventional Vehicles, and comprised cars, utilities, 15-cwt, 2-ton and 3-ton trucks.

In July, 1940, after leaving most of its equipment on the Dunkirk beaches, the British Government placed orders in Canada for an initial supply of 7000 vehicles. The Canadian forces themselves also needed equipment, as did Australia, India, New Zealand, South Africa, etc. Canada had declared war on Germany on 10 September, 1939, four days after the first British troops had landed in France.

Modified Conventional types were shipped out in large numbers, notably for use in the North African campaigns. The development of CMP vehicles was stepped up and eventually consisted of a range of chassis types which was, if only in terms of standardisation, among the best in the world. These CMP vehicles were built by General Motors (Chevrolet) and Ford, and the main standard commercial components used were the well-proven Chevrolet 216 six-cylinder engine ('Stovebolt Six') and the equally famous Ford V8 engine, both with their standard gearboxes. Driven front axles were patterned on the American Marmon-Herrington design, using standard-type differentials. The first CMP vehicles came to Britain with the Canadian forces in December, 1939.

By 1941 Canada was the main supplier of automotive equipment to the Empire, turning out during that year 189,178 military MT vehicles, the majority being 4×2 types. In 1942 the total figure rose to 199,542 plus 12,987 AFVs, the country's highest yearly total during the war. The total number of MT vehicles produced in Canada between 1939 and 1945 was 815,729 plus 50,663 AFVs. In addition, Canada produced countless other items of military hardware.

The design, development and pre-production testing of the multitude of vehicle types (despite the limited number of basic chassis) was carried out by the Army Engineering Design Branch in close co-operation with the industry. There were many special requirements to be met, especially for Lend-Lease orders.

Following repeated criticism of the cramped position in which the driver was placed in the early CMP cabs (copied from British design), a new, much roomier cab, known as No. 13, was designed and introduced in 1942 for all CMP vehicles. The new cab had reverse-slope windscreens, to reduce glare and also to prevent the piling up of snow, slush, etc.

The fact that only a few factories turned out over 815,000 MT vehicles in five years was a very creditable achievement. The high degree of standardisation throughout the CMP range, from 8-cwt 4×2 up to 3-ton 6×6, was an achievement of equal importance. What made these CMP vehicles particularly

Typical RCASC MT column just before the war

interesting was that they combined British WO design features (forward control, large single tyres, etc.) with American engineering (engines, transmissions, axles, etc.) and mass-production techniques.

The fact that numbers of CMP vehicles are still in use after 25 years proves that the quality was excellent and the design sound.

Concentrating on mass production of vehicles from 8-cwt to 3-ton capacity meant that the Canadian Army had to procure its lighter and heavier vehicles from elsewhere, mainly the USA (Jeeps, Diamond T, FWD, Mack, Federal, etc.).

In Australia the picture was rather different. The Australian Army, owing to the country's large area and sparse population, was a very mobile one. Until after World War II Australia depended entirely on the supply of motor vehicle chassis from overseas, notably Canada. These chassis were assembled at a number of plants on the mainland and fitted with locally produced cabs and bodywork. Military vehicles turned out by these plants were referred to as Local Pattern (later: Australian Pattern) to distinguish them from overseas military pattern vehicles which were also used. Most Australian Pattern vehicles produced early in the war were militarised commercial types (4×2, 4×4, 6×6).

General Motors products were the most numerous, followed by Ford, Chrysler (Dodge, Fargo, etc.) and IHC. Some conventional-type Bedfords were turned out by GM-H for the RAAF during 1939–40 and a small number of AEC Matador tractors were assembled in Sydney. The only CMP vehicle of this period was the 1940–41 Ford 3-ton 4×4. CMP vehicles were officially known in Australia as CWO (Canadian War Office) vehicles but popularly they were (and still are) known as 'Blitz' models. After 1942 most GM and Ford trucks were 4×4 CMP, Dodge and IHC continuing as modified commercial models. Despite their Canadian chassis, Australian 'Blitz' trucks were easily recognisable as being Australian Pattern, not only by their bodywork, but because of small but unmistakable cab variations (the GM-Holden-built CMP models, for example, had rectangular ventilator doors in the sides of the front shell). During the early part of the war, large numbers of civilian vehicles, new and used, were impressed, but despite this the supply of vehicles lagged behind the demand until about 1942. Although Australian Pattern vehicles were sent to Egypt for use by the AIF (Australian Imperial Force), the AIF in the Middle East still had to rely on British sources for much of their vehicle requirements. In 1942, the year of the expected Japanese invasion of the Australian mainland, all available vehicles were used to get the army mobile. In addition to Australian Pattern vehicles there were British, Canadian, Indian and US types on issue to units. Many of these were brought back from the Middle East by returning AIF units; others were 'refugee cargo' or direct Lend-Lease from the USA. In 1943, when the threat of a Japanese invasion disappeared and it became apparent that the fighting in the SW Pacific Area would be confined to the islands north of Australia, the need for MT declined sharply. A new 'Tropical Scale' was introduced which among other things reduced MT in fighting formations to a minimum. Also many types of unit became obsolete, armoured units being particularly affected. The most widely used vehicle from 1943 was the US 'Jeep', which was often the only type of vehicle capable of being operated anywhere near the scene of fighting in the SWPA jungles. Many 'Jeeps' were adapted for special roles, varying from mobile bath unit to mobile cinema.

Australian Pattern vehicles included tracked carriers, armoured cars, tanks and trailers.

A feature of this period was the construction and operation of a strategic supply road, almost 1000 miles long, linking the isolated Darwin base in the north with the more populated southern portion of Australia. Many US Mack NR diesel

Ford of Canada's products range in 1941

CMP Cab No. 11

CMP Cab No. 12

CMP Cab No. 13

CMP Cab No. 43

10-ton 6×4 units saw service on this North–South Road and also on the East–West Road. Other types used included Australian Pattern 7-ton semi-trailers, towed by Ford, Chevrolet and International truck-tractors. The North–South Road and the East-West Road, incidentally, became known as the Stuart Highway and Barkley Highway respectively.

Vehicle Types
Like their British counterparts, Commonwealth military vehicles were divided into 'A' and 'B' categories. 'A' vehicles were armoured fighting vehicle types; 'B' vehicles were the other types, varying from motorcycles to tank transporters.

Army MT Basic Chassis Types These, too, were generally similar to the British. Canadian chassis types were designated in several ways, namely by manufacturer's type symbol, DMS model code and a simplified model designation which at first was used only in maintenance manuals and parts lists, but which later became the most widely used in the Services. The latter system has been used in this book, but a listing of the various alternatives is on page 217, covering the principal types. Chevrolet CMP trucks had the same manufacturer's type symbol throughout the war, whereas Ford used different numbers to indicate 1940–41 models (early-type cab) and 1942–45 models (late-type cab with reverse sloping windscreen).

Cab Types Cabs were identified by a number as follows: first digit indicated pattern (1—CMP, 2—Modified Conventional, 3—Modified Conventional COE, 4—open); second digit indicated modifications, as follows:

11	original design	
12	alligator hood	
13	reverse-slope windscreen	
21	mod. conv., closed	
31	mod. conv., COE, closed	
41	No. 11 but cowl only (India)	
42	No. 12 but top removed	
43	No. 13 but top removed	
43S	No. 43 but extra wide (4 seats)	

Body Types Bodies were identified by a code with three or four digits, as follows. First or first and second (letters): type of vehicle (Series 1—bodies mounted on 8-cwt and HU chassis, etc.). Middle digit (figure) represented basic type of body. The last digit or two digits (figures) represented a variation of the basic body. For examples see section on 8-cwt 4×4 Heavy Utility vehicles. Complete listing of body-type designations is prohibitive for space reasons.

Differences between Ford and GM CMP Vehicles To identify the various CMP 1942–45 vehicles by make, here follows a list of their characteristic differences (see also illustrations). Most of these features also apply to 1940–41 production.

	Chevrolet[1]	Ford
Engine, type	6 in-line	V-8
Axles (driven), type	banjo	split
Radiator grille mesh, pattern	diagonal	square
Radiator grille badge, type	Chevrolet	Ford (oval)
Radiator guard springs, type	double leaf	laminated leaf
Radiator overflow tank, position	exposed[2]	concealed
Horn, position	concealed (on inlet manifold)	exposed (opposite steering box)
Steering wheel, type (4-spoke)	wood-rim[3]	black hard rubber
Pre-1943 dash instruments[4]	GM commercial	Ford commercial

[1] GM (C60X 6×6) as Chevrolet but GMC engine, hence radiator grille slightly protruding; usually square GM badge.
[2] Concealed from 1943 (under left-hand floor board).
[3] Prior to 1943 black hard rubber.
[4] From 1943 universal circular instruments were used instead of the rectangular cluster type inherited from 1940–41 commercial trucks.

Colours
Depending on area of operation, Canadian military vehicles (overseas) were painted khaki (matt green), light stone or olive drab. Vehicles for domestic use were usually painted 'No. 2 Brown' or olive drab. In addition various camouflage colours were used to suit local requirements. Australian vehicles from 1939 to about the end of 1941 were coloured a medium shade of khaki. Vehicles for service in the Middle East were painted a sandy yellow colour. From late 1941 to about mid-1944

vehicles were turned out in a camouflage pattern of khaki and dust colour (somewhat darker than desert yellow). This also applied to US Lend-Lease vehicles assembled in Australia. From about mid-1944 vehicles were painted a dark khaki or 'camouflage green' (a peculiar mud colour).

Registration Numbers

Canadian Army vehicles used the same numbering system as the British except that the letter C (Canada) appeared before the type prefix letter. Australian vehicles had various types of registration numbers. From 1939 vehicles carried number plates prefixed with a red 'C' (Commonwealth) or 'DD' (Defence Department). AIF vehicles had special number-plates prefixed AIF and a red letter denoting type of vehicle (V for Van, L for Lorry, T for Tank, etc.). Later, the use of number-plates was discontinued and the number and prefixes were painted on the bonnet and at the rear, or in equivalent positions. Sometimes the prefixes were not used, however, and British vehicles supplied to the AIF in the Middle East retained their original number with often, but not always, the AIF prefix, until the vehicle was repainted. After repainting, such vehicles were sometimes given Australian serial numbers while others retained their original numbers without prefix. Indian Army vehicles registration numbers were prefixed with a broad WD-type arrow.

Who's Who in the British Commonwealth Motor Industry (1939–45)

Canada

Chrysler Corporation of Canada Ltd, Windsor, Ontario (Dodge cars and trucks).

Ford Motor Company of Canada Ltd, Windsor, Ontario (Ford cars and trucks, CMP and armoured vehicles).

The Four Wheel Drive Auto Co, Kitchener, Ontario (basically US-type FWD 4×4 trucks and tractors for semi-trailers).

General Motors Products of Canada Ltd, Oshawa, Ontario (Chevrolet, GMC and Maple Leaf trucks, Chevrolet and Buick cars, CMP and armoured vehicles).

Note: there were also some small specialist firms who produced limited numbers of automotive vehicles such as Bombardier Snowmobile Ltd in Valcourt, Quebec, and Farand & Delorme Ltd.

Australia

General Motors–Holden's Ltd, Melbourne (also Sydney, Brisbane, Adelaide, Perth) (assembly of Canadian and British General Motors conventional, modified conventional and CMP chassis, manufacture of cabs and bodies).

Ford Motor Company of Australia Pty Ltd, Melbourne, Victoria (assembly of North American Ford cars and trucks, manufacture of bodywork).

Chrysler Australia Ltd, Adelaide (assembly of North American Chrysler Corp. cars and trucks, incl. Dodge, DeSoto, Plymouth, Fargo).

International Harvester Company of Australia Pty Ltd, Melbourne (assembly of North American IHC chassis, manufacture of bodywork).

India

Ford Motor Company of India, Ltd, a subsidiary of Ford Motor Co. of Canada Ltd (assembly of Canadian Ford vehicles, manufacture of bodywork).

Note: armoured hulls for the rear-engined Ford Wheeled Carriers were produced by the East Indian Railway Workshops and the Tata Iron and Steel Co. Ford India's total war production included 64,216 4×2 vehicles, 45,213 4×4 vehicles, 3088 6×4 vehicles, 11,614 civilian type and 9876 miscellaneous vehicles.

South Africa

Ford Motor Company of South Africa (Pty) Ltd, Port Elizabeth (assembly of North American Ford chassis, manufacture of bodywork, Marmon-Herrington front-wheel drive conversions).

Note: Ford South Africa's war production included 31,336 trucks, 1643 cars and utilities (station wagons), 1890 miscellaneous vehicles.

General Motors of South Africa Ltd, Port Elizabeth (assembly of North American GM chassis).

New Zealand

Ford Motor Company of New Zealand Ltd, Lower Hutt (assembly of North American Ford vehicles).

Note: Ford New Zealand, during World War II, turned out 1162 CMP vehicles, 3611 modified conventional vehicles and 427 other vehicles including 164 staff cars.

CANADIAN MILITARY TRUCK CHASSIS TYPES
MILITARY PATTERN & MODIFIED CONVENTIONAL

TYPE			GENERAL MOTORS (Chevrolet, GM)			FORD				CHRYSLER (Dodge)		
Payload	Drive	wb[1]	A	B	C	A	B1	B2[7]	C	A	B	C
8-cwt	4×2	101	C8	8420	C8421-M	F8	C011DF	C291DF	F8421-M	—	—	—
HU	4×4	101	C8A	8445	CHU441-M	—	—	—	—	—	—	—
8-cwt	4×4	101[2]	C8AX	8448	CHU441-M	F8A	C011DQ	—	F8441-M	D8A	T212	D8446-M
15-cwt	4×2	101[3]	C15	8421	C15421-M	F15	C101WF	C291WF	F15421-M	D15	T222	D15428-C
15-cwt	4×4	101[4]	C15A	8444	C15441-M	F15A	C011WQF	C291WQF	F15441-M	D3/4APT	T236	—
15-cwt	4×4	101	C15TA	8449	C15441-M	—	—	—	—	—	—	—
30-cwt	4×2	134	—	1533X2	C30424-C	—	—	—	—	—	—	—
30-cwt	4×4	134	C30	8441	C30444-M	F30	C01QF	C29QF	F30444-M	—	—	—
2-ton	4×2	134	—	—	—	FS40S	—	21T	F40424-C	—	—	—
2-ton	4×2	158[5]	—	—	—	F40L	—	218T	F40428-C	—	T118G	D40420-C
3-ton	4×2	134[6]	—	—	—	FC60ST	—	C39TFS	F60424-C	D60S	T110L6	D60426-C
3-ton	4×2	134	—	—	—	FC60S	—	C29TF	F60424-C	—	—	—
3-ton	4×2	134	—	—	—	F602S	—	C29WFS	F60424-M	—	—	—
3-ton	4×2	158	—	—	—	F602L	—	C298WFS	F60428-M	—	—	—
3-ton	4×2	160	CC60L/X2	1543X2	C60420-C	FC60L	C098TFS	C298TFS	F60428-C	D60L	T110L5	D60420-C
3-ton	4×2	160	CC60L/X3	1542X3	C60424-C	—	—	—	—	D60L/D	T110L9	D60420-C
FAT	4×4	101	CGT	8440	C60441-M	FGT	C011QF	C291QF	F60441-M	—	—	—
Bofors SP	4×4	134	—	—	—	F60B	—	C39QB	F60444-S	—	—	—
3-ton	4×4	115	—	—	—	F60T	—	C395Q	F60445-M	—	—	—
3-ton	4×4	134	C60S	8442	C60444-M	F60S	C01QF	C29QF	F60444-M	—	—	—
3-ton	4×4	158	C60L	8443	C60448-M	F60L	C018QF	C298QF	F60448-M	—	—	—
3-ton	6×4	160	—	—	—	F60H	C010QF	C290QF	F60640-M	—	—	—
3-ton	6×6	160	C60X	8660	C60660-M	—	—	—	—	—	—	—

[1] wb for GM models (Ford CMP ¼ in longer). [2] Dodge 116. [3] Dodge 128. [4] Dodge (¾-ton WC) 98. [5] Dodge 160. [6] Dodge 136; also T110LS (2 DT). [7] 'F' not used after 1943.

A Simplified Chassis Model Designation (C—Chevrolet & GM, D—Dodge, F—Ford; 8, 15, 30, 60—payload in cwt, where applicable; suffix letter(s) to indicate front-wheel drive, lwb, etc., where necessary). This code was standardised early in 1944.

B Manufacturer's Chassis Type Symbol (B1 is Ford 1940–41, B2 is Ford 1942–45; first figure in Ford model symbol indicates model's year of introduction: 0-1940, 1-1941, etc.).

C DMS Chassis Model Code. Early code used by Canadian Department of Munitions and Supply until superseded by the simplified code (A). The first digit indicated the manufacturer (as under A). The second and third digits indicated load capacity in cwt (as under A). The fourth and fifth digit indicated the number of wheels and the number of driven wheels respectively (e.g. 64 for a 6×4). The sixth digit indicated the wheelbase (1—101 in, 4—134 in, 8—158 in, 0—160 in). Digit to right of dash indicated pattern (M—CMP, C—Commercial, S—Special).

BRITISH COMMONWEALTH

MOTORCYCLES and CARS

Motorcycles used by the Canadian forces comprised British Matchless (G3L), Norton (solo and w/SC), and US Harley-Davidson WLC and Indian 340 and 640 (solo and w/SC). The Australian forces had BSA, Harley-Davidson and Indian solo machines and Indian and Norton combinations.

Passenger cars were mainly products of General Motors (Chevrolet, Buick), Ford (Ford, Mercury) and Chrysler (Plymouth). The Canadian Army used some British Austin 10 HP 2-seaters also. The Australian forces had a variety of American and European (British, French, German) cars before standardising on Chevrolet and Ford.* US 'Jeeps' were used extensively.

*For additional coverage see *American Cars of the 1940s* (Olyslager Auto Library/Warne).

Motorcycle, Solo, Heavy (Harley-Davidson WLC) V-2-cyl., 18 bhp, 3F, wb 57½ in, 88×36×42 in, 550 lb approx. 45·12 CID engine. Tyres 4.00–18. Model WLC was built in the USA expressly for the Dept of National Defense, Canada. Some had pillion seat or special gear ratios for sidecar use. 1943.

Car, Light, Sedan, 4×2 (Ford C21AS) V-8-cyl., 95 bhp, 3F1R, wb 114 in, 195×73×68 in, 3660 lb. Mil. version of 1942 sedan, fitted with special bumpers, tyres, lighting, roof luggage-rack, etc. Used by Staff officers of certain Canadian formation headquarters.

Car, Light, Sedan, 4×2 (Plymouth 114 DS) 6-cyl., 85 bhp, 3F1R, wb 117 in, 199×73×68 in, 3200 lb. Militarised 1941 Model P11 Plymouth 5-passenger sedan, featuring 9.00-13 sand tyres and other modifications to meet Canadian service requirements. Used as staff car. LHD.

BRITISH COMMONWEALTH

HEAVY UTILITIES
4×2

There were two classes of heavy utilities: civilian type station wagons with varying degrees of modifications and military pattern models. The former were mainly Canadian-built Fords (four-door five- and seven-seaters, some of which were converted to soft-top tourer types) as well as Chevrolet, Dodge and Fargo carryalls (station wagons on light commercial chassis). The military pattern models consisted of special bodywork on CMP chassis, e.g. Trucks, 6-str, Utility (Aust.) (Chevrolet C15 and Ford F15 chassis). In the Canadian Army the 4×2 models were superseded by the Chevrolet C8A 'HUP' which was a four-wheel drive CMP vehicle (see page 222).

Station Wagon/Heavy Utility, 4×2 (Ford C21AS) V-8-cyl., 95 bhp, 3F1R, wb 114 in, 194×79×72 in, 4004 lb. 6.00–16 tyres. Crew seven incl. driver. Equipped with roof rack, blackout curtains, rifle clips, map container, first-aid kit, POW cans, etc. Basically 1942 station wagon.

Station Wagon/Heavy Utility, 4×2 (Ford C11ADF) V-8-cyl., 95 bhp, 3F1R, wb 114¼ in, 194×79×72 in, 4230 lb. 9.00–13 tyres. Crew five incl. driver. Two roof hatches. Full-floating truck-type rear axle with open propeller shaft and semi-elliptic springs (as C011DF). Used by British also. 1941.

Car, Heavy Utility, 4×2 (Ford C011DF–F8) V-8-cyl., 95 bhp, 4F1R, wb 101 in, 162×78×78 in, 4032 lb. 9.00–13 tyres. Crew six incl. driver. British-built body (Mulliner, Stewart & Ardern) on 8-cwt CMP chassis. Used mainly by RAF. Rear seats folding to provide cargo space.

BRITISH COMMONWEALTH

TRUCKS, 8-CWT
4×2 and 4×4

Makes and Models: Chevrolet C8 (4×2), C8AX (4×4). Dodge T212 (D8A) (4×4). Ford F8 (4×2), F8A (4×4).

General Data: The Canadian Chevrolet (except C8A) and Ford 8-cwt models were generally similar to their British counterparts (Ford, Humber, Morris-Commercial) and were designed to the same basic specifications. The Dodge T212 was basically identical to the US Dodge ½-ton 4×4 but featured right-hand drive and a British type pickup body. The Ford and Chevrolet had the early 1940–41 type CMP (Canadian Military Pattern) cab. These cabs were of all-steel construction and had side curtains to the doors with 'Monsento' transparent inserts. The top was removable at half cab. 1941 models (Cab No. 12) were similar to 1940 (Cab No. 11) but had an alligator-type engine hood. A hook arrangement allowed the (rear-hinged) doors to be left partly open. The Indian Army was probably the only user of the Ford F8A (4×4), relatively few of which seem to have been made. This model had 9.25–16 tyres. The Chevrolet C8A (4×4) was almost exclusively produced with Heavy Utility body (see next section) but it was also available as Chassis with cab for special bodies. This was the No. 13 cab with reverse-sloping windscreen. Bodies on the above chassis included Personnel/GS and Wireless. Early models had the same body as their British counterparts, although only in the case of the wireless body could the superstructure with tilt be used separately off the vehicle as a shelter. Later-type bodies were longer. The Chevrolet C8A differed from the heavier CMP types in having narrower front wings. The towing lugs on the front bumper were hinged horizontally (vertically on all heavier models). Like the 15-cwt 4×4 models, it had a single-speed transfer case. From 1st July, 1944, the square hatch in the cab roof of No. 13 cabs was replaced in production by a circular 'hip ring'.

Vehicle shown (typical): Truck, 8-cwt, 4×2, Wireless (Chevrolet C8).

Technical Data:
Engine: Chevrolet 216 6-cylinder, I-I-W-F, 216·5 cu in, 85 bhp @ 3400 rpm.
Transmission: 4F1R.
Brakes: hydraulic.
Tyres: 9.00–13.
Wheelbase: 101 in.
Overall l×w×h: 166×83×90 in.
Weight: 5348 lb.
Note: 1940 model shown (cab No. 11). Chorehorse battery-charging unit between cab and body (not shown).

Truck, 8-cwt, 4×2, Wireless (Ford F8) V-8-cyl., 95 bhp, wb 101¼ in, 166×83×90 in, 5292 lb. 1941 model with No. 12 cab featuring alligator-type hood. Used by British Army in Egypt. Note petrol-engine-driven battery-charging generator (chorehorse) behind cab. Tarpaulin removed.

Truck, 8 cwt, 4×2, Wireless (Chevrolet C8) 6-cyl., 85 bhp, 4F1R, wb 101 in, 166×83×90 in, 5350 lb. Second body type for 8-cwt Wireless and Personnel types with No. 12 cab. Equipped to carry No. 19 wireless set. Seating for four in rear body. Spare wheel and battery-charging set inside.

Truck, 8-cwt, 4×4, Personnel (Dodge T212, D8A) 6-cyl., 74 bhp, 4F1R×1, wb 116 in, 172×76×83 in, 4939 lb. 9·25-16 tyres. Similar to US Army Dodge ½-ton 4×4 but slightly smaller engine (b×s $3\frac{3}{8} \times 4\frac{1}{16}$ vs $3\frac{1}{4} \times 4\frac{5}{8}$). Rhd. British WO type body, wheels and tyres.

Truck, 8-cwt, 4×4, GS (Chevrolet C8AX) 6-cyl., 85 bhp, 4F1R×1, wb 101 in, 165×80×90 in approx. GVW 7500 lb. Tyres 9.25–16. Supplied by GM of Canada in small quantity only in CKD form. Shown in use with Royal New Zealand Air Force. Chassis also used for prototype APT truck.

BRITISH COMMONWEALTH

TRUCKS, 8-CWT 4×4
HEAVY UTILITY

Makes and Models: Chevrolet C8A HUA, HUP, HUW, etc.

General Data: This type was introduced in 1942 and was made only by General Motors of Canada Ltd. It was produced, with minor changes, until 1945. Although basically an 8-cwt chassis, it was usually referred to as Heavy Utility and in 1945 it was redesignated 15-cwt 4×4. All models had an all-steel box-shaped body, the forward end of which was the No. 13 CMP cab without the rear panel. The rear body was modified according to role and the following versions existed:

1C1: Personnel carrier (HUP), five doors, windows all round, six seats.
1C2: Wireless (HUW), five doors, no rear quarter windows, five seats.
1C3: Ambulance (HUA), five doors, two stretchers, four seats, no rear quarter windows.
1C4: not built but intended as 1C1 fitted out as office.
1C5: as 1C3 but 'articised' (Stewart-Warner gasoline-burning heater, etc.).
1C6: not built (except prototype) but intended as 1C3 with tyre carrier on right-hand side, replacing one side door (pneumatic instead of 'Run Flat' tyres).
1C7: as 1C1 but spare wheel on right-hand side, four doors.
1C8: as 1C2 but tyre carrier as 1C7, four doors.
1C9: as 1C7 but rear seats replaced by computer equipment (also used as Cipher Office and Paymaster Office).
1C10: as 1C7 but equipped with machinery and special fitments for wireless repair.
1C11: Staff car; as 1C7 but sliding glass windows all round; special interior finish and equipment.

Some types carried penthouse and poles for side or rear extension. The Australian Army used a 6-str utility body on the Canadian 15-cwt 4×4 and 4×2 chassis.

Vehicle shown (typical): Truck, Heavy Utility, 4×4, Personnel (Chevrolet C8A, Body 1C7)

Technical Data:
Engine: Chevrolet 216 6-cylinder, I-I-W-F, 216·5 cu in, 85 bhp @ 3400 rpm.
Transmission: 4F1R×1.
Brakes: hydraulic.
Tyres: 9.25–16.
Wheelbase: 101 in.
Overall l×w×h: 163×79×90 in.
Weight: 6200 lb.
Note: 1944–45 model with hip ring and lifting sling flanges on wheel hubs. Radiator guard removed. Also known as 'HUP'.

Truck, Heavy Utility, 4×4, Ambulance (Chevrolet C8A, Body 1C5) Chassis as Personnel version. Curtain at rear of driver. Racks for two stretchers on left. Medicine lockers over rear wheel arches. Bench-type seat on rh locker for two passengers. Sliding glass windows in rear compartment (not on 1C3).

Truck, Heavy Utility, 4×4, Wireless (Chevrolet C8A, Body 1C2) Chassis as Personnel version. Rear compartment with sliding windows, designed to accommodate No. 9 or 19 wireless set and amplifier, chorehorse, three operator seats and other equipment. Note aerial bracket on body side.

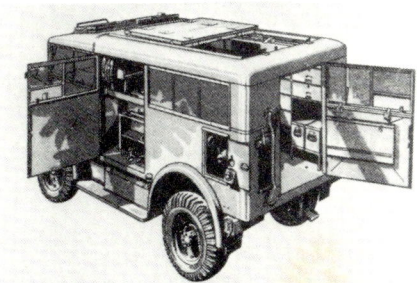

Truck, Heavy Utility, 4×4, Machinery 'ZL' (Chevrolet C8A, Body 1C10) Chassis as Personnel version. Used by RCEME for wireless repairs. Known as 'Z light'. Equipped with generating set, air-cooled transformer, power panel, test panels, valve test sets, work benches, shelves, drawers, 2 chairs, etc.

Truck, Heavy Utility, 4×4, Staff Car (Chevrolet C8A, Body 1C11) Limited production, 1945. Sides, roof and doors insulated with pressed board, floor covered with linoleum, hot-water heater, sliding roof in centre (at rear on other models), four seats and map table in rear compartment. Exterior door handles.

BRITISH COMMONWEALTH

TRUCKS, 12- and 15-CWT and 1-TON 4×2

Makes and Models: *Australian types: Note:* all models officially known as Vans. 12-cwt Utility GS: Chevrolet, DeSoto, Dodge, Ford, Plymouth. 12-cwt Panel Van: Chevrolet, Ford. 15-cwt Utility GS: Chevrolet, Dodge, Fargo, Ford, GMC, International. 15-cwt Coupé Utility: Chevrolet, Dodge, Fargo, Ford, GMC, International. 15-cwt Panel Van: Chevrolet, GMC. 15-cwt Wireless: Ford. 1-ton Utility GS: Chevrolet, Dodge, Fargo, Ford, International. 1-ton GS and Office: Chevrolet, Dodge, Fargo, Ford, GMC, International. 1-ton Panel Van: Ford. 1-ton Battery Staff, Wireless (Aust.): Ford.
Canadian Modified Conventional types: 15-cwt GS: Chevrolet, Dodge D15. 15-cwt Water Tank: Dodge D15.
Canadian Military Pattern types: 15-cwt GS and various other body styles: Chevrolet C15, Ford F15.

General Data: The Australian types in this category were all adaptations or modifications of commercial-type vehicles which were in production during 1939–41. These vehicles were of North American origin and assembled in Australia. General Motors products were assembled by General Motors-Holden's Ltd and fitted with GM-H's own cab which was similar but not identical to the North American type. The Coupé Utility was and still is a type peculiar to Australia. Canadian Military Pattern vehicles in the 15-cwt 4×2 class were used by the British, Canadian, Australian, Indian and South African forces, and later by several more. The Ford CMP model of 1940–41 was also used with open cab (early British WD style), notably by the Indian Army in North Africa. There was a wide range of body options, especially for the CMP vehicles, most of which were also fitted on 4×4 chassis (see next section). Chevrolet trucks assembled and bodied in India were popularly known as 'Bombay Chevs'.

Vehicle shown (typical): Van, 12-cwt, 4×2, GS (Aust.) (Dodge D20)

Technical Data:
Engine: Dodge 6-cylinder, I-L-W-F, 230 cu in, 90 bhp @ 3000 rpm (or 201·3 cu in, 82 bhp @ 3600 rpm).
Transmission: 3F1R.
Brakes: hydraulic.
Tyres: 6.00-16.
Wheelbase: 117½ in.
Overall l×w×h: 195×80×81(74) in.
Weight: 3584 lb.
Note: Tarpaulin and superstructure removed. Similar body on DeSoto and Plymouth chassis. 1941 style car front end. In background: IHC coupé utility and trucks.

Van, 12-cwt, 4×2, GS (Aust.) (Ford 01C) V-8-cyl., 85 bhp, 3F1R, wb 112 in, 3136 lb. Australian coupé utility type body on 1940 Ford Commercial chassis. Oversize tyres. Later production had spring bumper, roof hatch, etc. Some had superstructure and canvas cover.

Van, 15-cwt, 4×2, GS (Aust.) (Ford, modified) V-8-cyl., 85 bhp, 3F1R, wb 112 in, approx. 3100 lb. Experimental modification of 1940/41 Australian Ford coupé utility. The obvious aim was to militarise a commercial vehicle, thus simplifying body-damage repair jobs in the field.

Van, 1-ton, 4×2, Battery Staff, Wireless (Aust.) Type GA (Ford 01Y) V-8-cyl., 95 bhp, 4F1R, wb 122 in, 1940 model chassis. Modified commercial-type cab with roadster-type folding top. Gun Position Officers' vehicle. Tarpaulin with roll-up sides. Later converted into GS truck. 7.00–17 tyres.

Truck, 15-cwt, 4×2, GS (Indian) (Chevrolet 1311X3) 6 cyl., 78 bhp, 3F1R, wb 113$\frac{1}{2}$ in, 3192 lb. Troop carrier based on Canadian-built 1939 (and later) $\frac{1}{2}$-ton commercial chassis. Used in Egypt and North Africa by British Indian Army and LRDG. 9.00–13 sand tyres. Similar bodywork on Ford O1C chassis.

Van, 15-cwt, 4×2, GS (Aust.) (Chevrolet 1300 Series) 6 cyl., 78 bhp, 3F1R, wb 113½ in. 1940 style commercial ½-ton chassis with Australian (GM-Holden's) cab and Utility GS body. Typical and common Australian vehicle, used in Middle East, Malaya, etc. 7.50–16 tyres.

Van, 1-ton, 4×2, GS (Aust.) (International K3) 6-cyl., 82·4 bhp, 3F1R, wb 130 in. 7.00–16 (2DT) tyres. Australian steel cab with canvas top. Canvas roll-up blind at rear. Removable side screens on 'half doors'. Wooden body reinforced with channel iron.

Truck, 15-cwt, 4×2, GS (Dodge D15) 6-cyl., 95 bhp, 4F1R, wb 128¼ in, 201×85×90(75) in, 5264 lb. 9.00–16 tyres. Standard pattern all-steel GS body. All-steel commercial type cab but top detachable. Widely used by British Army.

Truck, 15-cwt, 4×2, Water Tank (Dodge D15) 6-cyl., 95 bhp, 4F1R, wb 128¼ in, 204×78×84(75) in, 6804 lb. 200-gal. tank body, generally similar to that fitted on Bedford MWC. Pump driven from PTO. Tarpaulin could be fitted for camouflage purposes.

227

Truck, 15-cwt, 4×2, GS (Ford F15) V-8-cyl., 95 bhp, 4F1R, wb 101¼ in, 170×86×90 in. 5870 lb. 9.00–16 tyres. Later types had pneumatic instead of Runflat tyres and spare wheel between cab and body. 1941 model No. 12 cab shown.

Truck, 15-cwt, 4×2, Personnel (Chevrolet C15) 6-cyl., 85 bhp, 4F1R, wb 101 in, 170×86×93(88) in, 6059 lb. Similar to GS but adapted for personnel carrying (six in rear body) and fitted with tilt. 1940 model No. 11 cab shown, recognisable by single-piece radiator grille. Introduced in 1939.

Truck, 15-cwt, 4×2, AA, No. 1 Mk II and III (Ford F15) V-8-cyl., 95 bhp, 4F1R, wb 101¼ in, 173×84×86(81) in, 6272 lb. Platform body equipped to take 20-mm AA gun. Gun is loaded onto vehicle by small hand winch up tubular ramps. Also on Chevrolet chassis. See also page 155.

Truck, 15-cwt, 4×2, Van (Chevrolet C15) 6-cyl., 85 bhp, 4F1R, wb 101 in, 170×86½×95 in, 7457 lb. Early-type GS body with tubular superstructure and tarpaulin on 1942 type CMP chassis. Cab No. 13. Similar body also for GS, AT Tractor, AA and AT, Personnel, Cable Layer.

BRITISH COMMONWEALTH

TRUCKS, 15-CWT 4×4

Makes and Models: Chevrolet C15A. Chevrolet C15A-APT (pilot only). Dodge D3/4 APT and APT/WP. Ford F15A. GM of Canada C15TA (armoured, see page 254).

General Data: Of the above models the CMP Chevrolet C15A and Ford F15A (early and later pattern) were the most numerous. They were similar to the C15 and F15 but had a driven front axle and a single-speed transfer case. As a result these vehicles were slightly higher than their 4×2 counterparts. These CMP vehicles were also assembled in other countries of the Commonwealth. The Canadian range of bodies available for the Chevrolet C15A included GS, Van, Battery Charging, Office, Wireless, Cable Layer, Personnel, Water Tank, AT Tractor and AA. The Ford F15A range included GS, Van, Cable Layer, Personnel, AT Tractor, AA, Machinery, etc. A pilot model for an airportable C15A was produced but did not enter production. Instead, the US Dodge Weapons Carrier was produced by the Canadian Chrysler factory. It differed from the parent vehicle only in minor detail (US model T214: bore and stroke $3\frac{1}{4} \times 4\frac{5}{8}$ in, displacement 230·2 cu. in, CR 6·7:1, Zenith carb., 10-in clutch; Canadian T236: $3\frac{7}{16} \times 4\frac{1}{4}$ in., 236·6 cu. in, 6·8:1, Ball & Ball (Carter) and 11-in resp.). The British counterpart of this vehicle was the Austin K7. The C15A assembled in Australia had a GM-Holden cab which differed from the Canadian cab in having oblong ventilator panels on the sides of the front shell. CMP Ford cabs produced in Australia had the same pattern ventilator panels as the original Canadian type but if fitted with the square roof hatch they differed from Canadian Ford and all Chevrolet CMP vehicles in having two separate grab handles in front of the hatch. On all other models there was a full-width single-piece grab rail (see illustrations). CMP cabs were also produced without a roof hatch or hip-ring. Some CMP models assembled in Australia (Chevrolet and Ford) has a canvas-top cab. The GM C15TA was an armoured 15-cwt truck, powered by a GMC 270 engine.

Vehicle shown (typical): Truck, 15-cwt, 4×4, Office (Chevrolet C15A)

Technical Data:
Engine: Chevrolet 216, 6-cylinder, I-I-W-F, 216·5 cu in, 85 bhp @ 3400 rpm.
Transmission: 4F1R×1.
Brakes: hydraulic.
Tyres: 9.00–16.
Wheelbase: 101 in.
Overall l×w×h: 171×88×97(90) in.
Weight: 9772 lb.
Note: GS body fitted for office role. Tarpaulin (removed in picture) of penthouse type.

Truck, 15-cwt, 4×4, GS (Ford F15A) V-8-cyl., 95 bhp, 4F1R×1, wb 101¼ in, 170×86×90 in, 5870 lb. One of the earliest CMP vehicles, featuring wood-bolted body with wheel arches and sheet-metal rear wings. Hinged sides. Some early-production F15As had GM banjo-type driving axles.

Truck, 15-cwt, 4×4, Van (Ford F15A) 6-cyl., 95 bhp, 4F1R×1, wb 101¼ in, 166×86½×92¼ in, 9533 lb. Std GS body (2B1) with bows and tarpaulin. No. 12 cab (which differed from No. 11 on left in having alligator-type bonnet). 'Run Flat' tyres (no spare tyre carrier). Also on Chevrolet C15A chassis.

Truck, ¾-ton, 4×4, GS, APT (Dodge D3/4APT T236) 6-cyl., 92 bhp, 4F1R×1, wb 98 in, 182×83(later 77)×85(67) in, 5900 lb. Almost identical to US Weapons Carrier (shown) but wider gap between spare wheel and body (as on later US types). APT/WP was wadeproofed. Left-hand drive. 236·6 CID engine. See also General Data.

Truck, 15-cwt, 4×4, FFW (Chevrolet C15A) 6-cyl., 85 bhp, 4F1R×1, wb 101 in, 171×88×102 in, 8255 lb. GS type body, fitted for wireless. Designed to carry No. 19 WT set (alternatively No. 9, 11 or 22). Late type shown. Note additional battery box below left-hand door.

Truck, 15-cwt, 4×4, Machinery 'KL' (Ford F15A) V-8-cyl., 95 bhp, 4F1R×1, wb 101¼ in, 168×86×94½ in, 11,630 lb. Used for electric welding in the field. 300-amp. welder driven by (separate) Ford V8 engine, portable grinder, welder's table and screen, etc.

Truck, 15-cwt, 4×4, Water Tank (Chevrolet C15A) 6-cyl., 85 bhp, 4F1R×1, wb 101 in, 172×85×90 in, 6743 lb. 200-gal. (Imp.) tank, used to provide filtered and sterilised water for troops in the field. Superstructure and tarpaulin provided for camouflage. Also with No. 12 cab.

Truck, 15-cwt, 4×4, Wireless, House Type (Ford F15A) V-8-cyl., 95 bhp, 4F1R, wb 101¼ in, 175½×85×114 in, 9670 lb. Usually on Chevrolet chassis. Designed to carry No. 9, 11 or 19 WT set. Two removable 300-watt charging generators. Early type shown.

Truck, 15-cwt, 4×4, Wireless, House Type (Chevrolet C15A) 6-cyl., 85 bhp, 4F1R×1, wb 101 in, 175½×85×114 in, 9772 lb. Pilot model for latest type. Spare wheel recessed in body side. Composite body construction. Permanent roof aerial eliminated. Redesigned interior.

Truck, 15-cwt, 4×4, GS (Aust.) No. 2 (Ford F15A) V-8-cyl., 95 bhp, 4F1R×1, wb 101¼ in, 171×88×96 in, 7220 lb. Australian GS body on imported Ford CMP chassis. Early-style cab with canvas top over steel frame. Also as FFW. No. 1 had wooden body.

Truck, 15-cwt, 4×4, Office (Aust.) No. 2 (Chevrolet C15A) 6-cyl., 85 bhp, 4F1R×1, wb 101 in, 171×88×96 in, approx. 7500 lb. Early-style cab (later models—all-steel with roof hatch). GS type body fitted out as office. Tent equipment provided. Valances on rear wheel arches later omitted.

Truck, 15-cwt, 4×4, Signals Office (Aust.) No. 2 (Chevrolet C15A) 6-cyl., 85 bhp, 4F1R×1, wb 101 in, 171×88×96 in, approx. 7500 lb. GS body equipped for signal office role. There was also a *Truck, 15-cwt, Wireless Signals (Aust.)*, which had full-length all-steel body, sliding windows.

Truck, 15-cwt, 4×4, Fire Tender (Aust.) (Chevrolet C15A) 6-cyl., 85 bhp, 4F1R×1, wb 101 in. Towed 500-gpm 'Jupiter' fire-pump unit on GM-H trailer. Similar body also fitted on C15 4×2 chassis. All-steel GM-H cab shown.

BRITISH COMMONWEALTH

TRUCKS, 30-CWT
4×2 and 4×4

Makes and Models: *Australian 1937–42 4×2 types:* Chevrolet, Dodge, Fargo, Ford, GMC, International. *Australian 1939-41 4×4 types:* Ford/Marmon-Herrington. *Canadian types:* Chevrolet and Maple Leaf (4×2 and 4×4). Chevrolet C30 (4×4). Ford F30S, F30 (4×4).

General Data: The Australian types in this category were adaptations or military modifications of standard commercial chassis. Many were fitted with oversize tyres, single rear. Various body styles were used, including Ambulance, GS, Water Tank and House type. Late-production Fords had spring-type radiator guard and square cab-roof hatch. All were basically North American chassis, assembled in Australia and fitted with Australian bodywork. The British Indian Army used large numbers of Canadian and US Chevrolets (commercial 1½-ton), often fitted with oversize tyres, local bodywork (usually GS) and open cab. Canadian Military Pattern 4×4 types in this class were produced from 1940 until 1943, following British War Office specifications. Like their British-built counterparts their production was discontinued in the interest of vehicle standardisation. The following bodies were among those fitted on these CMP chassis. *Chevrolet C30:* GS, Office, Cable Layer, LAA tractor, Stores, Wireless, 2-pdr Gun Portee. *Ford F30:* GS, Ambulance, Winch truck, Derrick, LAA tractor. The Chevrolet Gun Portee was later converted into an AT (17-pdr) tractor. Most other types were superseded by 3-ton 4×4 models. The 30-cwt C30 and F30 differed from the swb 3-ton C60S and F60S mainly in tyre size, spring ratings and brakes, which were not power-assisted. The Ford Motor Co. of Canada produced an experimental 30-cwt 4×4 compressor truck on a 101-in wb chassis. Very early Ford 30-cwt 4×4 trucks were designated F30S. The 'S' indicating short wheelbase, it was later omitted since there was no long-wheelbase version of this model. These early production models had GM banjo-type axles. The CMP driven front axle was first designed by Marmon-Herrington in 1939.

Vehicle shown (typical): Truck, 30-cwt, 4×4, Office (Chevrolet C30)

Technical Data:
Engine: Chevrolet 216 6-cylinder, I-I-W-F, 216·5 cu. in, 85 bhp @ 3400 rpm.
Transmission: 4F1R×2.
Brakes: hydraulic.
Tyres: 10.50–16.
Wheelbase: 134 in.
Overall l×w×h: 199×88×110 in.
Weight: 9441 lb; GVW 12600 lb.
Note: No. 11 cab (1940) and early type body with hinged side doors. Superseded by Lindsay body with sliding doors and pullman-type windows. Equipped with 12-volt generator (chorehorse). Hinged luggage platform at rear.

Truck, 30-cwt, 4×4, LAA Tractor (Chevrolet C30) 6-cyl., 85 bhp, 4F1R×2, wb 134 in, 210×88×101(88) in, 10,400 lb. Early type with No. 12 cab. 10-ft open-end body. Lockers and bins for ammunition and personal, vehicle and gun equipment. Fitted with winch.

Truck, 30-cwt, 4×4, LAA Tractor (Chevrolet C30) 6-cyl., 85 bhp, 4F1R×2, wb 134 in, 210×88×101(88) in, 10,400 lb. Late type with No. 13 cab. Also on Ford chassis. Crew nine. One crew member sat in centre at rear, facing the rear, controlling the cable-operated Bofors gun brakes.

Truck, 30-cwt, 4×4, Gun Tractor (Maple Leaf) 6-cyl., 80 bhp, 4F1R×2, wb 134 in. The Maple Leaf was a product of GM Products of Canada Ltd and could be described as a heavy-duty Chevrolet. This unit was produced for Australia, where it was fitted with GM-H cab and fixed-side body.

Truck, 30-cwt, 4×4, Gun Tractor (Ford/Marmon-Herrington) V-8-cyl., 100 bhp, 4F1R×2, wb 134 in. Officially known as 'Tractor, Artillery (Aust.), No. 3' or '3A'. Basically 1940 North American Ford V-8 truck, converted to 4×4 with Marmon-Herrington front-wheel drive kit. Soft-top cab.

Truck, 30-cwt, 4×2, Ambulance, Indian Army Type (Aust.) No. 1 (Ford 01T) V-8-cyl., 85 bhp, 4F1R, wb 134¼ in. Four stretchers. Open at rear with canvas roll-down flap. Red Cross insignias often appeared much larger but were painted out for a short period in Malaya. Tyres 9.00–18.

Truck, 30-cwt, 4×2, Ambulance, Indian Army Type (Aust.) No. 1 (Chevrolet 1500 Series) 6-cyl., 85 bhp, 4F1R, wb 134½ in. Standard body type (not unlike that used in WW I), also fitted on Dodge, Ford, GMC and International. Officially: *Lorry, Ambulance, Indian Army Type (Aust.).*

Truck, 30-cwt, 4×2, Water Tank (Aust.) No. 2 (Chevrolet 1500 Series) 6-cyl., 85 bhp, 4F1R, wb 134½ in. 200-gal. tank. Standard body type, also on Ford, GMC and IHC commercial-type chassis. Note spare wheel in box behind cab and wing cut-outs to accommodate oversize (9.00–18) tyres.

Truck, 30-cwt, 4×2, GS (Aust.) (Ford 01T) V-8-cyl., 85 bhp, 4F1R, wb 134¼ in, 238×84×105(77) in. Militarised commercial 1940 type, widely used until about 1943. Similar body on Chevrolet, Dodge and GMC chassis. Various wb lengths and tyre sizes.

Truck, 30-cwt, 4×2, GS (Chevrolet 1533×2) 6-cyl., 85 bhp, 4F1R, wb 134½ in, 219×80×58½ in, 6450 lb. Tyres 10.50–16. Welded body (No. 4B1, by Gotfredson), ordered for ammunition carrying in Middle East theatre of war, 1941. Used extensively by 'Long Range Desert Group'.

Truck, 30-cwt, 4×4, 2-pdr, AT Portee (Chevrolet CGT) 6-cyl., 85 bhp, 4F1R×2, wb 101 in, 180×86×112 in, 8836 lb. Chassis similar to CGT gun tractor, using 3-tonner axles, etc. Winch omitted. Gun loaded by hand winch, pointing forward. Vehicles were later converted into AT tractors (17-pdr).

Truck, 30-cwt, 4×4, Wireless, House Type (Chevrolet C30) 6-cyl., 85 bhp, 4F1R×2, wb 134 in, 209×86¼×114 in, 9764 lb. Designed to carry No. 9, 11 or 19 WT set. Two removable 300-watt generator charging sets or single 500-watt unit. Roof aerials. Some were arcticised or winterised.

Truck, 30-cwt, 4×4, Ambulance (Ford F30) V-8-cyl., 95 bhp, 4F1R×2, wb 134¼ in, 207×90¼×110 in, 10,838 lb. Canadian steel-panelled body. Four stretchers. Upper stretchers laid on raising gears. Equipped with electric fan, ventilators, water tank. Superseded by 3-ton type.

BRITISH COMMONWEALTH

TRUCKS, 3-TON 4×2

Makes and Models: *Australian 1939–42 types:* Bedford, Chevrolet, Dodge, Ford, GMC, International, Maple Leaf. *Canadian Modified Conventional types:* Chevrolet CC60L/X2, CC60L/X3. Dodge D60S, D60L, D60L/D (T110 Series). Dodge T120, T130 (COE). Ford FC60S, FC60L. *Canadian Military Pattern types:* Ford F602S, F602L.

General Data: The 3-ton payload class was represented by large numbers of the above models as well as purely civilian-type vehicles (many impressed) which, because of their great variety, have not been listed. Although many of the above types were originally intended as interim types to be superseded by 4×4 types, most remained in use throughout the war. Most of the Australian types were basically North American vehicles assembled and militarised in Australia. Many of these were fitted with oversize tyres for desert use. Bodywork was Australian. The Bedfords were of British origin and used mainly by the RAAF (Armoured Tender, Fire Tender, Slide-Tilt, etc.). Canadian Modified Conventional types were militarised commercial trucks, usually with oversize tyres (single rear). They were fitted with left-hand drive only when built for domestic use. The only CMP vehicles in this class were the Ford F602S (swb) and F602L (lwb). They were similar to the F60S and F60L (4×4) except: tubular front axle, no transfer case, two-speed rear axle. Bodywork on the above chassis was of a wide variety, including Ambulance, Breakdown, Dump, GS, Office, Stores, Workshop, etc. The Canadian Maple Leafs were basically US-type Chevrolets with certain 'heavy duty' features. The 1940/41 Maple Leaf 1600 Series (2½-ton civilian rating) was similar to the contemporary GMC 9600 but had the Chevrolet 216 engine; the 3-ton 1600H Series Maple Leaf was similar to the GMC 9600H, both powered by the GMC 248 engine. Bus chassis were supplied by Chrysler (Dodge), Ford (e.g. Model C294T, 194-in wb, for use in Middle East), General Motors and International (Australia).

Vehicle shown (typical): Truck, 3-ton, 4×2, Dump (Dodge T110L-6, D60S)

Technical Data:
Engine: Dodge 6-cylinder, I-L-W-F, 236·6 cu in, 95 bhp @ 3600 rpm.
Transmission: 4F1R; two-speed rear axle.
Brakes: hydraulic (hydrovac).
Tyres: 8.25–20(2DT), later jobs 10.50–16.
Wheelbase: 136 in.
Overall l×w×h: 204½×84×83 in.
Weight: 7640 lb.
Note: hydraulically actuated steel dump body. Chrysler Corp. of Canada produced total of 180,816 military Dodge trucks during 1939–45.

Truck, 3-ton, 4×2, GS (Dodge T110L-5, D60L) 6-cyl., 95 bhp, 4F1R, wb 160 in, 258×90×124(84) in, 7865 lb. 10.50–16 tyres. Split-top cab. T110L-4 similar, but dual rear tyres. 12-ft all-steel (shown) or composite body. Eaton two-speed rear axle. Also used by British Army.

Truck, 3-ton, 4×2, GS, COE (Dodge T130) 6-cyl., 95 bhp, 5F1R, wb 159 in, 265×85$\frac{3}{4}$×124(94), 7685 lb. Canadian lhd model (for domestic use). 14-ft stake body with tilt. Similar type later used by Rhodesian Army. A 105-in wb version was produced as truck-tractor.

Truck, 3-ton, 4×2, GS (Chevrolet 1543×2) 6-cyl., 78 bhp, 4F1R, wb 160 in, 253×88×123(84) in, 6832 lb. 32×6(2DT) tyres. Two-speed rear axle. Military all-steel GS body. Basically 1941–42 commercial model, produced in Canada. Used by British RAF.

Truck, 3-ton, 4×2, GS (Chevrolet CC60L/X2) 6-cyl., 85 bhp, 4F1R, wb 160 in, 259×88×123(84) in, 7308 lb. 10.50–16 tyres, split-top cab with opening windscreen, rhd (for overseas use), spring bumper, etc. Two-speed rear axle. CC60L/X3 had single-speed axle with dual tyres.

Truck, 3-ton, 4×2, GS (Aust.) (Maple Leaf 1600 H) 6-cyl., 88 bhp, 4F1R, wb 175¾ in. 1940-style Canadian chassis, assembled and modified in Australia. Fitted with sleeping berth behind cab. Vehicle shown is partly stripped, used on Australian North–South Road.

Truck, 3-ton, 4×2, GS (Aust.) (Chevrolet 1500, 1600 Series) 6-cyl., 85 bhp, 4F1R, wb 160 in. Canadian chassis (1941–42), assembled and modified in Australia. 10.50–20 tyres. Standard Aust. GS type, but in this case front end militarised (experimental).

Truck, 3-ton, 4×2, Ambulance, Conventional (Aust.) (Chevrolet 1500 Series) 6-cyl., 78 bhp, 4F1R, wb 133 in. Australian four-stretcher body using standard GM-H cab doors. Used by Australian Imperial Force and RAAF. Similar body on Bedford chassis for RAAF.

Truck, 2½-ton, 4×2, Aircraft Refueller (Maple Leaf 1600 Series) 6-cyl., 78 bhp, 4F1R, wb 131½ in. Basically Canadian heavy-duty Chevrolet, 1939 style, produced for RCAF for domestic use. Left-hand drive. Swinging-boom type refuelling body.

Truck, 3-ton, 4×2, Breakdown (Aust.), LP3 (Maple Leaf 1600 Series) 6-cyl., 78 bhp, 4F1R, wb 157¾ in. Australian-assembled RHD Canadian Chevrolet Maple Leaf 2½-ton commercial chassis, fitted with GM-H cab, Australian military pattern bodywork and manually operated crane. 1940 model shown.

Truck, 3-ton, 4×2, Aircraft Refueller (Aust.) (GMC 9600 Series) 6-cyl., 80 bhp, 4F1R, wb 157¾ in. Australian-assembled RHD Canadian GMC 2½-ton truck with GM-H cab. 1942 model shown. Basically as Maple Leaf 1600 Series but 224 vs 216·5 CID engine. Similar types on 1942 Ford and International. Used by RAAF.

Truck, 3-ton, 4×2, Breakdown (Ford 098T) V-8-cyl., 95 bhp, 4F1R, wb 158 in, 228×85×98 in, 7392 lb. Canadian Ford, 1940, with 10.50–16 tyres and Holmes wrecker body. Two swinging booms with two hand-powered winches. Similar body also on Chevrolet chassis.

Truck, 3-ton, 4×2, GS (Ford FC60L) V-8-cyl., 95 bhp, 4F1R, wb 158 in, 261×88×122 in, 7925 lb. Modified conventional Canadian 1942 type. 10.50–16 tyres. Two-speed rear axle. Widely used in North Africa. Device at rear of body (both sides) is mg mounting.

Truck, 3-ton, 4×2, GS (Ford F602L) V-8-cyl., 95 bhp, 4F1R, wb 158¼ in, 243×90×118(78) in, 7280 lb. Tubular front axle, two-speed rear axle. This CMP chassis superseded the Modified Conventional Pattern 4×2 type towards the end of the war. Also with 300-gal. petrol tanker body (British Army).

Truck, 3-ton, 4×2, Ambulance (Ford F602L) V-8-cyl., 95 bhp, 4F1R, wb 158¼ in, 229×90×110 in, 12,992 lb (laden, with four stretcher cases). Chassis as shown on left but with 10.50-16 (vs 10.50-20) tyres and special ambulance springing. Stretcher-raising gear as in Austin K2. Used by British.

Truck, 3-ton, 4×2, GS (Ford F602S) V-8-cyl., 95 bhp, 4F1R, wb 134¼ in, 204×90×87 in, 7845 lb. Load carrier with No. 43 soft-top cab and 10-ft welded or bolted (KD) steel body with flat tarpaulin. Designed on 3-ton chassis but actual payload restricted by 10.50-16 'Run Flat' tyres.

Truck, 3-ton, 4×2, Bomb Crane (International KS5) 6-cyl., 82·4 bhp, 4F1R, wb 159 in. 7.00-20 tyres. Cab with canvas top on wooden frame. One of several types of Australian IHC trucks. This type used by RAAF, others (mainly GS) by AIF, US Army, etc. Some had 10.50-18 tyres.

BRITISH COMMONWEALTH

TRUCKS, 3-TON 4×4

Makes and Models: Chevrolet C60S, C60L. Ford F60S, F60L, F60T, F60B. FWD HAR.

General Data: Of the 815,729 'B' vehicles produced in Canada during World War II, 345,831 were 4×4 types and of these a very high percentage were in the 3-ton payload class. Many thousands were exported to other Commonwealth countries, notably Australia and India, where they were often fitted with locally built bodywork. The number of basic chassis types was small, as can be seen above. The type designation suffixes had the following meaning: S = swb (134 in), L = lwb (158 in), T = Tractor (wb 115 in), B = Bofors SP carriage (wb 134 in). The C60L and F60L were the most numerous, and the following were amongst the wide range of Canadian bodies fitted on these chassis. *C60L:* GS, Stores, Dental, Machinery (various types), Breakdown (wrecker), Office, Petrol Tanker, Command HP and LP, Signal Construction, etc. *F60L:* GS, Stores, Portee, Breakdown (wrecker), Machinery (various types), Ambulance, Cable Layer, etc. The swb chassis (C60S and F60S) were used with GS, Derrick, Dump, Stores, Wireless, Breakdown and other bodies. Australian bodies on CMP 3-ton 4×4 chassis were also of a wide variety and included GS (various types), Ambulance, AA SP, Bacteriological, Breakdown, Compressor, Garage, Office, Portee, Stores, Signal Office, Water Tank, etc.

Note: most of the above types were produced with the early and the later types of cab. The basic components of the 4×4 3-tonner were also used for several other types, including gun tractors, armoured cars and 6-wheeled trucks. The first 4×4 Fords had GM banjo-type driving axles but Ford split-type axle conversion kits were made available for all models for installation in the field if and when axle replacement or overhaul was necessary. The FWD HAR had a 10-ft Budd-type steel body and a winch between cab and body. It had a 156-in wb and 9.00–20 (2DT) tyres.

Vehicle shown (typical): Truck, 3-ton, 4×4, GS (Ford F60L)

Technical Data:
Engine: Ford V8, 8-cylinder, V-L-W-F, 239 cu in, 95 bhp @ 3600 rpm.
Transmission: 4F1R×2.
Brakes: hydraulic (hydrovac).
Tyres: 10.50–20.
Wheelbase: $158\frac{1}{4}$ in.
Overall l×w×h: $244\frac{1}{2}$×88×120(89) in.
Weight: approx. 8950 lb.
Note: No. 12 cab, early QMG 12-ft steel body (model 5C1).

Truck, 3-ton, 4×4, Light D Machinery (Ford F60L) V-8-cyl., 95 bhp, 4F1R×2, wb 158¼ in, 241×88×120 in. Also on C60L. GS body modified for workshop role. Equipped with three steel work benches, paint and welding equipment, compressor, battery charger, etc.

Truck, 3-ton, 4×4, Stores (Chevrolet C60L) 6-cyl., 85 bhp, 4F1R×2, wb 158 in, 241×88×120 in. Also on F60L. GS body modified for stores role. Superstructure covered with wire screen. Screen doors above tail gate. Steel furniture incl. desk.

Truck, 3-ton, 4×4, GS (Ford F60L) V-8-cyl., 95 bhp, 4F1R×2, wb 158¼ in, 244×90×120(75) in, 9695 lb. 1942 (No. 13) model cab shown. Most common body style on this chassis and on C60L. Many body variations existed, incl. types with folding troop seats.

Truck, 3-ton, 4×4, Dump (Chevrolet C60S) 6-cyl., 85 bhp, 4F1R×2, wb 134 in, 207×84×89 in, 10,055 lb. Also on Ford F60S. Gar Wood hoist driven from PTO, with 60° angle lift. All-steel body with hinged tail gate. Also with No. 12 cab.

243

Truck, 3-ton, 4×4, Stores (Chevrolet C60S) 6-cyl., 85 bhp, 4F1R×2, wb 134 in, 204×88×117(90) in, 9255 lb. 10-ft GS body, modified for stores role. Steel counters each side of body with storage underneath. Writing desk. Wire screening and tarpaulin. Also on F60S.

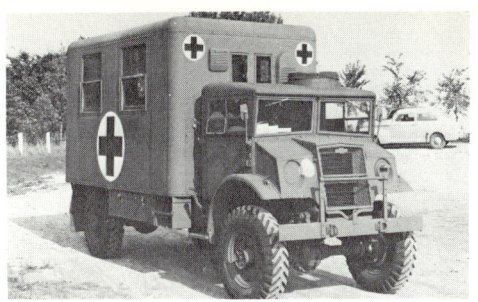

Truck, 3-ton, 4×4, Dental (Chevrolet C60L) 6-cyl., 85 bhp, 4F1R×2, wb 158 in, 238×91×125 in, 10,750 lb. Equipped to perform dental services in the field. 12-ft Lindsay house type body with rear entrance. Accommodation for two dental operators. One dental chair.

Truck, 3-ton, 4×4, Ambulance (Chevrolet C60L) 6-cyl., 85 bhp, 4F1R×2, wb 158 in, 229×90×118 in, approx. 14,000 lb. Four-stretcher field ambulance. Usually on Ford F60L chassis. Early production had 30-cwt type steering U-joints and steering, 10.50–16 tyres and no brake booster.

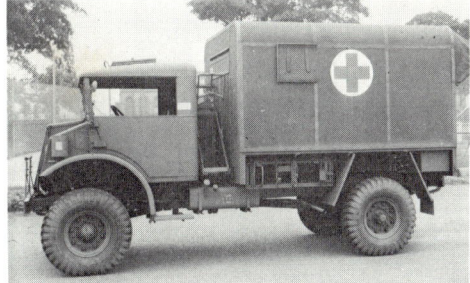

Truck, 3-ton, 4×4, Ambulance (Aust.) (Chevrolet C60S) 6-cyl., 85 bhp, 4F1R×2, wb 134 in, 206×90×108 in, 8836 lb. Australian composite body construction. Four stretchers. 10.50–16 tyres. Earlier production had canvas-top cab.

244

Truck, 3-ton, 4×4, Office (Chevrolet C60L) 6-cyl., 85 bhp, 4F1R×2, wb 158 in, 228×91×124 in, 11,920 lb. 10.50–16 tyres (10.50–20 on early production). 10-ft Lindsay body with two sliding doors. Fully equipped for general office work. Heating, lighting, 12-volt chorehorse.

Truck, 3-ton, 4×4, Breakdown (Chevrolet C60L) 6-cyl., 85 bhp, 4F1R×2, wb 158 in, 245×96×108 in, approx. 14,150 lb. Gar Wood twin swinging-boom wrecker. Also with similar equipment by Holmes. Later models had steel bodywork on short-wheelbase C60S chassis.

Truck, 3-ton, 4×4, Gasoline Tank (Chevrolet C60L) 6-cyl., 85 bhp, 4F1R×2, wb 158 in, 230×89×91 in, 9400 lb. 800-gal. (Imp.) capacity, divided into two compartments of 400 gal. Two rotary hand pumps with capacity of 12 gpm. Tarpaulin for camouflage.

Truck, 3-ton, 4×4, Signal Construction (Chevrolet C60S) 6-cyl., 85 bhp, 4F1R×2, wb 134 in, 218×95×106 in, 13,320 lb. Used by RC Sigs. to erect cable on poles after having been paid out along proposed route by line construction companies. Special body. Fitted with power winch.

245

Truck, 3-ton, 4×4, Bofors, SP (Ford F60S) V-8-cyl., 95 bhp, 4F1R×2, wb 134¼ in. Bofors AA guns were mounted on 1940 model chassis by Leyland Motors' plant at Kingston, Surrey, for Canadian Army. Leyland also devised and fitted chassis stabilisers and electrical control for laying the gun.

Truck, 3-ton, 4×4, 75-mm Gun, SP (Ford F60L) V-8-cyl., 95 bhp, 4F1R×2, wb 158¼ in. A number of Ford trucks were converted in 1942 in North Africa and fitted with the famous 'Seventy-Five' on a revolving mount. Front end was armoured and tilt was carried to make vehicle look like GS truck.

Truck, 3-ton, 4×4, Bofors, SP (Aust.) (Ford F60L) V-8-cyl., 95 bhp, 4F1R×2, wb 158¼ in. One of several types of self-propelled AA guns. This is an Australian-built unit on F60L chassis. Ford Canada produced F60B with 134-in wb and special open cab seating four (Cab No. 43S).

Truck, 3-ton, 4×4, AT Portee and Fire (6-pdr) (Chevrolet C60L) 6-cyl., 85 bhp, 4F1R×2, wb 158 in, 224×90×129(93) in, gross wt approx. 16,000 lb. Usually on Ford F60L chassis. No. 43 cab. Gun could be operated from vehicle (forward) or off vehicle. Later converted to open-cab GS.

BRITISH COMMONWEALTH

TRUCKS, 3-TON
6×4 and 6×6

Makes and Models: *Australian types:* Ford/Marmon-Herrington 01T, 11T, 296T (6×6).
Canadian Conventional types: Ford/Marmon-Herrington MM5-6 (6×6). Ford/Welles-Thornton 098T (6×4).
Canadian Military Pattern types: GM C60X (6×6) (designated Chevrolet in Australia). Ford F60H (6×4).

General Data: All the above chassis were derived from standard two-axle vehicles. The Ford/M.-H. 6×6 was basically a commercial-type Ford truck, fitted with two extra driving axles. Such M.-H. conversions (6×6 and 4×4) were popular, particularly in Australia and South Africa. The Thornton rear bogie was another commercial-type modification, consisting of a twin driving axle rear bogie with central power divider (available for several of the popular North American trucks). The Ford F60H, with early and late type cab, was a standardised modification of the F60L. The rearmost axle was non-driving (trailing axle). The General Motors C60X was basically a Chevrolet CMP C60L but fitted with a twin driving axle rear bogie, much like that of the US GMC CCKW-352/353. It also had the more powerful GMC 270 engine and to accommodate this in the CMP front end the radiator grille was brought forward approx. one inch. It usually had a GM nameplate, although in Australia a Chevrolet badge was often fitted. The F60H and C60X chassis were used mainly for house-type bodies (workshop, stores), sometimes for other types such as wrecker and bridging equipment. Pilot models for new Canadian 3-ton 6×6 chassis (150½-in wb, GS and FAT) were produced by Chrysler and General Motors towards the end of the war but no quantity production followed. The following types were used by the Canadian Army based on a special lwb Diamond T 4-ton 6×6 chassis: Crane (revolving type) Mark VI (Coles) and VII (Bay City), Folding Boat Equipment (FBE), GS, Machinery 'H', 'M' and 'RE 25-KW', and Pontoon. The wheelbase of these model 975(A) chassis was 201 (BC 52) in and at least 1118 units were supplied. The Canadians also used British Albion and Leyland 3-ton 6×4 trucks.

Vehicle shown (typical): Truck, 3-ton, 6×4, Workshop (Ford F60H)

Technical Data:
Engine: Ford 8-cylinder, V-L-W-F, 239 cu in, 95 bhp @ 3600 rpm.
Transmission: 4F1R×2.
Brakes: hydraulic (hydrovac).
Tyres: 10.50–20 (earlier models 10.50–16).
Wheelbase: 160¼ in, BC 48 in.
Overall l×w×h: 251¼×94×134 in.
Weight (loaded): 15,710 lb.
Note: rearmost axle is not driven; early-type workshop body of composite construction.

Truck, 3-ton, 6×4, Breakdown Gantry (Ford F60H) V-8-cyl., 95 bhp, 4F1R×2, wb 160¼ in, 250×90×133 in. British body on early Canadian chassis with No. 11 cab. Similar body on Austin, Crossley, Guy, Leyland, Dodge. Max. lift 2½ tons.

Truck, 3-ton, 6×4, Aircraft Refueller (Ford/Welles-Thornton 098T) V-8-cyl., 95 bhp, 4F1R, wb 153 in. Capacity 1000 gal. Swinging booms. Basically Canadian-built Ford, 1940 style, converted with Thornton rear bogie. Produced for RCAF for domestic use (lhd).

Truck, 3-ton, 6×6, Crash Tender (Ford/Marmon-Herrington MM5-6) V-8-cyl., 95 bhp, 4F1R×2, wb 134 in. Tyres 9.00–16. 1940 style. Converted to 6×6 by Marmon-Herrington for RCAF (domestic use). Similar units supplied to Dutch East Indies, using 1941 chassis.

Truck, 3-ton, 6×6, Breakdown (Aust.) No. 3A (Ford/Marmon-Herrington 296T) V-8-cyl., 95 bhp, 4F1R×2, wb 156 in, 290×98×118 in. 10.50–18 tyres. 1942 front end (No. 3: 1941). Flat-floor body with bench and lockers for tools and equipment. Also with soft-top cab. Aust.-type spring bumper.

Truck, 3-ton, 6×4, Breakdown (Ford F60H) V-8-cyl., 95 bhp, 4F1R×2, wb 160¼ in. 10.50–16 tyres. Very early Canadian model with No. 11 cab (single-piece radiator grille). Holmes twin swinging-boom power wrecker equipment. Delivered early 1941.

Truck, 3-ton, 6×4, Breakdown (Ford F60H) V-8-cyl., 95 bhp, 4F1R×2, wb 160¼ in. 10.50–20 tyres. Holmes power wrecker on 1942-type chassis. Telescoping brace legs to relieve frame when doing side pull with boom. Limited production.

Truck, 3-ton, 6×6, Breakdown (Aust.) No. 4 (GM C60X) GMC 270 6-cyl., 100 bhp, 4F1R×2, wb 160 (BC 52) in. 10.50–20 tyres. Canadian chassis, Australian bodywork. Fitted with 15,000-lb power winch, two-speed hand-operated winch and very comprehensive equipment.

Truck, 3-ton, 6×6, MGO Stores (GM C60X) GMC 270 6-cyl., 100 bhp, 4F1R×2, wb 160 (BC 52) in, 245×90×126 in, 15,960 lb. Similar bodies used for Machinery and Workshop. Drop sides could be lowered and used as tables or workbenches. Also supplied to Russia (see page 259).

Truck, 3-ton, 6×6, Aircraft Refueller (Aust.) (GM C60X) GMC 270 6-cyl., 100 bhp, 4F1R×2, wb 160 (BC 52) in. 900-gal. capacity. Used by RAAF. Cab with modified doorways, usually with easily removed canvas-on-steel frame doors. These doors were also fitted on other Australian CMP-type trucks.

Vehicle, Armoured Command (Aust.) (Ford/Marmon-Herrington 6×6) V-8-cyl., 95 bhp, 4F1R×2, wb 156 in. Limited production in 1942/3. Double doors and spare wheel at back. Camouflage was khaki and dust colour (dark sandy yellow).

Truck, 3-ton, 6×4, Folding Boat Equipment (Ford F60H) V-8-cyl., 95 bhp, 4F1R×2, wb 160¼ in. Used in FBE platoons of RCASC bridging company to carry three folding boats. Hand winches in the four corner stanchions. Usually on Diamond T 4-ton 6×6 201-in wb chassis.

Truck, 3-ton, 6×6, GS, APT, w/Winch (GM/Chrysler) Chrysler 8-cyl., 140 bhp, 4F1R×2, wb 150½ in, 257½×88×113 in, 13,290 lb. Five pilot models built, incl. FAT. GS body (by Cusson Frères) could be split for loading in Dakota C47 aircraft. 323·5 CID L-head straight-8 engine.

BRITISH COMMONWEALTH

TRACTORS and SEMI-TRACK VEHICLES

Makes and Models: *Tractors:* Chevrolet, Maple Leaf (4×2, for Aust. S-T). Chevrolet C60S, C60L (4×4, for Aust. S-T). Chevrolet CGT (4×4 FAT, Canadian and Aust. types). Dodge T110-L5 (4×2, for Aust. S-T). Dodge T130 (4×2, for Canadian S-T). Ford 01T, 11T, 118T (4×2, for Aust. S-T). Ford FC60ST (4×2, for 1500-gal. S-T). Ford F60T (4×4, for 6-ton S-T). Ford F60S, F60L (4×4 for Aust. S-T). Ford FGT (4×4 FAT, Canadian and Aust. types). Ford/Marmon-Herrington 01T, 11T (4×4 and 6×6, Aust. Artillery tractors and tractors for S-T). FWD (see below). International K (4×2 Aust. LAA). International K5, KS5, K6, KS6, K7 (4×2, for Aust. S-T). International/Farmall, various industrial types (4×2, RAAF).
Note: for LAA tractors 4×4 see Trucks, 30-cwt, 4×4 (pages 232/233).
Semi-track vehicles: Chevrolet, Ford, International (Aust. experimental types). Bombardier Snowmobile. General Motors Snowmobile.

General Data: With the exception of the Dodge T130, Ford FC60ST and F60T, all the tractors for semi-trailers were Australian types. The Chevrolet CGT and Ford FGT were CMP gun tractors and were basically 101-in wb versions of the 3-ton 4×4, fitted with winch. In 1941 the CGT chassis was also used for the 2-pdr AT Portee (later 17-pdr AT tractor) and in Australia the CGT and FGT were fitted with Australian Pattern artillery tractor bodies (Nos. 8 and 9). Tractors, Artillery (Aust.) Nos. 1, 2(A), 3(A), 4(A) used Ford/M.-H. 4×4 134-in wb chassis, Nos. 5 and 7 used Ford/M.-H. 6×6 156-in chassis, all of 1940-41 pattern. Nos. 2A, 3A and 4A had a power winch; all had open-type cab. The Ford/M.-H. 4×4 types (see page 233) were used to haul the 18-pdr, 4·5 howitzer and the 25-pdr and their respective ammunition trailers (limbers); the 6×6 types towed the 6-in howitzer and 3-in AA gun. Both were also used as tractor for semi-trailer, the 6×6 for a tank transporter. Two 4×4 tractors were built by FWD in the USA for the Canadian (and British) forces, namely the HAR 136-in wb 3½-ton for 6-ton S-T and the SU-COE 144-in wb 4-ton for 10-ton S-T (incl. low-loader).

Vehicle shown (typical): Tractor, 4×4, Field Artillery (Ford FGT)

Technical Data:
Engine: Ford V8, 8-cylinder, V-L-W-F, 239 cu in, 95 bhp @ 3600 rpm.
Transmission: 4F1R×2.
Brakes: hydraulic (hydrovac).
Tyres: 10.50–20.
Wheelbase: 101¼ in.
Overall l×w×h: 170×88×90 in.
Weight: 8876 lb.
Note: 1940 type shown; believed first CMP type to go into series production.

Tractor, 4×4, Field Artillery (Ford FGT) V-8-cyl., 95 bhp, 4F1R×2, wb 101¼ in, 170½×88×90 in, 8880 lb. Second type, featuring canvas-covered opening in roof and two small doors in rear deck. Also on Chevrolet CGT chassis. Crew six incl. driver.

Tractor, 4×4, Field Artillery (Chevrolet CGT) 6-cyl., 85 bhp, 4F1R×2, wb 101 in, 169×90×96 in, 10,800 lb. Third type, also on FGT, with open-backed std cab. Fourth type had spare wheel on rear deck; fifth was winterised for −20°F. All types had power winch with front and rear fairleads.

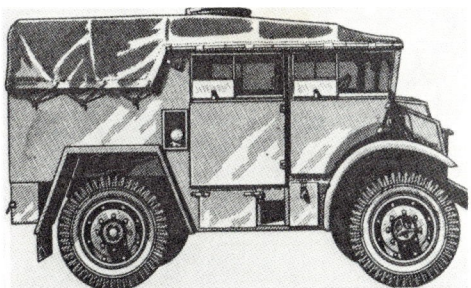

Tractor, 4×4, Field Artillery (Chevrolet CGT) 6-cyl., 85 bhp, 4F1R×2, wb 101 in, 169×83×97½ in, 10,560 lb. Sixth (latest) type, featuring open-top hull with tubular superstructure and tarpaulin (like latest MCC FAT). Two hip rings. Few produced (1945).

Tractor, 4×4, Artillery (Aust.) No. 8 (Chevrolet CGT) 6-cyl., 85 bhp, 4F1R×2, wb 101 in. Also on FGT. Superseded No. 3 and 3A. Used without limber. No. 9 basically similar but without outside lockers. No. 6 was on C60S chassis and had LAA body.

Tractor, 4×2, No. 4 (Ford 118T) with Semi-Trailer, 7-ton, GS (Aust.) No. 3 (McGrath) V-8-cyl., 95 bhp, 4F1R, wb 158 in. Some had cabs with open doorways with steel or canvas top. Used extensively on Australia's North-South road (Stuart Highway) and East-West road (Barkley Highway).

Tractor, 4×2, No. 5 (IHC K7) with Semi-Trailer, 7-ton, GS (Aust.) No. 3 (McGrath) 6-cyl., 125 bhp, 5F1R, wb 158 in. Early type S-T. One of several type cabs used on Aust. Internationals. Pictures show trucks being 'piggy-backed' as return load on North-South road, 1942.

Tractor, 4×4 (Ford F60T) with Semi-Trailer, 6-ton, GS V-8-cyl., 95 bhp, 4F1R×2, wb 115 in, 170½×84×91 in, 7168 lb. S-T: 220×84×134(88) in, 5824 lb. Built mainly for British in 1943–44 (3000 units). Australians employed F60S and L as Tractors, 4×4, No. 1.

Tractor, 4×4 (FWD HAR) with Semi-Trailer, 6-ton, GS Waukesha 6-cyl., 95 bhp, 5F1R×1, wb 136 in. 1200 produced in USA. Canadian 'fifth wheel'. Tyres 10.50–20. S-T could also be used as full-trailer (with dolly). Canadians also had FWD SU-COE tractors (144-in wb) for 10-ton S-Ts.

Truck, 3-ton, Half-Tracked, Chassis (Aust.) (Chevrolet) 6-cyl., 85 bhp, 4F1R. Experimental model utilising Universal Carrier tracks on US-pattern bogie. Truck's original rear springs were used as well as rear axle which was moved forward.

Truck, 3-ton, Half-Tracked, GS (Aust.) (Ford 218T) V-8-cyl., 95 bhp, 4F1R. Australian 1942 Ford GS truck experimentally fitted with Universal Carrier track bogies at rear. The result was very similar to the German Ford 'Maultier' (Mule) conversion.

Truck, 3-ton, Half-Tracked, Platform (Aust.) (International K6) Another experimental vehicle, utilising Universal Carrier track bogies. In this case the truck's rear axle, fitted with driving sprockets, has been moved rearward. Cab has canvas top and was one of several used by IHC in Australia.

Carriage, 40-mm Bofors Gun, Semi-Track (GM Snowmobile) Pilot model for snow-traversing vehicle produced by General Motors of Canada Ltd. Also produced with Personnel body, both in 1942–43. A load carrier was also projected but all development was discontinued in mid-1943. IWM photo MH6978.

BRITISH COMMONWEALTH

ARMOURED VEHICLES

Principal Types: *Australia:* Scout Car, Dingo. Armoured Car, Rover. Carriers, Universal, etc. Cruiser Tanks AC1 and 2 (Sentinel I, III).
Canada: Scout Cars, Lynx I and II (Ford 4×4). Light Reconnaissance Car, Otter I (GM 4×4). Armoured Car, Fox I (GM 4×4). Armoured Truck, C15TA (GM 4×4). Carriers, Universal, Windsor and other types (Ford). Cruiser Tanks, Ram I, II and derivations (Montreal Locomotive Works). Infantry Tank, Valentine (Canadian Pacific Railway Co.). SPs, Sexton I, II (Montreal Locomotive Works).
India: Armoured Carriers, Wheeled, Mks I–IV (Ford 4×4). Armoured Trucks (Ford 4×4).
South Africa: Armoured Cars, Marmon-Herrington Mk I (Ford, 4×2), Mks II–V (Ford, 4×4), Mk VI (8×8).

General Data: In addition to the above there were countless experimental and limited-production AFVs, wheeled and tracked, incl. SPs on the Canadian Lynx and Fox. The Canadian Ford 3-ton 4×4 chassis was used for several types of armoured vehicles and a special rear-engined version was designed and produced for armoured bodywork. This rear-engined chassis was made in three versions (CO11QRF, C191QRF, C291QRF) and was supplied mainly to India as a basis for the Indian Pattern wheeled carriers. A rather similar chassis but with 81-in wb was used for the Canadian Lynx scout car. Marmon-Herrington front-wheel drive conversion kits were used to convert standard Ford V8 truck chassis for armoured-car use, especially in South Africa and Australia. Universal (Bren) Carriers to British design, or modified, were produced in Australia, Canada and New Zealand. Canada turned out 1420 Valentine tanks (1940–43), 1390 of which were supplied to Russia. It was an infantry tank of British design. Also produced were almost 2000 Ram tanks and 250 Grizzly Is. These were Canadian versions of the American M3 and M4 medium tanks. Canada produced 38,032 tanks and carriers, 2577 other tracked AFVs, 10,054 wheeled AFVs. The Ram chassis was also used for several other AFVs (Sexton SP, AOP, ARV, APC, etc.). Australia produced a small number of their Sentinel Cruiser tanks, powered by triple Cadillac V8 engines.

Vehicle shown (typical): Truck, 15-cwt, 4×4, Armoured (GM C15TA)

Technical Data:
Engine: GMC 270 6-cylinder, I-I-W-F, 269·5 cu in, 106 bhp @ 3000 rpm.
Transmission: 4F1R×2.
Brakes: hydraulic (hydrovac).
Tyres: 10.50–16.
Wheelbase: 101 in.
Overall l×w×h: 187×92×91 in.
Weight: 10,030 lb.
Note: used as APC (seating eight), load carrier, and armoured ambulance; introduced in 1943 to replace US White Scout Car M3A1. Still in use in South Vietnam, 1969.

Car, Scout, Mk III*, Lynx I (Ford SC, 4×4) V-8-cyl., 95 bhp, 4F1R×1, wb 81 in, $140\frac{3}{4} \times 72\frac{1}{2} \times 65\frac{1}{4}$ in, 7830 lb. Canadian equivalent of British Daimler 'Dingo'. IHC-built hull on conventional chassis with rear engine. First 336 hulls had full-length engine cover, etc. (like earlier Mk III).

Car, Scout, Ford II, Lynx II (Ford C29SR, 4×4) Like Mk III* the Lynx II had heavier axles and springs and smaller steering ratio than Lynx I, Mk III. Sand channels (for desert use) carried at rear. Dimensions 144×73×70 in, 9370 lb. No armoured roof. Shown is last vehicle produced.

Car, Light Reconnaissance, Mk I, Otter I (GM RAC, 4×4) 6-cyl., 106 bhp, 4F1R×1, wb 81 in, 177×84×96 in, 9070 lb. Welded hull (Hamilton Bridge Co.) with open-top turret for Bren mg. Front port for Boys rifle. Crew three (driver, commander, gunner). 1761 produced.

Car, Armoured, Universal (GM, 4×4, 'Caplad') Experimental armoured vehicle produced in Canada towards the end of the war as projected universal vehicle (Car, Armoured, LAD, Ambulance, Demolition). Based on Fox armoured car. Similar 'Caplad' built in Britain, using Ford CMP running gear (q.v.).

Carrier, Armoured, Wheeled, I.P., Mk IIA (Ford C011QRF, 4×4) V-8-cyl., 95 bhp, 4F1R×2, wb 101¼ in, 186×90×78 in, 12,770 lb. Indian hull on Canadian rear-engined chassis. 10.50–16 tyres, 69-in track. Mk IIC onwards: 10.50–20 tyres and 78-in track. 4655 built (six marks). IWM.

Car, Armoured, Mk 1, Fox I (GM AC-MK1, 4×4) 6-cyl., 106 bhp, 4F1R×2, wb 101 in, 176¼×89¼×97 in, 15,083 lb. Canadian equivalent of Humber armoured car on GM rear-engined chassis. Hull and turret made by Hamilton Bridge Co. Crew four. Approx. 200 built, 1942. 10.50–20 tyres. GMC 270 engine.

Car, Scout (Aust.), Dingo (Ford/M.-H., 4×4) V-8-cyl., 85 bhp, 4F1R×2, wb 110 in, 180×82×73 in. Australian armoured body on shortened Ford truck chassis with M.-H. front-wheel drive. 10.50-18 Run Flat tyres. Crew two. Armament one Bren or Lewis mg, one TSMG.

Car, Armoured, Marmon-Herrington Mk II (Ford/M.-H., 4×4) V-8-cyl., 85 bhp, 4F1R×2, wb 134 in, wt 6 tons. Modified and fitted with captured Italian Breda 20-mm gun. Double rear door. Mk III: generally similar but 117-in wb. Mk IV, IVF: rear-engined. Mk I: Ford, 4×2. IWM Photo E2873.

Carrier, Universal, Mark I* and II* (Ford C01UC) V-8-cyl., 85 bhp, 4F1R, 148×83×63 in, 7910 lb (gross wt 9774 lb). 33,988 produced by Ford Canada to British design. Crew three. Derivations: C21UCG (tank hunter with 2-pdr Gun), C21UCM (3-in Mortar), C31UCW (Welsh Guards Stowage).

Carrier, Windsor, Mark I* (Ford C49WC) V-8-cyl., 95 bhp, 4F1R×2 (two-speed rear axle), 172¼×83×57 in, 9350 lb. Canadian design, could serve 4·2-in mortar platoon and 6-pdr AT gun crew in five different roles, incl. tractor. Crew, depending on role, two to five.

Snowmobile, Canadian, Armoured, Mk I (Farand & Delorme) Cadillac V-8-cyl., 125 bhp, 4F1R (Hydramatic), 154×101×58 in, 9400 lb. Also known as *Car, Tracked, Armoured*. Role: Light Reconnaissance car with crew of two, No. 19 W/T set, Bren and Sten gun, etc. 16 Run Flat tyres (4.50–16); 35-in tracks.

Jeep, Tracked, Canadian, Mk II (Marmon-Herrington) Willys MB 4-cyl., 54 bhp, 3F1R, 111×65×48 in. Mk I had open top hull, engine at rear. Canadian design, produced by M.-H. and Willys (TJ). Also known as '*Jeep tank*'. Six Mk IIs produced (five Mk Is). Crew two or three. Intended for airborne use.

U.S.S.R.

Until the early 1930s Russia had no automotive industry of great importance and for its civilian as well as military motor transport needs the country relied to a large extent on imports from several countries, notably Britain, Germany, Italy and the USA.

In 1924 the first Soviet truck was introduced in the form of a 1½-tonner designated the AMO-F-15. Its 35-bhp 4-cyl. engine was patterned on the contemporary Fiat 1·5-litre, hence the F-15 suffix. Some touring cars (phaetons) were also built on this same chassis.

The first Soviet car to go into series production was the NAMI-1. Introduced in 1927, it had been designed by the Soviet institute for motor vehicle and engine research in Moscow (NAMI) and was produced, until 1932, in the Spartak factory which, like the AMO factory, was located in Moscow. The NAMI-1 was a tourer with 20-bhp twin-cylinder air-cooled engine and some 300 were made.

In 1933 the name AMO was changed to ZIS, in honour of Stalin (ZIS—Zavod Imeny Stalin, factory named after Stalin) and in 1956 this was changed to ZIL, the last letter of which stands for Likhachev, a one-time director. ZIS became one of the two most important motor vehicle manufacturing plants in Russia. The other one was, and is, GAZ.

The GAZ factory was built in 1931–32 in Gorkiy with American assistance and is also known as the Molotov factory (ZIM—Zavod Imeny Molotov). The tooling was said to have been acquired from Germany when the German Ford operations were moved from Berlin to Cologne to start production of the Ford Model B. With the initial help of American Ford technicians the first GAZ-A ('Russki-Ford' Model A) and GAZ-AA (Model AA truck) came off the assembly line in 1932. The following years saw the addition of the GAZ-AAA, a six-wheeled version of the AA, and other derivatives, all of which bore a close resemblance to US Ford products. From 1936, for example, the Model A four-cylinder engine was used in what looked like a 1933–34 US Ford sedan body shell with a distinctive GAZ radiator grille (the GAZ-M-1) and from 1940 the same body shell was used with a new rounded grille but with a six-cylinder engine (GAZ-11-73). Other factories which came into being during the 1930s included KIM, JA(A), JAG, JAS, and PAZ. These letter combinations were usually abbreviations indicating the product and the location of the plant. Later 'makes' include UAZ, MAZ, JAAZ, KRAZ, etc. In most instances the first or first two letters represent the location (G—Gorkiy, M—Minsk, KR-Kremenchug, etc.), the middle letter the product (A—Automobile), whereas the letter Z stands for Zavod, translated as plant or factory (works).

During the first Five-year Plan, started in 1928, Russia produced some 50,000 trucks. During the second this number was quadrupled. The Soviet motor industry has always been characterised by its emphasis on truck production. During 1937, for example, ZIS produced 70,000 trucks, compared with 10,000 cars. Today about three-quarters of all vehicles produced are trucks.

At present, GAZ/ZIM and ZIL manufacture about two-thirds of all the wheeled vehicles produced in the USSR and, in addition, they turn out other products and supply components to other plants.

Tractor and truck production also started in the early 1930s. US-type track-laying tractors were built in Chelyabinsk, Kharkov (Komintern), Kirov and Stalingrad.

World War II started for Russia when the Germans invaded the country on 22 June, 1941. From that year until June, 1945, Russia received considerable quantities of war material under Lend-Lease from the USA, Canada and Great Britain. The USA alone, during that period, shipped to the Soviet Union 16,651,000 tons of material, valued at $9,119,204,000. This included over 400,000 trucks plus many thousands of other military vehicles such as artillery prime movers, tanks, tank transporters, scout cars, half-track vehicles, etc. About a quarter of the trucks were 2½-ton 6×4 and 6×6 Studebakers, the remainder consisting of 'Jeeps', Dodge Weapons Carriers, Chevrolet and Dodge 4×4 1½-tonners, GMCs, Macks, etc. Canada supplied Valentine tanks and GM trucks, England trucks (Albion, Austin, Bedford and Ford), motorcycles, carriers, etc. At one time more than half the Red Army's supplies on some sectors of the front were carried in Lend-Lease trucks, without which the spectacular Soviet advances of 1943–45 (following the equally spectacular retreat of 1941) would have been greatly hampered, if not held up altogether. The 1400-mile sweep of the Soviets from Stalingrad through Poland to Berlin and Torgau on the River Elbe, where on 25 April, 1945, their 58th Guards Division met with the US First Army, was made possible by what was probably the greatest array of automotive equipment ever assembled. In addition to their own and the Lend-Lease vehicles, the Soviet used large numbers of captured enemy transport.

It was not surprising that the second generation of Soviet tactical military transport vehicles should be so similar to American equipment. The GAZ-67B 'Jeep', first introduced in 1943, was patterned on the US $\frac{1}{4}$-ton 4×4, the GAZ-63 on the US 1$\frac{1}{2}$-ton 4×4, the ZIS-151 on the US 2$\frac{1}{2}$-ton 6×6, etc.

It is also interesting to note that the Soviet M-72 motorcycle with sidecar is almost identical to the German WWII BMW, and the GAZ-69A, which superseded the GAZ-67B in the mid-1950s, is similar in general layout to the *Wehrmacht*'s *'Kubel'*-type field car.

Road conditions in Russia were far from ideal

GAZ-AAA trucks lined up in Red Square

Lend-Lease Canadian and US GMC trucks

USSR

CARS and MOTORCYCLES

Makes and Models: *Cars and Field Cars:* GAZ-A, 1932–36 (Phaeton, 4×2). GAZ-M-1, 1936–41 (Sedan, 4×2). GAZ-11-40, 1940 (Phaeton, 4×2). GAZ-11-73, 1940–48 (Sedan, 4×2). GAZ-61, 1941–48 (Sedan, 4×4). GAZ-64/67/67B, 1942–53 (Field Cars, 4×4). KIM-10, 1940–41 (2-door Sedan, 4×2). ZIS-101, 1936–41 (6-seater Sedan, 4×2). ZIS-102, 1939–41 (6-seater Phaeton, 4×2).
Note: Russian spelling of the above designations: ГАЗ (GAZ), КИМ (KIM) and ЗИС (ZIS).
Motorcycles: AM600; IZ-8, -9, -12; L-8, -300, -600; M-72; etc.

General Data: Except for the GAZ-67 and 67B the above cars were basically civilian types, used by the Red Army as Command and/or Staff cars. The GAZ-61 was a four-wheel drive version of the civilian GAZ-11-73 sedan (1940–48). The GAZ-67 and 67B were patterned on the American 'Jeep', large numbers of which were supplied during the war by the USA via Lend-Lease (Bantam and Willys). From early 1942 till the end of 1943 over 20,000 US 'Jeeps' were delivered. In addition, Russia received numbers of the Ford GPA amphibious 'Jeep', which was later copied and produced as the MAV (GAZ-46). The GAZ-67 and 67B had the old Ford Model A four-cylinder side-valve engine and in Model A fashion had a dash-mounted fuel tank. Front springs were twin $\frac{1}{4}$-elliptic. Although reliable and a good off-road performer, it was a rather crude vehicle. It was in production from 1943 until 1953 and was eventually replaced by the more modern and much better looking GAZ-69A. The GAZ-67B saw extensive combat service in Korea. Like most Russian cars of the period, the ZIS-101 and 102 were very American in design and appearance. Both had a 5·75 litre straight-eight engine developing 110 bhp. The GAZ-A and -M-1 also appeared as half-track cars and some of the latter (VM Pikap) had pickup bodywork. In addition to their own motorcycles the Red Army used British solo machines, including BSA, Matchless, Velocette, etc. Captured German machines were also used and the BMW R-71 sidecar combination was copied in 1942 and is still in use, designated M-72. The M-72 was also used as a solo machine.

Vehicle shown (typical): Field Car, 4-seater, 4×4 (GAZ-67B)

Technical Data:
Engine: GAZ-M1, 4-cylinder, I-L-W-F, 3280 cc (98·43 × 107·95 mm), 54 bhp @ 2800 rpm. CR 4·6: 1.
Transmission: 4F1R × 1.
Brakes: hydraulic.
Tyres: 6.50–16 or 7.00–16.
Wheelbase: 2100 mm.
Overall l × w × h: 3350 × 1690 × 1700 mm.
Weight: 1320 kg. *Payload:* 400 kg.
Note: Main difference between 67B and earlier 67 was in wider wheel track, viz. 1445 and 1250 mm resp. Prototype (1942) was designated GAZ-64.

Car, 5-seater, 4×2, Phaeton (GAZ-11-40) 6-cyl., 76 bhp, 3F1R, wb 2845 mm, 4655×1770×1800 mm, 1400 kg. Tyres 7.00–16. 3·48-litre (82×110 mm) SV engine. CR 5·6:1. Track, front and rear 1440 mm. Used in Soviet Army as command car. Radius of action *c.* 400 km. Speed 120 km/h.

Car, 5-seater, 4×2, Sedan (GAZ-M-1) 4-cyl., 50 bhp, 3F1R, wb 2845 mm, 4625×1770×1775 mm, 1370 kg. Max. speed 100 km/h. Tyres 7.00–16. 3·28-litre GAZ M-1 (Model A) engine with 4·6:1 CR. Produced 1936–41. Body similar to American Ford 1933/34. Also produced as half-track (VM 'Pikap').

Car, 5-seater, 4×4, Sedan (GAZ-61) 6-cyl., 85 bhp, 4F1R×1, wb 2855 mm, 4670×1770×1905 mm, 1650 kg. Tyres 7.00–16. Cubic capacity 3·48 litres (82×110 mm). CR 6·5:1. Max. speed 100 km/h. Produced during 1941–48 as 4×4 version of GAZ-11-73 sedan. Superseded by GAZ M-72, 1955.

Motorcycles, Solo Various types were used. *Shown on left:* wheeled cavalry, motorcycles equipped with machine guns; *right:* Red Army despatch rider on conventional type, fitted with pillion seat. Soviet motorcycle types varied from 124-cc two-strokes to 746-cc four-stroke twins (M-72).

USSR

TRUCKS, ½- to 5-TON
4×2 and 4×4

Makes and Models: GAZ-AA, 1932–38 (1½-ton, 4×2). GAZ-MM,* 1938–48 (1½-ton, 4×2). GAZ-42, 1939–42 (1¼-ton, 4×2, w/gas producer). GAZ-44,* 1939–42 (1¼-ton, 4×2). GAZ-55,* 1938–45 (Ambulance, 4×2). GAZ-410,* 1936–46 (Dump, 1¼-ton, 4×2). GAZ-M-415,* 1939–42 (Pickup, 0·4-ton, 4×2). JAG-6, 1936–42 (5-ton, 4×2). JAS-3, 1936–41 (Dump, 4-ton, 4×2). ZIS-5, 1933–44 (3-ton, 4×2). ZIS-11, 1935–41 (Fire Fighter, 4×2). ZIS-16C, 1939–45 (Ambulance, 4×2). ZIS-21, 1939–44 (2½-ton, 4×2, w/gas producer). ZIS-30, 1941 (2½-ton, 4×2, w/gas cylinders). ZIS-32, 1941 (2½-ton, 4×4). Ural-ZIS-5, 1944–55 (3-ton, 4×2).

Note: Russian spelling of the above designations: ГАЗ (GAZ), ЯГ (JAG), ЯС (JAS), ЗИС (ZIS) and Урал-ЗИС (Ural-ZIS).

Bodywork: cargo, unless stated otherwise. Some chassis, notably the ZIS-5, were also used for special bodywork, e.g. bridging equipment (Pontoon), air compressor (KS-200), snow-plough (DAK-5, D-151), tanker, searchlight, etc.

General Data: The first Russian trucks were the AMO-F-15 1½-tonner of 1924–31 and the 3-ton JA-3 of 1925–28. Large-scale production, however, started in 1932 with the GAZ-AA, which was the Russian-built American Ford Model AA truck of 1930. It was in production, in modified form (GAZ-MM) and in many variations, until after the war. The other mass-produced Russian truck of that period was the ZIS. The basic model was the ZIS-5 but many variations existed. Of the basic products, GAZ and ZIS, there were versions with gas producer, compressed gas, dump body, etc., as well as six-wheelers and half-tracks. Following the Russian retreat in 1941 a new ZIS plant was built in the Urals. Production started there in 1943/44 and the vehicles were known as Ural or Ural-ZIS. JAG and JAS trucks were generally of heavier construction than the GAZ and ZIS products but much less numerous. Many American, British and Canadian 4×2 and 4×4 trucks (Lend-Lease) were also used, as well as captured German and Finnish vehicles.

* Based on GAZ-AA but more powerful M-1 engine (50 bhp @ 2800 rpm vs 40 @ 2200).

Vehicle shown (typical): Truck, 3-ton, 4×2, Searchlight (ZIS-5).

Technical Data:
Engine: ZIS-5 6-cylinder, I-L-W-F, 5550 cc (101·6×114·3 mm), 73 bhp @ 2300 rpm.
Transmission: 4F1R.
Brakes: mechanical.
Tyres: 34×7 (2DT).
Wheelbase: 3810 mm.
Overall l×w×h: 6060×2235×2160 mm.
Weight: 3100 kg.
Note: Also known as 'Stalin' truck. For cargo truck version see illustration on page 259. Other versions: ZIS-21 (with gas producer), ZIS-30 (with gas cylinders).

Truck, 3-ton, 4×2, Cargo (Ural-ZIS-5) 6-cyl., 76 bhp, 4F1R, wb 3810 mm, 6060×2235×2160 mm, 3100 kg. Similar to ZIS-5 but engine had higher CR (5·3 vs 4·6:1). Produced in Ural factory of ZIS (see General Data). Flat-top wings were also fitted on certain ZIS-5 trucks.

Truck, 1½-ton, 4×2, Cargo (GAZ-MM) 4-cyl., 50 bhp, 4F1R, wb 3340 mm, 5335×2040×1970 mm, 1800 kg. Tyres 6.50-20 (2DT). Known as 'Russki-Ford'. Basically as GAZ-AA but 50- vs 40-bhp engine (4·6 vs 4·22:1 CR). Also with conventional (rounded) wings.

Truck, 5-ton, 4×2, Cargo (JAG-6 or YaG-6) ZIS 6-cyl., 73 bhp, 4F1R, wb 4200 mm, 6500×2500×2550 mm, 4930 kg. Tyres 40×8. Superseded JAG-4 (1934–36) which had similar appearance, as did JAS-3 (dump truck). Max. speed 40 km/h. Produced at Yaroslavl.

Truck, 2½-ton, 4×4, Cargo (ZIS-32) 6-cyl., 73 (later 88) bhp, 4F1R×2, wb 3810 mm, 6060×2215×2330 mm, 3680 kg. Tyres 36×8 (2DT). Payload on made roads 3 tons. Developed from ZIS-5. Max. speed 60–65 km/h. Superseded by GAZ-63 series (1946).

USSR

TRUCKS, 1½- to 8-TON
6×4

Makes and Models: GAZ-AAA, 1933–45 (1½-ton, 6×4) (ГАЗ-AAA). GAZ-05-193, 1936–45 (Ambulance, 6×4) (ГАЗ-05-193). JAG-10, 1932–41 (8-ton, 6×4) (ЯГ-10). ZIS-6, 1934–45 (2½-ton, 6×4) (3ИC-6).
Bodywork: cargo or special equipment.

General Data: Most common of the above were the GAZ-AAA and the ZIS-6, both modifications of standard 4×2 trucks, namely the GAZ-AA and the ZIS-5 respectively. Both were used not only with standard cargo bodies but also with special equipment. The GAZ-AAA, for example, was widely used as a mobile mount for an anti-aircraft weapon (Quadruple 7·62-mm Maxim water-cooled mg) and the ZIS-6 as a mobile rocket launcher (BM-13 'Katuscha'). The latter was popularly known as 'Stalin Organ' and was equipped to launch 16 rockets. The firing salvo for all 16 projectiles was 8 to 10 seconds. The GAZ-AAA chassis was also used as the basis for the BA-10 armoured car. A large proportion of the vehicles in this class, however, consisted of Lend-Lease trucks, notably the US Studebaker 2½-ton 6×6 and 6×4. These, too, were used with various types of bodywork. Other Lend-Lease vehicles were the US Studebaker 5-ton 6×4 (as truck-tractor with semi-trailer), the US GMC 6×6 and the Canadian GM (C60X) 6×6. Studebaker alone supplied over 100,000 6×6 and 6×4 trucks for the Soviet Army. The post-war ZIS/ZIL 6×6 chassis was clearly patterned on the American type 2½-ton 6×6, its front end and cab resembling the International Harvester K series. American Marmon-Herrington 6×6 trucks (dump trucks and wreckers) were also used by the Soviet Union. Most of the American Lend-Lease trucks reached the Red Army via the Persian Gulf. They arrived in CKD form and were assembled at the port of entry, mainly at Khorramshahr. They were then driven northward across Iran (Persia) to the USSR. In October, 1942, the US Government bought the Ford Motor Co's tyre plant and shipped all the equipment to Russia where it was set up for an annual output of at least a million military truck tyres.

Vehicle shown (typical): Truck, 1½-ton, 6×4, Cargo (GAZ-AAA).

Technical Data:
Engine: GAZ-M1 4-cylinder, I-L-W-F, 3280 cc (98·43×107·95 mm), 50 bhp @ 2800 rpm. CR 4·6:1.
Transmission: 4F1R×2.
Brakes: mechanical.
Tyres: 6.50–20 (4DT).
Wheelbase: 3200 (BC 940) mm.
Overall l×w×h: 5335×2040×1970 mm.
Weight: 2475 kg.
Note: payload on made roads 2½ tons. Max. speed 65 km/h. Also with wrecker (breakdown) equipment. Early production had GAZ-A engine with 4·22 CR, developing 40 bhp @ 2200 rpm.

Truck, 1½-ton, 6×4, Ambulance (GAZ-05-193) 4-cyl., 50 bhp, 4F1R×2, wb 3200 (BC 940) mm. Weight 3140 kg. Based on the GAZ-AAA chassis, produced from 1936 to 1945. Accommodation for nine. Maximum speed 65 km/h. Fuel consumption 27 l/100 km.

Truck, 2½-ton, 6×4, Cargo (ZIS-6) 6-cyl., 73 bhp, 4F1R×2, wb 3900 (BC 1080) mm, 6060×2235×2160 mm, 4230 kg. Tyres 34×7 (4DT). Payload on made roads 4 tons. Developed from ZIS-5. Max. speed 55 km/h. Also used as searchlight carrier, rocket launcher, etc.

Truck, 8-ton, 6×4, AA Gun (JAG-10 or YaG-10) Hercules YXC 6-cyl., 93·5 bhp, 4F1R×2. American 7020-cc (111×120·6 mm) L-head engine with 4·6:1 CR. Tyres 40×8. Track, front/rear 1448/1829 mm. Max. speed 40 km/h. Also cargo truck version. Produced at Yaroslavl and known as 'Yagio'.

Truck, 2½-ton, 6×4, Rocket Launcher (Studebaker US6×4) 6-cyl., 87 bhp, 5F1R×2, wb 4115 (BC 1177) mm. Typical Soviet application of Lend-Lease Studebaker truck. Similar launcher on other chassis (ZIS-6, British Ford WOT8, etc.). Popularly known as 'Stalin Organ' or 'Stalin Organ Pipes'.

USSR

TRUCKS
HALF-TRACK and 8×8

Makes and Models: *GAZ-based half-track:* GAZ/V and VM 'Vezdekhods' ($\frac{1}{2}$-ton, 1936/37). GAZ-AA/Kégresse (1$\frac{1}{2}$-ton). GAZ-AA/Nati-3 (1$\frac{1}{2}$-ton). GAZ-60 (1$\frac{1}{4}$-ton, 1938–42). *ZIS-based half-track:* ZIS-22 (2$\frac{1}{4}$-ton, 1938–42). ZIS-33 (2-ton). ZIS-42 (2$\frac{1}{4}$-ton, 1942–44). *8×8:* JAG-12 (12-ton, 1932–41).

General Data: Already during the First World War vehicles with tracks instead of rear wheels made their appearance in the Russian Army. This was largely the work of the Frenchman Kégresse, then manager of the Imperial garages, who, probably inspired by the American Lombard half-track tractor which was used by the Russian forces, developed a successful half-track conversion for the Czar's cars. Some British Austin armoured cars were subsequently converted and during the 1930s several half-tracked trucks appeared. During the war Russia received over 1200 armoured half-tracks from the USA under Lend-Lease, as well as an unarmoured truck version, produced by Autocar. Soviet-built half-track trucks were basically modifications of conventional 4×2 types, consequently they were underpowered and lacked front-wheel drive which made steering difficult, especially in mud and snow conditions. In 1942 the Germans carried out an investigation into the variety of semi-track vehicles used by the Soviet Army. Comprehensive details were subsequently published about six models based on GAZ and two based on ZIS chassis, ranging from passenger cars to 2$\frac{1}{4}$-ton load carriers/prime movers. Although rather crude by Western standards, this type of vehicle seems to have been used in relatively large numbers. The half-track principle was not further developed after 1945 when similar (or better) cross-country performance could be obtained from multi-wheeled all-wheel drive vehicles. A good example of this is the BTR-152 armoured personnel carrier which is in many respects similar to the WW II US half-track APC but on a 6×6 truck chassis.

Vehicle shown (typical): Truck, 2$\frac{1}{4}$-ton, Half-Track, Cargo (ZIS-42)

Technical Data:
Engine: ZIS-5, 6-cylinder, I-L-W-F, 5550 cc (101·6×114·3 mm), 73 bhp @ 2300 rpm. CR 4·6:1.
Transmission: 4F1R×2.
Brakes: mechanical.
Tyres: 34×7 or 36×8 (front).
Wheelbase: 3810 mm.
Overall l×w×h: 6095×2360×2950 mm.
Weight: 5250 kg.
Note: Max. speed 35 km/h. Later production had ZIS-16 engine (88 bhp, CR 5·7:1, max. speed 40 km/h). Somua-type tracks.

Truck, ½-ton, Half-Track, Cargo (GAZ/VM 'Pikap') GAZ-M1 4-cyl., 50 bhp, 3F1R×2, wb 2845 mm, 5020×2254×1825 mm, 2460 kg. Nati-type bogie on modified GAZ-M-1 car (q.v.). Payload 680 kg or 500 kg plus two men. Also as 5-pass. sedan. Known as 'Vezdekhods' (go-anywhere).

Truck, 2½-ton, Half-Track, Cargo (ZIS-33) 6-cyl., 73 bhp, 4F1R. 34×7 tyres. Modification of standard ZIS-5 truck, employing half-track conversion but retaining original rubber-tyred twin rear wheels. Weight 4585 kg. Track ground pressure 0·46 kg/cm². For US Autocar half-track truck see p. 119.

Truck, 1¼-ton, Half-Track, Cargo (GAZ-60) 4-cyl., 50 bhp, 4F1R×2, wb 3340 mm (to BC), 5335×2360×2100 mm, 3520 kg. Max. speed 35 km/h. Fuel consumption 57–96 l/100 km. Developed from GAZ-AA. Ground pressure of the tracks 0·21 kg/cm².

Truck, 12-ton, 8×8, Cargo (JAG-12 or YaG-12) Continental 6-cyl., 93 bhp. CR 4·6:1. 7·02-litre (111×120·6 mm) engine, developing 38 mkg of torque. Heaviest and most powerful Russian truck of the 1930s. Believed limited production. Speed 40 km/h. Fuel consumption 52 l/100 km.

USSR

TRACTORS
FULL-TRACK

Makes and Models: *Conventional type (driver at rear):* ATZ, ChTZ-60, DT-54, GB-58, KD-35, S-60, S-65 (SG-65), S-80, STZ-Nati, etc.
Truck type (normal-control type cab): Stalin (various), Ya-12 (diesel), Ya-13 (petrol), etc.
COE truck type (forward-control cab): KT-12, STZ-5-2TB, etc.

General Data: During the 1930s the Soviet Army and motor industry developed a wide range of full-track light, medium and heavy tractors for agricultural and military use. The military versions were used in large numbers for gun hauling. In fact, practically all heavy artillery hauling was performed by full-track prime movers. The obvious advantages were that this type of tractor was best suited for local Russian operating conditions (mud, snow) plus the fact that many, if not most, Russian soldiers were used to or familiar with operating this type of machinery. Apart from many which were very similar to agricultural types, several models had a truck-like appearance, featuring a closed cab and cargo-type body. Some were fitted with a gas producer, using solid fuel. By the end of the war the Ya-12 and Ya-13(F) had been developed and these remained in use until long after the war. They were light tractors with conventionally mounted closed cab and personnel/cargo bodies. These models were, in turn, superseded by the post-war M2 and AT-L. The heavy types were superseded in 1950 by the AT-T, the medium in about 1954 by the AT-S. In addition to towing field pieces, AA guns, heavy mortars, etc., the conventional types were also used in a number of other roles, such as bulldozers, cranes, etc. Numbers of tracked artillery tractors, such as the International Harvester TD-14, were also supplied to Russia by the USA via Lend-Lease. Tractors without a cab were sometimes fitted with cabs taken off disabled captured enemy trucks (Opel Blitz and others) as field modifications. Light artillery pieces were usually towed by wheeled trucks, such as the ZIS-5 and US Lend-Lease types, and half-tracks.

Vehicle shown (typical): Tractor, Artillery, Full-track (Stalinets-65, SG-65)

Technical Data:
Engine: 6-cylinder diesel, I-I-W-F.
Transmission: 5F1R.
Steering: controlled differential.
Brakes: mechanical.
Tracks: steel link.
Overall l×w×h: 4086×2416×2803 mm.
Weight: 11,200 kg.
Note: Widely used military version of agricultural-type tractor. Wording on radiator surround, top and sides: СТАЛИНЕЦ (STALINETS), bottom: ДИЗЕЛЬ (DIESEL).

Tractor, Artillery, Full-Track (Stalin) Large prime mover-cum-load carrier, used to tow heavy artillery and carry crew, ammunition and equipment. Note large number of track rollers and 'scissors'-type coil-spring suspension. German designation: *Stalin 607(r)*.

Tractor, Artillery, Full-Track (Komintern) Rather similar in overall dimensions to Stalin shown on left, but as cab is mounted further back, rear body is consequently shorter. Suspension units are similar to those used on the T24 tank of 1929. Tows 152-mm gun-howitzer M1937.

Tractor, Artillery, Full-Track (STZ-5-2TB) Cab-over-engine prime mover/load carrier. Oscillating bogies with horizontally mounted coil springs (as on STZ-Nati and some other agricultural-type crawler tractors). Germans used captured vehicles as *Artillerie Schlepper CT3-601(r)*.

Tractor, Artillery, Full-Track (STZ Komsomolets) Used as tractor, personnel and ammunition carrier, etc. $3450 \times 1859 \times 1400$ (min.) mm, 4200 kg. Six folding seats, back to back, over engine. Armour cab with ball-mounted mg. German designation: *l.gp. Art.S.630(r)*. Shown minus tilt.

USSR

COMBAT VEHICLES ARMOURED
WHEELED and TRACKED

Principal Models: *Armoured cars:* GAZ BA-10 and -32 (6×4, from 1932); BA-20 (4×2, 1937); BA-64 (4×4, from 1943).
Light tanks: T-40 (1940); T-50 (1940); T-60 (1941); T-70 (1942); T-80 (1942).
Medium tanks: T-34/76 A, B, C (1939–43); T-34/85 (from 1943).
Heavy tanks: KVI, IA, B, C (1939–43); JSI, II, III (1943–45).
SPs: SU-37 (1943); SU-76 (1942); SU-85 (1943); SU-100 (1944); SU-122 (1941); JSU-122 (1945); JSU-152 (1943).

General Data: The principal Soviet armoured cars of the 1930s were the four-wheeled Bronieford, BA-20(M), and the six-wheeled BA-10 and -32, all built on Russian Ford (GAZ) chassis. Of the six-wheelers, an amphibious version (BAZ) also existed. In 1943 a new light armoured car was put into production based on the newly developed GAZ light 4×4 chassis. This model, the BA-64, remained in use until long after the war. Like so many Russian vehicles, the design of these armoured cars was rather unsophisticated but suitable for mass-production by unskilled labour. This also applied to the famous and legendary T-34 medium tank, which surprised the Germans. Designed in 1937 the 32-ton T-34 was in production in 1939, in service in 1940, and in action against the Germans when Hitler invaded Russia in June, 1941. Originally it had a long 76-mm gun, but from 1943 a revised turret with 85-mm gun was fitted. This T-34/85, as it was designated, first went into action in Rumania against Manteuffel's *Gross Deutschland Panzer Division* in May, 1944. The Russians' heavy tank was the Joseph Stalin (JS or IS), a redesigned version of the 48-ton KV. The KV was produced in the Kirov plant in Leningrad from 1939, the JS I and II in the Urals in 1943–44. The revised JS III first went into action in Poland in 1945. The Soviets also employed large numbers of mobile armoured assault and support weapons. These were known as SUs (*Samochodnaja Ustanovka*, Self-propelled gun) and were built on contemporary tank chassis.

Vehicle shown (typical): Car, Armoured, Light, 4×4, BA-64 (GAZ)

Technical Data:
Engine: GAZ-M1 4-cylinder, I-L-W-F, 3280 cc, 50 bhp @ 2800 rpm.
Transmission: 4F1R×1.
Brakes: hydraulic.
Tyres: 6.50 or 7.00–16.
Wheelbase: 2130 mm.
Overall l×w×h: 3660×1530×1900 mm.
Combat weight: 2400 kg.
Note: Developed in 1942/43, based on GAZ-67B chassis. Main armament: one 76·2-mm mg or 14·5-mm AT rifle. Armour 6–10 mm. Crew two. Later also referred to as 'Bobby'.

Car, Armoured, 6×4, BA-32 (GAZ) GAZ-M1 4-cyl., 50 bhp, 4F1R×2, wb 3400 mm, 4650×2100×2430 mm, approx. 6 tons. Based on GAZ-AAA chassis. This variant was armed with one 45-mm gun and two 7·62-mm mgs. Crew four. Similar to BA-10. Both were widely used in WW II.

Tank, Medium, T-34 V-12-cyl. diesel, 500 bhp, 5920×3000×2400 mm, 26–28 tons. One of the most successful tanks of World War II. Christie-type coil-spring suspension. 76·2-mm gun (later 85-mm). Note sloping armour and low silhouette. Some early models had all-metal bogie wheels.

Tank, Heavy, JS (Joseph Stalin) V-12-cyl. diesel, 600 bhp, 6770×3070×2730 mm, 44–46 tons. Heavy cruiser tank with 122-mm gun as main armament. Torsion-bar suspension. JS I and II (shown) were much alike; JS III had redesigned turret and glacis plate. SPs on this chassis were known as JSUs.

Howitzer, 122-mm, Self-propelled, SU-122 V-12-cyl. diesel, 500 bhp, 6950×3000×2150 mm, 31 tons. Introduced in 1941/42, based on T-34 tank chassis. Maximum speed 55 km/h. On similar chassis: SU-85 (85-mm gun, 1943) and SU-100 (100-mm gun, 1944).

GERMANY

Unlike the sections which deal with the Allied forces, the German section of this book contains many pre-war types, simply because a very large proportion of the *Wehrmacht*'s vehicles were of pre-war manufacture or design. One aspect in which the German section differs from all the others is that it includes a number of 'foreign' vehicles which were captured or confiscated or produced in occupied territories, notably in Austria, Czechoslovakia and France, as well as some which were taken over from or used by their Axis allies (Hungary, Italy, Rumania).

Another aspect that needs some attention is the fact that although motorised *'Blitzkrieg'* and *'Deutsche Wehrmacht'* were, at least in the eyes of the public at large, almost synonymous, this does not imply that the Nazis were equipped with an abundance of tactical-type motor vehicles when they started World War II. In fact, a high proportion of their transport of troops and supplies was carried out with confiscated/impressed civilian vehicles. Horses were also used in enormous quantities right throughout the war, mainly behind the front lines and in the occupied territories.

When the writer's home town in the north of the Netherlands was occupied early in May, 1940, the majority of invading German troops were on horseback, and from then until the liberation in April, 1945, the civilian-type vehicles used by the Nazis outnumbered the real military types easily by something like a hundred to one. The former category included not only commandeered vehicles of Dutch civilian origin (anything from bicycles to buses) but also many of Austrian, British, Czech, French, Italian and US origin. Admittedly this was not a typical front line sector but we lived in a garrison town and did see our share of tactical types, even including columns of the Skoda *Ostradschlepper* and the like.

The vehicles with which the *Wehrmacht*'s 'elite divisions' were equipped were invariably of a high standard and were built specifically for military tactical use. However, production of these types, many of which were of elaborate and sophisticated design, was sufficient to equip only a comparatively small portion of Hitler's war machine. The remainder were of a wide variety and became even more varied as the war drew on.

The purely military vehicles were, as mentioned before, of quite elaborate design and this reflected the national thinking that was behind the German rearmament programme which started in 1932. It must have been a paradise for vehicle designers. What other army would produce beautifully designed tourer-type cars with independent suspension and other refinements just to transport four soldiers comfortably, with their kit and equipment? What army would spend a small fortune on every one of thousands of highly sophisticated semi-track artillery tractors providing comfortable theatre-type seating for every member of the crew of a dozen or so? This was preparation for war and somehow was interconnected with Goebbels' propaganda machine. The photographs, many in full colour, of this equipment in such magazines as *Signal*, *Wehrmacht* and *Motor-Schau* looked superb and must have impressed millions.

It was not very long before a German general decided that these fantastic and costly developments had to come to a halt. This officer, General von Schell, devised a programme, known as the '*Schell Programm*' which, for a start, reduced the total number of German civilian motor vehicles of all types then in production. He reduced the number of truck models (all classes) from 113 to 30, cars from 52 to 19, and motorcycles from 150 to 30. Several so-called '*Einheits*' types, standardised military cars and trucks, were axed. The development of some special-purpose vehicles (outside the AFV sphere) was abandoned or curtailed, and the *Wehrmacht* was persuaded to be content with a minimum number of reliable basic types which would meet most, if not all, transport requirements. The following new load classes were subsequently standardised to form a universal truck fleet: $1\frac{1}{2}$-, 3-, $4\frac{1}{2}$- and $6\frac{1}{2}$-ton, each with a basic 4×2 (*S-Typ*) version and a 4×4 (*A-Typ*) modification, the latter being chiefly for military use. 'S', incidentally, meant standard/conventional, 'A' indicated '*Allradantrieb*' or all-wheel drive. The various types were produced by a number of different firms in Germany and Austria, and in some cases one make of vehicle which had proved to be of good quality was produced by more than one firm (example: the successful Opel Blitz 3-ton *S-Typ* was produced by Opel and Daimler-Benz). Within this framework there was plenty of room to manœuvre. For example the 1½-ton *A-Typ* was used for bodywork in at least three different categories, namely passenger-carrying, light truck, and ambulance; the 3-ton *S-Typ* was also used for '*Maultier*' semi-track trucks, etc.

The mass-produced Volkswagen '*Kübelwagen*' was another example of a universal, simple and reliable vehicle which superseded much more expensive machinery.

This system worked reasonably well, although they still managed to devise more than a hundred different versions of one basic truck with a universal house-type body (with such typical and cumbersome designations as *Kfz.305/89 Röntgenschirmbilddunkelkammerkraftwagen* and *Kfz.305/98 Sauerstoffgerätinstandsetzungskraftwagen*), and it never satisfied the numerical need for vehicles.

The many ever-changing front lines, the vast occupied territories, and the immensely long supply routes into Russia called for large numbers of motor vehicles of various sorts. Moreover, not every type of vehicle was suitable for every type of combat area. The problems of making the same basic vehicles work satisfactorily under winter conditions on the Eastern front and almost the exact opposite in North Africa were enormous.

Once the British and Canadian war industries had got into top gear in 1941, and much more so when by 1942 the gigantic US automotive industry began to go all-out on producing war materials and the Soviets began receiving huge quantities of additional equipment from their Allies, under Lend-Lease, there was no hope left for the *Third Reich*. Their industry was totally incapable of supplying enough material to enable them to retain the territorial gains they had made during their swift and successful *Blitz* campaigns, especially in Russia and Africa.

Clearly, but inexplicably, the Nazi leaders had expected the war to be of short duration and their victory to be certain. When the turning-point came they changed from the offensive to the defensive with little or no choice but to utilise whatever automotive equipment they could lay their hands on, whether suitable or makeshift. The general results were chaotic.

Added problems, of course, were those of material shortages and interruptions in production as a result of sabotage, bombing, etc. In connection with this and with production capacity proper, it was decided during the early part of the war to engage as many 'foreign' plants as possible, notably in Austria, Czechoslovakia and France, not only for the production of certain vehicles for which these factories were set up at the time of occupation but also for additional output of '*Schell Programm*' and other types of vehicles.

One of the results of material shortage was the substitution of steel truck cabs by a standardised universal *Ersatz* type. This cab, called '*Wehrmacht-Einheitsfahrerhaus*', was made of pressed cardboard panels on timber framing and was used on the majority of new vehicles produced in 1944–45 (they even remained in production after the war, until 1947).

Special mention should be made here of the unique family of German standardised semi-track vehicles. Well over 25,000 were produced. Their origins can be traced back to the late 1920s, although World War I had already seen some interesting German vehicles of this type. By 1936 the various models had begun to take the general shape they would retain until their production tailed off as war progressed. Their size varied from the small NSU-built '*Kettenkrad*' motorcycle tractor to the huge Famo 18-ton '*Zugkraftwagen*' prime mover. Between these two extremes, which had lengths of 9 ft and 27 ft respectively, and turned the scales at 1235 and 15,470 kg (empty), there was a whole range of models in many different versions, including armoured. Except the little '*Kettenkrad*', which had an Opel car engine, all standardised models had a Maybach power plant, 6-cyl. or V-12, varying in output from 90 to 230 bhp. The V-12 engines were equipped with an inertia-type starter. Features in common to all models included: needle roller-bearing tracks with detachable rubber track pads, torsion bar suspension on all but the front wheels (except early 8-ton), large double and overlapping rubber-tyred bogie wheels which carried the top run of the track, rubber-padded sprockets driving through rollers instead of the more usual teeth, and '*Cletrac*'-type geared differential steering combined with normal steering of the front wheels for road work. Many similar things can be said about German armoured fighting vehicles but this is prohibitive for space reasons. In this directory only the better-known types of armoured vehicles are listed.

One characteristic item common to all motor vehicles officially taken into German government service was the Notek *Tarnscheinwerfer* or black-out lamp. Its shape was not unlike a flattened *Wehrmacht* helmet and it was fitted on the left front wing or in an equivalent position. It was part of the so-called *Kfz.-Nachtmarschgerät*, developed by the army in co-operation with Nova-Technik GmbH. The complete equipment consisted of the (detachable) black-out lamp, a distance/stop/tail light unit and a special switch. The rear light unit had two pairs of two square openings in a row (which could be blanked off by a hinged lid), by which means the distance from the vehicle could be determined during convoy driving at night. If the four squares appeared to the following driver as one light area, he was too far behind (300–35 metres); if they looked like two lights the distance was correct (35–25 metres); if all four squares could be identified he was too close (25–0 metres). This equipment is still in use in some countries, notably Czechoslovakia and East Germany, where it was continued after the war.

Towards the end of the war many German Army vehicles had a metal seat mounted on the right-hand front wing for an aircraft observer and main roads were lined with drive-in dug-outs at regular intervals into which the whole vehicle could be driven in case of air attacks.

As it was, German Army motorisation started off very professionally with high hopes and ideals during the 1930s, but it ended in a complete shambles. Of course, Hitler's War Machine should never have got on the move, but unfortunately it did, and there is now no denying the fact that from the military automotive history point of view 'Germany 1932–45' is an extremely interesting period, which on purely technical and historical grounds ought to be given more attention. Unfortunately, only relatively few specimens of genuine *Wehrmacht* vehicles remain, usually in the hands of firms who specialise in letting them out on hire to the film industry. Probably the largest collections are owned by Mr Oliver of Egham in England and the firm of Bapty & Co. in London. Of all types that have survived, the Volkswagen *'Kubelwagen'* is the most numerous, and many of these are still in daily use in Eastern Europe.

Vehicle Types

At the outbreak of war these were the following main classifications for *Wehrmacht* motor vehicles:

Motorcycles (*Krad*): Light (up to 350 cc), medium (350–500 cc), heavy (over 500 cc, 2×1, 3×1, 3×2).

Cars (*Pkw.(o)*): light (up to 1500 cc), medium (1500–3000 cc), heavy (over 3000 cc).

Field Cars (*gl.Pkw*): as Cars, but military-type bodywork, open (4×2, 4×4).

Std. Cars (*E.Pkw*): light, medium, heavy (4×4).

Ambulances (*Krkw.*): one basic type, on chassis of heavy car or light truck (4×2, 4×4, 6×4).

Buses (*Kom.*): light (up to 15 seats), medium (up to 30 seats), heavy (over 30 seats) (4×2).

Trucks (*Lkw.(o)*): light (up to 2 tons), medium ($2\frac{1}{2}$ to 4 tons), heavy (over 4 tons) (4×2).

Cross-country Trucks (*gl.Lkw.*): as Trucks, but 6×4, 4×4, 6×6.

Std. Trucks (*E.Lkw.*): light (6×6).

Tractors, Wheeled (*Rd.Schlp.(o)*): light (up to 25 bhp), medium (up to 40 bhp).

Tractors, Tracked (*Kett.Schlp.(o)*): light (up to 25 bhp), medium (up to 40 bhp), heavy (over 40 bhp).

Prime Movers, Semi-track (*Zgkw*): light (up to 3-ton towed load), medium (5 to 8 tons), heavy (12 to 18 tons).

Armoured Cars (*Pz.Spähw.*): light (4×4, heavy (6×4, 8×8).

APCs (*Schtz.Pz.Wg.*): light (*Sd.Kfz. 250* series), medium (*Sd.Kfz.251* series).

AFVs (*Pz.Kpfw.*, *Pz.Jäg.*, etc.): tanks, SP artillery, etc.

Note: for later additional types (*Schell Programm*), etc., see preceding text.

Military Designation Symbols

German military vehicles, if intended for any specific role, were given a *Kfz.* (*Kraftfahrzeug*—motor vehicle) number to indicate their role, regardless of the make of the vehicle's chassis. If the whole vehicle was of a special design, evolved by the army itself, it was given a *Sd.Kfz.* number (*Sonderkraftfahrzeug*—special motor vehicle). The use and allocation of *Kfz.* numbers was discontinued in 1943 when it was no longer practical.

For reasons of space, it is impossible to give a complete list of *Kfz.* and *Sd.Kfz.* numbers, but the following are the main categories. It must be emphasised that there are several exceptions, just as there are many inconsistencies and variations in official German nomenclature.

Kfz. Numbers and Chassis Types on which Bodywork is mounted:
Kfz.1–10: l.Pkw.; Kfz.11–20: m.Pkw.; Kfz.21–30: s.Pkw.; Kfz.31–40: l.Lkw. or s.Pkw.; Kfz.41–50: m.Lkw.; Kfz.51–60: s.Lkw.; Kfz.61–70: l.gl.Lkw. or s.Pkw.; Kfz.71–80: m.gl.Lkw. or l.gl.Lkw.; Kfz.81–90: l.gl.Lkw. or s.Pkw.

Note: chassis may be commercial (o), cross-country (gl.) or standard military (E).

Sd.Kfz.1–100: unarmoured special vehicles; Sd.Kfz.101 upwards: armoured special vehicles.

Most German military vehicles had their 'vital statistics' painted on the driver's door or in an equivalent position. The information usually appeared in a 'panel' surrounded with a broken line. Examples:

```
Kfz.le.Pkw.  Kfz.1(o)
Leergew.:  970 kg
Nutzlast:  420 kg
Verl.Kl.:  II
```
(Tatra 57K *'Kübel'*)

```
Kfz.m.Lkw.(V3000S/III)
Leergew.:  2,85 to.
Nutzlast:  3,–to.
Verl.Kl.:  Ia
```
(Ford medium truck)

indicating abbreviated vehicle nomenclature, weight (empty), payload, and loadclass for railway transport.

Sometimes the broken-line surround was omitted and only weights were shown.

German Nomenclature

German vehicle nomenclature was not always consistent and sometimes there were several 'official' designations for the same vehicle. It is important for the user of this book to be familiar with some of the basic terminology and some of the widely used abbreviations. The following glossary gives the most common abbreviations and their nearest English translation:

E. or *Einh.*	*Einheits* = standardised design (various mfrs.)
Fgst.	*Fahrgestell* = chassis
Flak	*Fliegerabwehrkanone* = AA gun
Gep., gp.	*gepanzert* = armoured
gl.	*geländegängig* = designed for off-road operation
gf.	*geländefähig* = suitable for off-road operation
Kfz.	*Kraftfahrzeug* = Motor vehicle
Kom.	*Kraftomnibus* = (Motor) bus
Krkw.	*Krankenkraftwagen* = Ambulance
Kw.	*Kraftwagen* = (Motor) vehicle
l. or *le.*	*leicht* = light
Lkw.	*Lastkraftwagen* = Truck
m. or *mittl.*	*mittler* = medium
NSKK	*Nationalsozialistisches Kraftfahrkorps* = Nat. Soc. Motor Corps
(o)	*handelsüblich* = commercially available
Pak	*Panzerabwehrkanone* = AT gun
Pkw.	*Personenkraftwagen* = Passenger car
Pz.Jäg.	*Panzerjäger* = SP AT gun, 'tank hunter'
Pz.Kpfw.	*Panzerkampwagen* = Tank
Pz.Spähw.	*Panzerspähwagen* = Armoured Reconnaissance car
s. or *schw.*	*schwer* = heavy
schf.	*schwimmfähig* = amphibious
schg.	*schienengängig* = for use on railroads
Schlp.	*Schlepper* = Tractor
Schtz.Pz.Wg.	*Schützenpanzerwagen* = Infantry armoured vehicle (APC)
Sfl.	*Selbstfahrlafette* = SP gun carriage
Sd.Kfz.	*Sonderkraftfahrzeug* = special motor vehicle
Zgkw.	*Zugkraftwagen* = Prime Mover

'Peace' · War · Peace (IWM NA26347)

Colours

Except in Africa where a yellow sand colour was used, German military vehicles were first universally painted a dark bluish-grey. In June, 1943, it was announced that the 'Africa Painting' would gradually be adopted for the whole of the *Kriegslieferungsprogramm* (programme of supply for war purposes), including those supplied to State authorities, industry and for export. Subsequently, all new vehicles were painted the yellow sand (or dried mud) colour and vehicles in use were gradually repainted. This *Einheits* colour was also used for most other kinds of painted equipment. Subsequently, most if not all vehicles were additionally painted with camouflage patterns, usually in brownish-red and green, sprayed over the basic colour. White and other colours were sometimes used to suit local conditions (snow, etc.). While the Germans were assured of their air superiority they tied a red, white and black swastika flag to the top of their vehicles as an identification to their own aircraft. This, however, was only the case during the first few years of the war.

By 1945 many *Wehrmacht* cars and trucks began to carry slogans, or rather warnings, painted large on the sides of their bodywork. Typical, almost classics, were *'Auch das Mundwerk Verdunkeln'* (blackout, also for the mouth) and *'Feind hört mit'* (the enemy is listening in). The latter warning was also displayed on radio equipment.

Registration Numbers:

The registration numbers of German military vehicles were generally shown on number-plates at front and rear. On certain vehicles like *Zgkw, Schtz.Pz.Wg.* and many of the early *gl.Pkw.* and *Lkw.* the rear number appeared on either side of the rear or rear side quarters of the bodywork, as shown in some of the illustrations. The registration number was prefixed by two or three letters, as follows:

WH for *Wehrmacht, Heer* (Army).
WL for *Wehrmacht, Luftwaffe* (Air Force).
WM for *Wehrmacht, (Kriegs)Marine* (Navy).
Pol for *Polizei*, including militarised police units.
SS for *SS* (*Schutzstaffel*), including *Gestapo*.
OT for *Organisation Todt* (semi-military).
RW for *Reichswehr* (the German Army in accordance with the Treaty of Versailles, until early 1935).

Who's Who in the German Automotive Industry (1935–45)

Note: only the major motor transport vehicle producers are listed, incl. the main plants in occupied territories.

Adler	Adler-Werke vorm. Heinrich Kleyer AG, Frankfurt/Main.
Audi	Auto-Union AG, Werk Audi, Zwickau i.Sa.
Austro-Daimler	Steyr-Daimler-Puch AG, Wien (A).
Auto-Union	Auto-Union AG, Chemnitz & Siegmar-Schönau.
BMW	Bayerische Motoren Werke AG, München, Eisenach & Spandau.
Borgward	Hansa-Lloyd & Goliath Werke AG (later Carl F. W. Borgward GmbH), Bremen.
Büssing-NAG	Büssing-NAG, Vereinigte Nutzkw. AG, Braunschweig, Leipzig & Berlin-Oberschöneweide.
Citroën	SA Citroën, Paris (F).
Daimler-Benz	(see Mercedes-Benz).
Demag	Demag AG, Wetter/Ruhr.
DKW	Auto-Union AG, Werk DKW, Zschopau i.Sa.
Famo	Fahrzeug- und Motorenwerke GmbH, Breslau 6.
Faun	Faun-Werke Nürnberg, Schnaittach.
Fiat	Fiat SpA, Turin (I).
Ford	Ford-Werke AG, Köln/Rhein.
Framo	Framo-Werke GmbH, Hainichen i.Sa.
Fross-Büssing	A. Fross-Büssing KG, Wien (A)
Graf & Stift	Graf & Stift Automobilfabriks AG, Wien-Döbling (A).
Hanomag	Hannoversche Maschinenbau AG, Hannover-Linden.
Hansa-Lloyd/Goliath	(see Borgward)
Henschel	Henschel & Sohn GmbH, Kassel & Berlin.
Horch	Auto-Union AG, Werk Horch, Zschopau & Zwickau i.Sa.
Kaelble	Motorenfabrik Kaelble, Backnang.
Klöckner-Humboldt-Deutz	Klöckner-Humboldt-Deutz AG, Köln & Ulm/Donau.
Krauss-Maffei	Krauss-Maffei AG, München-Allach.
Krupp	Friedrich Krupp AG, Essen.
Laffly	Éts. Laffly, Asnières (F).
Lanz	Heinrich Lanz AG, Mannheim.
Magirus	C. D. Magirus AG, Ulm/Donau (KHD).
MAN	Maschinenfabrik Augsburg-Nurnberg AG, Augsburg, Nürnberg & Gustavsburg.
Matford	SA Ford, Asnières (F).
Maybach	Maybach Motorenbau GmbH, Friedrichshafen/Bodensee.
Mercedes-Benz	Daimler-Benz AG, Stuttgart-Untertürkheim, Gaggenau, Berlin–Marienfelde & Sindelfingen.
NSU	NSU AG, Neckarsulm.
ÖAF	Österreichische Automobil-Fabriks-AG, Wien (A).
OM	OM SpA, Milan (I).
Opel	Adam Opel AG, Rüsselsheim/Main & Brandenburg/Havel.
Peugeot	SA Automobiles Peugeot, Sochaux (F).
Phänomen	Phänomen-Werke Gustav Hiller AG, Zittau.
Porsche	Dr.Ing.h.c. F. Porsche KG, Stuttgart-Zuffenhausen.
Praga	Böhmisch-Märische Maschinenfabriken AG, Praha (ČS).
Renault	Usines Renault, Billancourt, Seine (F).
Saurer	Österreichische Saurerwerke, Wien (A).
Skoda	Skoda-Werke, Plzen & AG vorm. Skodawerke, Praha (ČS).
Somua	Somua-Werke, Saint-Quen, Seine (F).
Steyr	Steyr-Daimler-Puch AG, Steyr, Oberdonau & Wien (A).
Stoewer	Stoewer-Werke AG, Stettin-Neutorney.
Tatra	Ringhoffer-Tatra-Werke AG, Praha & Kolin (ČS).
Tempo	Vidal & Sohn Tempo-Werke GmbH, Hamburg.
Trippel	Trippel Werke GmbH, Molsheim, Elsass.
Unic	SA des Automobiles Unic, Puteaux (F).
Volkswagen (KdF)	Volkswagenwerk GmbH, Wolfsburg.
Vomag	Vomag-Maschinenfabrik-AG, Plauen i.Vo.
Wanderer	Auto-Union AG, Werk Wanderer, Siegmar i.Sa.
Zündapp	Zündapp-Werke GmbH, Nürnberg.

GERMANY

MOTORCYCLES

Types, Makes and Models: *Motorcycles, Light (under 350 cc) (l.Krad (o))*: BMW R35; DKW RT125; Horex; NSU 251 OS(L); Phänomen AHO 1·125; Steyr-Daimler-Puch 125; Triumph BD250W; etc.
Motorcycles, Medium (350–500 cc) (m.Krad (o)): DKW NZ350; Triumph B350 and S350; Victoria KR35 SN/WH; etc.
Motorcycles, Heavy (over 500 cc) (s.Krad (o)), Motorcycles, Heavy, w/Sidecar (s.Krad (o) mit Beiwagen (E)), Motorcycles, Heavy, w/Sidecar, 3×2 (s.Krad mit Beiwagenantrieb (E)): BMW R12 and KS600 (solo and w/SC); BMW R75 (3×2); FN M12 (3×2; Belgian); Gnome & Rhône AX2 (3×2; French); NSU 601 OSL (solo and w/SC); Zündapp KS750 (3×2); etc.
Semi-Track, Light (Kettenkrad): NSU (see Semi-Track section).

General Data: *Krad* was the official abbreviation of *Kraftrad*. For sidecar the Germans used *Seitenwagen* or *Beiwagen für Krad*. Most sidecards were of a standardised *Einheits* type. Motorcycles were used in the German forces by all arms and in relatively large numbers. The makes and models listed above were the most common. In addition there were impressed and captured machines. In the Channel Islands, for example, the *Wehrmacht* used the ex-Belgian Army FN M12 (a 1000-cc shaft-drive 3×2 heavy sidecar combination, first introduced in 1938). The *Einheits* sidecar was suitable for the mounting of a machine-gun and could also carry a mortar. For some of its duties the motorcycle with sidecar was superseded by the VW *Kübelwagen*, even in the so-called *Kradschützen* units. The 3×2 combinations were used on all fronts and also in airborne operations. The BMW 3×1 combination was copied by the Soviet Army (M-72). The BMW R75 and Zündapp KS750 were specially built for heavy sidecar work but were occasionally used as solo machines. The NSU *Kettenkrad* could be described as a semi-track motorcycle and was first used in the landings at Crete in 1941. It was powered by an Opel 1·5-litre car engine.

Vehicle shown (typical): Motorcycle, Heavy, with Sidecar, 3×2 (BMW R75) (*schweres Kraftrad 750 cc mit Seitenwagen (angetrieben)*)

Technical Data:
Engine: BMW R750 2-cylinder, H-I-A-C, 745 cc, 26 bhp @ 4400 rpm.
Transmission: normal ratio: 4F1R; low ratio: 3F1R.
Brakes: hydraulic (rear and sidecar wheels).
Tyres: 4.50–16.
Wheelbase: 1444 mm.
Overall l×w×h: 2400×1730×1000 mm.
Weight: 400 kg.
Note: final drive with lockable differential. Produced 1940–44.

Motorcycle, Light (NSU 251OSL) 1-cyl., 10 bhp, 4F, wb 1320–1340 mm, 2040×780×950 mm, 144 kg. Basically a commercial (o) model. OHV engine of 241-cc capacity. Tyres 3.00–19. *Autobahn* speed 90 km/h (max. 100 km/h). Range 300–350 km. 35,000 produced during 1934–44.

Motorcycle, Heavy (BMW R12) 2-cyl., 18 bhp, 4F, wb 1380 mm, 2100×900×940 mm, 188 kg. Basically a commercial (o) model, often fitted with sidecar (E). 745-cc OHV engine. Shaft drive. Tyres 3.50–19. Max. speed 110 km/h. Production period 1935–41.

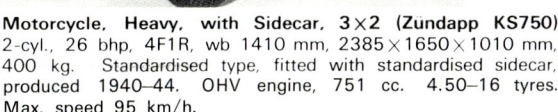

Motorcycle, Heavy, with Sidecar, 3×2 (Zündapp KS750) 2-cyl., 26 bhp, 4F1R, wb 1410 mm, 2385×1650×1010 mm, 400 kg. Standardised type, fitted with standardised sidecar, produced 1940–44. OHV engine, 751 cc. 4.50–16 tyres. Max. speed 95 km/h.

Motorcycle, Semi-Track (NSU, 'Kettenkrad') Opel Olympia 4-cyl., 36 bhp, 3F1R×2. This was one of the prototypes for the lightest semi-tracked tractor (see also relevant section). Note wire-spoke wheels and different body details. Another prototype had shorter track bogies with fewer wheels.

GERMANY

CARS, LIGHT
4×2 and 4×4

Types, Makes and Models: *Cars, Light, 4×2 (under 1500 cc) (l.Pkw. (o)):* Bianchi; BMW; Citroën; DKW; Fiat; Ford Eifel; Hanomag Garant, Kurier and Rekord; Opel P4, Kadett and Olympia; Peugeot; Renault; Röhr Junior; Steyr 50/55; Volkswagen 11 and 51 (KdF). etc. *Bodywork:* standard commercial (closed or open) or military pattern (open; see General Data).
Cars, Light, 4×2 (l.gl.Pkw.): Framo MW, 1936/37; Porsche 62, 1939; Skoda Popular 1100; Tatra 57K, 1940–43; Volkswagen (KdF) K1 Typ 82 (*Kübel*), 1939–45. *Bodywork:* military pattern (open).
Cars, Light, 4×4, (l.gl.Pkw. (o) and l.E.Pkw.): BMW El Pkw/325, 1938–40, and E1l, 1939–40. Hanomag El Pkw/20B, 1937–40. Stoewer El 1–4, 1937–40; R180 and R200 Spezial, 1938–40, and 40, 1939–43. Tempo G1200, 1936–39. *Bodywork:* military pattern (open).
Cars, Light, 4×4, Amphibious (l.schwf.Pkw.): Porsche/VW K2 Typ 128, 1940/41. Volkswagen (KdF) K2S Typ 166, 1942–44.

General Data: The chassis in the second category were made specifically for the fitting of military bodywork, but some of the chassis of the first category were also fitted with open military-pattern bodywork and cars thus converted became *l.gl.Pkw. Kfz.1, mit Fahrgestell der l.Pkw. (o)*, freely translated: light cross-country car with chassis of conventional light car. The (o) indicated that the car (or chassis) was of a type which was commercially available. The *Kfz.1* conversion consisted of a simple platform with two front seats, one rear seat (all of the *Kübel* or bucket type), a folding top, side curtains and little else. Later models were more sophisticated and featured steel doors. They were popularly known as *Kübelsitzer, Kübelwagen* or just *Kübel*. *Kfz.2, 2/40, 3* and *4* were basically similar to *Kfz.1* but were fitted out for specialist roles. The std light car chassis I (*l.E.Pkw.*) was intended to replace the multitude of models of the early and mid-1930s, but the mass-produced Volkswagen 82 was more successful in achieving at least part of this ideal.

Vehicle shown (typical): Car, Light, 4×2, Kfz.1 (Volkswagen (KdF) 82) (*leichter Personenkraftwagen K1 Typ 82*)

Technical Data:
Engine: VW Typ 1, 4-cylinder, H-I-A-R, 985 cc, 24 bhp @ 3000 rpm (from March, 1943: 1131 cc, 25 bhp @ 3000 rpm).
Transmission: 4F1R; limited-slip diff.
Brakes: mechanical.
Tyres: 5.25–16 (or oversize for desert use).
Wheelbase: 2400 mm.
Overall l×w×h: 3740×1600×1650 mm.
Weight: 685 kg.
Note: popularly known as *Kübelwagen*. Total produced about 52,000, in several versions. 'Step-down' rear axle final drive gear cases for improved ground clearance.

Car, Light, 4×2, Signals, Kfz.2 (BMW 303) 6-cyl., 30 bhp, 4F1R, wb 2400 mm, approx. 4100×1500×1500 mm, 1000 kg. Tyres 5.25–16. Officially: *Nachrichten-* or *Fernsprechkraftwagen*. Also on Hanomag (4/20 and Garant), Mercedes-Benz (170V), *l.E.Pkw* and Volkswagen (*Typ 82*) chassis.

Car, Light, 4×2, Siren (Opel P4) 4-cyl., 23 bhp, 4F1R, wb 2460 mm, approx. 4100×1500×2700 mm, 930 kg. Basically three-seater *Kfz.1* body, equipped with *Kraftfahrsirene*. Tyres 4.25–17. 1936–38. Also used as basis for dummy tanks (for training purposes) and with light truck body.

Car, Light, 4×2, Kfz.1 (BMW 303) 6-cyl., 30 bhp, 4F1R, wb 2400 mm, approx. 4100×1500×1600 mm, 1000 kg. One rear seat. 1933–34. 1173-cc OHV engine. Similar Model 315 (1934–36) had 34-bhp 1490-cc engine. Also on Hanomag and other chassis.

Car, 4×2, Small Repairs, Kfz.2/40 (Mercedes-Benz 170V) 4-cyl., 38 bhp, 4F1R, wb 2845 mm, approx. 4110×1580×1800 mm, 1260 kg. *Kfz.2/40* bodywork on *medium* car chassis. One rear seat. Three removable side doors. Rear locker lid formed work bench. Also used as *Kfz.2* (signals car).

Car, Light, 4×2 (Porsche 62) Porsche/VW E60 4-cyl., 24 bhp, 4F1R, wb 2400 mm, 3750×1550×1550 mm, 642 kg. Prototype for VW *Kübelwagen*. 5.00–18 tyres. Differed from later VW 82 in many details. Note recessed spare wheel, deeper body sides (lower sill panels).

Car, Light, 4×2, Sedan (Volkswagen) 4-cyl., 25 bhp, 4F1R, wb 2400 mm, 4200×1550×1650 mm. Shown with gas producer. Other variants: *Typ 82E* (Sedan on 82 chassis), *Typ 86* and *87* (*Kübel* 4×4), *Typ 92* (Sedan 4×4), *Typ 155* (half-track *Kübel*), etc.

Car, Light, 4×2, Kfz.1 (Tatra 1,3/57K) 4-cyl., 23 bhp, 4F1R, wb 2550 mm, 3980×1550×1690 mm, 970 kg. Produced by Ringhoffer-Tatra in Czechoslovakia for German Army. Based on T57 car chassis. Later also built for and used by Czech Army. Superseded T57a (1936-41), which had open sides.

Car, Light, 4×2, Kfz.1 (Skoda Popular 1100) 4-cyl., 32 bhp, 4F1R, wb 2485 mm, 4000×1500×1560 mm, 950 kg. Tyres 5.25–18. Derived from civilian passenger car. Produced during 1939–41. Special soft-top steel body with large locker at rear on backbone-type chassis with independent suspension.

Car, Light, 4×4, Amphibious (Porsche 128) 4-cyl., 24 bhp, 5F1R, wb 2400 mm, approx. 4225×1480×1620 mm, 950 kg. Pre-production model of VW 166 (shown in background). 150 produced in 1940 for field trials. Officially: *VW-Schwimmwagen A, gross.*

Car, Light, 4×4, Amphibious, Kfz.1/20, K2s (VW 166) 4-cyl., 25 bhp, 5F1R, wb 2000 mm, 3825×1480×1615 mm, 910 kg. Known as *Schwimmwagen* and *Kradschutzen-Ersatzwagen*. 14,265 produced, 1942–44. Hinged three-blade propeller. Speed in water 6 mph, on roads 50 mph.

Car, Light, 4×4, Radio, Kfz.2 (Stoewer 40) 4-cyl., 50 bhp, 5F1R×1, wb 2400 mm, 3850×1690×1900 mm, 1700 kg. Radio version of *l.Einheits-Pkw.* 6.00–18 tyres. Independent suspension all round. Front-wheel steering. 1997-cc Stoewer AW2 OHV engine. Also produced as three-door *Kfz.2/40.*

Car, Light, 4×4, AA Protection, Kfz.4 (BMW E.I.Pkw./325) 4-cyl., 45 bhp, 5F1R×1, wb 2400 mm, 3850×1690×1900 mm, 2200 kg. Twin machine guns on *l.Einheits-Pkw.* Earlier version was on *m.gl.Pkw.* chassis. Same *E.Pkw.* also produced by Hanomag and Stoewer. Four-wheel steering if required.

GERMANY

CARS, MEDIUM
4×2 and 4×4

Types, Makes and Models: *Cars, Medium, 4×2 (1500–3000 cc) (m.Pkw. (o)):* Adler Favorit, Standard, 3Gd; Aero 50; Audi; BMW 320, 326; Ford V8; Horch 830, 830B/Bk/Bl; Mercedes-Benz 170, 170V, 200, 260, 290, 320 and 340; Opel Super 6 and Kapitän; Skoda Superb 3000; Steyr 220, 250 and 630; Stoewer; Tatra 87; Wanderer W11, W23S, W24 and W26; etc. *Bodywork:* standard commercial (closed or open) or military pattern (open or closed).
Cars, Medium, 4×4 (m.gl.Pkw. and m.E.Pkw.): Auto-Union/Horch EFm, 1937–40, and 40, 1940–43. Laffly V15R and V15T, 1939–40. Latil M7T1, 1939–40. Mercedes-Benz 170 VL/W139, 1936, and G5/W152, 1937–41. Opel EFm, 1937–39. Tatra V809, 1940–42. *Bodywork:* military pattern (open or closed).
Cars, Medium, 4×4, Amphibious (m.schwf.Pkw.): Trippel SG6 (etc.), 1940–44.

General Data: The story of the medium car is much the same as that of the light types. From the late 1920s until the mid-1930s first the *Reichswehr* and later the *Wehrmacht* purchased civilian (*o*) type car chassis and fitted these with open *Kübel*-type bodywork, or, in case of the *Kfz.17*, a van-type body. About 1936/37 Adler, Auto-Union, and Daimler-Benz started production of special (modified conventional) car chassis which were fitted with special tyres, wings, open bodywork (with removable steel doors), etc. Generally speaking medium-type cars had an engine capacity of between 1500 and 3000 cc and/or seating capacity for max. five men, incl. driver. The 3000-cc limit was not always adhered to for foreign-made 'medium' cars. The standard medium car chassis was intended to be used for all roles in this class, with either open or closed bodywork. This chassis (*Einheitsfahrgestell für m.Pkw.*) was produced by Auto-Union/Horch and Opel (to a common military specification) and was used mainly in the *Kfz. 15, 16* and *17* roles. From about 1941 the 1½-ton 4×4 *A-Typ* truck chassis was also used for the *Kfz. 15* role. Confiscated and captured vehicles of roughly the required specification were also employed, as well as some Austrian, Czechoslovakian and French military types.

Vehicle shown (typical): Car, Medium, 4×2, Kfz.11 (Horch 830) (*mittlerer geländegängiger Personenkraftwagen (Kfz. 11) mit Fahrgestell des m.Pkw. (o)*)

Technical Data:
Engine: Horch 3-litre 8-cylinder, V-L-W-F, 2983 cc, 70 bhp @ 3600 rpm (830B fitted with 3·2- or 3·5-litre engine).
Transmission: 4F1R.
Brakes: hydraulic.
Tyres: 6.00–18.
Wheelbase: 3200 mm.
Overall l×w×h: 4800×1800×1850 mm.
Weight: 1850 kg.
Note: widely-used chassis for various types of open and closed bodywork, 1933–38.

Car, Medium, 4×2, Kfz.15 (Horch 830) V-8-cyl., 70 bhp, 4F1R (OD top), wb 3200 mm, 4800×1800×1850 mm, 1950 kg. *Nachrichten Kw.* (signalling service car), basically similar to *Kfz.11* on opposite page but featuring locker at rear, cable reel holders on front wings, etc.

Car, Medium, 4×2, Kfz.17 (Horch 830BI) V-8-cyl., 75 bhp, 4F1R, wb 3200 mm, 4800×1800×1850 mm approx., 2550 kg (gross). *Kleinfunk Kw.* (radio car) on *m.Pkw.(o)* chassis. Coachbuilt four-door body, largely made of wood. Later models on *m.E.Pkw.* 4×4 chassis.

Car, Medium, 4×2 (BMW 326) 6-cyl., 50 bhp, 4F1R, wb 2870 mm, 4600×1600×1650 mm, 1100 kg. 1971-cc (66×96 mm) OHV engine. Tyres 5.25/5.50–16 or –17. Five-seater four-door bodywork. Also two-door convertible version. Standard commercial types (*m.Pkw.(o)*), produced 1936–41.

Car, Medium, 4×2 (Opel Super 6) 6-cyl., 55 bhp, 4F1R, wb 2642 mm, 1215 kg. 2473-cc (80×82 mm) OHV engine. Tyres 5.50–16. Coil-spring ifs. Produced 1937–39. Superseded by Opel Kapitän which had same engine. Both were widely used by the *Wehrmacht* as *m.Pkw.(o)*.

Car, Medium, 4×2, Kfz. 11 (Wanderer W23S) 6-cyl., 60 bhp, 4F1R, wb 3100 mm, 5000×1800×2000 mm, 1650 kg. Late-type body (1937–39) featuring removable steel doors, but civilian-type front and rear wings. Similar bodywork also on Mercedes-Benz 170V.

Car, Medium, 4×2, Kfz.12 (Adler 3Gd) 6-cyl., 60 bhp, 4F1R, wb 3335 mm, 4800×1800×2000 mm, 2210 kg (gross). Special *Wehrmacht* model, introduced in 1938, with towing hook for light artillery pieces. 6.00–20 tyres. Believed still in production in 1943. Derived from Adler Diplomat.

Car, Medium, 4×2, Kfz.12 (Mercedes-Benz 230) 6-cyl., 78 bhp, 4F1R, wb 2880 mm, 4800×1800×2000 mm, 2200 kg (gross). 1764 produced during 1938–40 specially for military use. 6.50–20 tyres on special wheels. Steel doors (shown removed).

Car, Medium, 4×2, Kfz.15 (Mercedes-Benz 320) 6-cyl., 90 bhp, 4F1R. Similar to M-B 230 *Kfz.12* but slightly longer and fitted with rear equipment locker and towing hook. Few produced. Note cable reel holders on front wing (both sides).

Car, Medium, 4×4 (Mercedes-Benz 170VL/W139) 4-cyl., 40 bhp, 5F1R×1, wb 2525 mm, 4065×1570×1608 mm, 1720 kg (gross). Tyres 5.25–18. Four-wheel steering if required. 1697-cc (73·5×100 mm) SV engine. 42 supplied to *Heeres-Waffenamt*, 1936. Canvas doors.

Car, Medium, 4×4, (Mercedes-Benz G5/W152) 4-cyl., 45 bhp, 5F1R×1, wb 2530 mm, 3990×1680×1900 mm, 2150 kg (gross). Tyres 6.00–18. Four-wheel steering. 2006-cc (82×95 mm) SV engine. 320 produced, 1937–41. Also with steel doors and as convertible (*Kommandeurwagen*).

Car, Medium, 4×4, Kfz.12 (Auto-Union/Horch) V-8-cyl., 80 bhp, 4F1R×2, wb 3100 mm, 4700×1860×2070 mm, 2500 kg. Early *m.E.Pkw.* (*mittlere Einheits Pkw.*—medium standard universal car), shown minus doors. Later production had detail differences. Fitted with towing hook. Four seats.

Car, Medium, 4×4, Kfz.16 (Auto-Union/Horch) V-8-cyl., 80 bhp, 4F1R×2, wb 3100 mm, 4700×1860×2070 mm, 2500 kg. As *Kfz.12* on left but fitted out as *m.Messtrupp Kw.* (medium surveying vehicle) or *Kfz.16/1 Vorwarner Kw.* (advance spotting vehicle), both used by Artillery units. *Circa* 1937.

Car, Medium, 4×4, Kfz.15 (AU/Horch and Opel EFm)
Horch V-8, 80 bhp or Opel 6-cyl., 68 bhp, 4F1R×2, wb 3100 mm, 4740×1850×2050 mm, 2500 kg. 190–18 tyres. Later type (model 40) only by Horch. Widely used standardised universal type (*m.E.Pkw.*), also as Staff car (open and closed).

Car, Medium, 4×4, Kfz.15 (AU/Horch 40) V-8-cyl., 80 bhp, 4F1R×2, wb 3100 mm, 4700×1860×2070 mm, 3300 kg (gross). Late type (*m.E.Pkw.*) without stub axle-mounted spare wheels. Wider body, seating five instead of four. Rommel used this model personally in North Africa.

Car, Medium, 4×4 (AU/Horch and Opel EFm) *m.E.Pkw.* chassis with platform body and pedestal for telephotolens camera. Equipment lockers underneath. The *m.E.Pkw.* was heavy and demanded excessive maintenance, with almost 100 grease nipples requiring regular attention.

Car, Medium, 4×4, Kfz.17 (AU/Horch and Opel EFm) Standardised (*m.E.Pkw.*) chassis with closed body, used (with different fittings) for Radio, Telephone Exchange, and similar roles. Late production dispensed with stub axle-mounted support/spare wheels.

Car, Medium, 4×2, Kfz.15 (Chevrolet Standard FC) 6-cyl., 72 bhp, 3F1R, wb 2769 mm, 4725×1830×1800 mm approx. *Kfz.15* conversion of confiscated or captured 1936 American Chevrolet four-door sedan. Original doors were retained but cut down to waist level.

Car, Medium, 4×2, Kfz.15 (Chevrolet Master DeLuxe JA) 6-cyl., 85 bhp, 3F1R, wb 2851 mm, 4775×1830×1800 mm approx. Converted 1939 Chevrolet sedan. Main modifications: folding top and windscreen, enlarged wing cut-outs, rear locker, four small running boards/steps, Notek lights.

Carrier, Personnel, 4×2 (Chevrolet Master DeLuxe GA) Typical example of personnel carrier/raid car as used during 1944–45. Wooden body had lengthwise bench seating for six and open rear entrance. Chassis included 1936–40 Buick, Cadillac, Chevrolet, LaSalle, Ford, Mercury, Opel, etc.

Car, Medium, 4×2, Kfz.15 (Morris-Commercial PU) 6-cyl., 60 bhp, 4F1R, wb 2743 mm, 4250×2000×1900 mm approx. Tyres 9.00–13. German *Kfz.15* bodywork on ex-British 8-cwt truck chassis, one of many left behind in the retreat from France in 1940 (see also pages 134 and 152).

Car, Medium, 4×2, Personnel (Steyr 250) 4-cyl., 25 bhp, 4F1R×2, wb 2600 mm, approx. 4100×1680×1870 mm. Austrian-built *Kübelsitzer*. 1200 produced, 1938–40 (also 205 light trucks, Steyr 150, on same chassis). Note hinged side-curtain frame top rail.

Car, Medium, 4×2, Kfz.15 (Skoda Superb 3000) 6-cyl., 80 bhp, 4F1R. 3140-cc (80×104 mm) OHV engine with 6·2:1 CR. Hyd. brakes. Independent front and rear suspension with transversal leaf springs. Differential lock. Produced in Czechoslovakia during 1942–43.

Car, Medium, 4×4, Kfz.12 (Tatra V809) 4-cyl., 50 bhp, 4F1R×2, wb 2600 mm, 3720×1600×2000 mm. 2470-cc (85×109 mm) air-cooled 4-in-line OHV engine. Ground clearance 255 mm. A number of these saw service with Rommel's *Afrika Korps*. 1940.

Car, Medium, 4×4, Command (Tatra V809) Basically as *Kfz.12* shown on left but with *Tropen-Limousine* bodywork and air-cooled V-8-cyl. engine of 3980 cc (80×99 mm) cubic capacity, developing 70 bhp at 2500 rpm. 'V' in model designation stood for *Vojenský* (military). 1942.

Car, Medium, 4×4, Amphibious (Trippel SG6/38) Opel 6-cyl., 55 bhp, 5F1R×1, wb 2500 mm, 4450×2000×1850 mm, 1800 kg. Permissible load on roads 1000 kg, in water 2000 kg. First made in 1935 with 4-cyl. 2-litre Adler engine. 1938 model (first military version) with 2·5-litre engine shown.

Car, Medium, 4×4, Amphibious (Trippel SG6/41) Opel 6-cyl., 55 bhp, 5F1R×1, wb 2500 mm, 4825×1800×1910 mm, 1750 kg. Produced in Bugatti works in Molsheim, 1941–44. SG6 production totalled about 1000 units. Variants included Tatra V-8-engined SG7, 1943, and armoured E3, 1944.

Car, Medium, 4×2 (Tatra 87) V-8-cyl., 75 bhp, 4F1R, wb 2850 mm, 4740×1670×1500 mm, 1370 kg. Air-cooled 2960-cc (75×84 mm) OHC engine at rear. Tyres 6.50–16. Max. speed 160 km/h. Developed from T77 (1934–37) and produced from 1938 until 1950. Independent front and rear suspension.

Car, Snow (Tatra V855) Experimental over-snow car, produced in 1942 for the *Wehrmacht* for possible use on the Eastern front. Max. speed 80 km/h. Driven by roller at rear and air propeller. Steering by skis (4). 2960-cc air-cooled V-8-cyl. engine at rear. Derived from T87.

GERMANY

CARS, HEAVY
4×2, 4×4, 6×4

Types, Makes and Models: *Cars, Heavy, 4×2 (3-litre and over) (s.Pkw. (o)):* Horch; Mercedes-Benz; Opel Admiral; Phänomen Granit; Skoda Superb 3000; etc. *Bodywork:* std (closed or open) except Phänomen, which was light truck chassis with *Kübel*-type body. *Cars, Heavy, 4×4 (s.gl.Pkw. and s.E.Pkw.):* Auto-Union/Horch EFm, 1937–40; a, b, 1a, 1b, 1c, 1937–40, and 40, 1940–41. Ford EGa, EGb, EGd, 1939–40. Horch-Argus, 1933–35. Mercedes-Benz L1500A/L301, 1941–43. Phänomen Granit 1500A, 1940–44. Steyr 1500A, 1941–44. *Bodywork:* military (open, few closed).
Cars, Heavy, 6×4 (s.gl.Pkw.): Krupp L2H 143, 1937–41, from 1933. Skoda 3-litre, 1941–43. Steyr 640, 1937–41; Tatra T82, 1935–40, T93 (6×6), 1937–40. Mercedes-Benz G4/W31,

General Data: The 4×4 and 6×4 type chassis were especially built for military use. The AU/Horch and Ford *s.E.Pkw.* were standard universal chassis with either Horch or Ford V8 engine and were also used with ambulance bodywork and, in modified form, for front- and rear-engined armoured cars (*Sd.Kfz.221*, etc.). They were originally intended to replace various vehicles built previously on light truck chassis (6×4 and 4×2), but this never fully materialised. The *A-Typ* 1½-ton light truck chassis of the *'Schell Programm'*, being much simpler and easier to maintain, was selected instead. The Phänomen, Steyr and Tatra had air-cooled engines which had obvious advantages for use in North Africa and Russia. Towards the end of the war civilian car chassis (mainly American makes) with simple open six-seater rear bodies with back entrance were used as raid cars and personnel carriers. These conversions were carried out mainly in *HKP (Heereskraftfahrzeugpark)* workshops in occupied countries like the Netherlands. *HKP*-workshops were usually civilian garages, forced to work for the *Wehrmacht*. The Tatra T82 and T93 had originally been produced for the Czech Army. The 3000-cc minimum engine capacity limit was not always adhered to, particularly for foreign-made 'heavy' cars.

Vehicle shown (typical): Car, Heavy, 4×4, Kfz.69 (Auto-Union/Horch 1a, 1b) (*Protzkraftwagen (Kfz.69) mit Einheitsfahrgestell II für s.Pkw.*)

Technical Data:
Engine: AU/Horch V8-108 8-cylinder, V-L-W-F, 3823 cc, 81 bhp @ 3600 rpm.
Gearbox: 5F1R×1.
Brakes: hydraulic.
Tyres: 210–18.
Wheelbase: 3000 mm.
Overall l×w×h: 4850×2000×2040 mm.
Weight: 3150 kg.
Note: used to tow light artillery. Model 1a had optional four-wheel steering. *Einh. Fahrgestell I für s.Pkw.* had rear engine (used for armoured cars).

Car, Heavy, 4×4, Personnel, Kfz.70 (AU/Horch 40) Horch V8, 90 bhp, or Ford V8, 78 bhp, 5F1R×1, wb 3000 mm, 4850×2000×2040 mm, approx. 3000 kg. Late *s.E.Pkw.* without side support/spare wheels. Steering only on front wheels. Superseded by light 4×4 truck with similar body (MB, Steyr).

Car, Heavy, 4×4, Kfz.21 (AU/Horch) V-8-cyl., 80 bhp, 4F1R×2, wb 3100 mm, 4850×1860×2040 mm, 2630 kg. Six-seater *s.Pkw.* bodywork on standard chassis of *m.E.Pkw.* (medium std car). Similar to *Kfz.12* on early *m.E.Pkw.* chassis but with additional seats. Used by Pioneers and *Luftwaffe*.

Car, Heavy, 4×2, Convertible (Mercedes-Benz 320) 6-cyl., 78 bhp, 4F1R, wb 3300 mm. Tyres 6.50–17. 3207-cc (82.5×100 mm) SV engine. Used by *Kriegsmarine*. One of various types of Mercedes-Benz civilian cars which were used by German officers. IWM photo BU 6052.

Car, Heavy, 4×2, Limousine (Horch 850) 8-cyl., 105 bhp, 4F1R, wb 3750 mm. Tyres 7.00–17. 4946-cc (87×104 mm) overhead-camshaft engine. Illustration shows car with loud-speaker equipment addressing Soviet soldiers hidden in houses of Eastern Front town, 1941.

Car, Heavy, 4×4, Command, Kfz.21 (AU/Horch EFm)
V-8-cyl., 90 bhp, 4F1R×2, wb 3100 mm, 4800×1860×1930 mm, 3080 kg. *Kommandeurwagen* on *m.E.Pkw.* chassis. Four-door coach-built convertible. Used by high commanders in the field, incl. Guderian (shown) and Rommel.

Car, Heavy, 4×2, Command, Kfz.21 (Skoda Superb 3000)
6-cyl., 80 bhp, 4F1R. Same chassis as Skoda *Kfz.15* shown on page 290, but fitted with standard pattern four-door convertible *Kommandeurwagen* bodywork. Note rear locker with fuel can holders on either side.

Car, Heavy, 4×4, Command, Kfz.21 (Steyr 1500A/01)
V-8-cyl., 85 bhp, 4F1R×2, wb 3250 mm, 5080×1850×2100 mm, 3630 kg. *Kommandeurwagen* on light truck chassis. Used by General von Arnim and others. Heavy coach-built body. Max. speed up to 100 km/h. Produced 1941–44.

Car, Heavy, 4×4, Personnel (Steyr 1500A/02) V-8-cyl., 85 bhp, 4F1R×2, wb 3250 mm, 5080×2000×2320 mm, 2485 kg. Chassis produced by Steyr in Austria (12,450 units) and Auto-Union. 8-seater body. Tyres 190–20 or 7.25–20, or, for desert use, 270–16. Early production had different doors.

Car, Heavy, 4×4, Personnel (Mercedes-Benz L1500A/L301) 6-cyl., 60 bhp, 4F1R×2, wb 3000 mm, 4930×2050×2225 mm, 2670 kg. Daimler-Benz *Kfz.70 Mannschaftswagen* bodywork on light 4×4 truck chassis. 190–20 tyres, 'half duals' rear. Also used for *Kfz.15* and other roles.

Car, Heavy, 4×4, Personnel (Phänomen Granit 1500A) 4-cyl., 50 bhp, 4F1R×2, wb 3270 mm, 5490×1980×2160 mm, 2215 kg. *s.Pkw.* version of light 1½-ton 4×4 truck, produced for use in N. Africa. Special 270–16 desert tyres. Additional air cleaner. Produced in Zittau, Saxony (now DDR).

Car, Heavy, 6×4, Convertible (Krupp L2H143) 4-cyl., 52 bhp, 4F1R×2, wb 2905 (BC 910) mm, 5100×1920×1800 mm, approx. 3000 kg. One of various body types on this widely used truck chassis. This model, produced in 1940, had a conventional front end and no side support wheels.

Car, Heavy, 6×4, Convertible (Skoda 3-litre) 6-cyl., 80 bhp, 4F1R×2. 3140-cc (80×104 mm) engine as in Skoda Superb 3000. Six seats. Produced in limited numbers for *Wehrmacht*, 1941–43. During the 1930s Skoda supplied similar command cars for the Czech Army.

Car, Heavy, 6×4, Command (Steyr 640) 6-cyl., 55 bhp, 4F1R×2, wb 3030 (BC 1060) mm, approx. 4900×1700×1900 mm. Used by Austrian and German Army. Four-door *Kommandeurwagen* (command car) on light truck chassis. Also with truck and ambulance bodywork.

Car, Heavy, 6×6, Convertible (Tatra T93) V-8-cyl., 70 bhp, 4F1R×2, wb 3270 (BC 940) mm. Tyres 6.00–20. Air-cooled 3980-cc (80×99 mm) engine. Based on *Rumänian-Typ* 6×6 truck chassis. Produced during 1937–40, in 1940 also with full-length soft-top body.

Car, Heavy, 6×4, Convertible (Mercedes-Benz G4/W31) 8-cyl., 100 bhp, 4F1R×2, wb 3575 (BC 950) mm, 5400×1890 ×1800 mm, 3500 kg (gross). 57 produced 1933–34. Later produced with 110-bhp 5401-cc engine (1939). Extensively used by Hitler. Self-locking diffs in rear axles.

Car, Heavy, 6×4, Communications (Mercedes-Benz G4/W31) 8-cyl., 100 bhp, 4F1R×2, wb 3575 (BC 950) mm. Chassis similar to that of Convertible shown above. 5018-cc straight-eight engine. Used in Hitler's personal transport fleet for convenience of press reporters. Operated by *Reichspost*.

GERMANY

AMBULANCES
4×2, 4×4, 6×4

Makes and Models: *4×2 types:* Adler Diplomat, W61. Horch V8. Mercedes-Benz L(E)1100. Opel Blitz. Peugeot 202U/T2LY, DK5J/D6S. Phänomen Granit 25H; etc.
4×4 types: Auto-Union/Horch (*s.E.Pkw.*). Phänomen Granit 1500A.
6×4 types: Steyr 640/643.

General Data: All ambulances used by the German forces were designated *Kfz.31 Krankenkraftwagen (Krkw.)*. The chassis on which the ambulance body was built could originally be any light 4×2 truck (*l.Lkw. (o)*) and the Phänomen Granit chassis was the most common during the 1930s. By 1939–40 the standardised heavy car chassis (*s.E.Pkw.*) with ambulance body came to the troops. It was widely used in the North African operations. Many captured ambulances were also taken into service, varying from the British Austin K2 to the Soviet ZIS. The Austrian Army Steyr 6×4 ambulance was produced for the Germans until 1941. The Peugeot 202U and DK5J were also 'foreign' ambulances produced for the *Wehrmacht*. The semi-track *Sd.Kfz.251/8* was an armoured front-line ambulance. Ambulances are also referred to as *Sanitätskraftwagen* or, abbreviated, *Sanka*. In addition to the above, many of which served as front-line ambulances, use was made of converted buses for the conveyance of casualties. The *Wehrmacht* employed military and civilian-type buses. The latter were usually confiscated/captured buses and coaches. The former were known as *Wehrmachts-Omnibus* and a common model was the Opel 3·6-47 *Typ S*. This was a lwb version of the Opel Blitz *S-Typ* 3-ton truck. Its interior could be adapted for use as bus, command centre, or ambulance. Mobile operating theatres (*Operationswagen*) existed on Ford, MAN, Opel and other truck chassis.

Vehicle shown (typical): Ambulance, 6×4, Kfz.31 (Steyr 640/643) (*Krankenkraftwagen Kfz.31, mit Fahrgestell des l.Lkw. (o)*)

Technical Data:
Engine: Steyr 640 6-cylinder, I-I-W-F, 2260 cc, 55 bhp @ 3800 rpm.
Transmission: 4F1R×2.
Brakes: hydraulic.
Tyres: 7.00–18.
Wheelbase: 3030 mm, BC 1060 mm.
Overall l×w×h: 4880×1800×2560 mm.
Weight: 2885 kg.

298

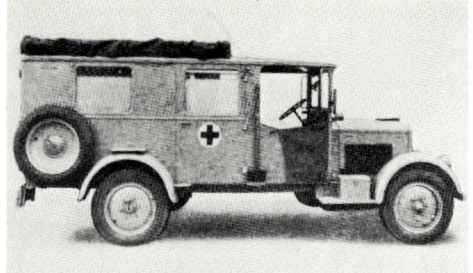

Ambulance, 4×2, Kfz.31 (Phänomen Granit 25H) 4-cyl., 37 bhp, 4F1R, wb 3600 mm, approx. 5400×2050×2300 mm, 2400 kg. Common *Wehrmacht* ambulance. Accommodation for four stretchers or eight sitting patients. Also with canvas-top cab and folding windscreen and with closed cab.

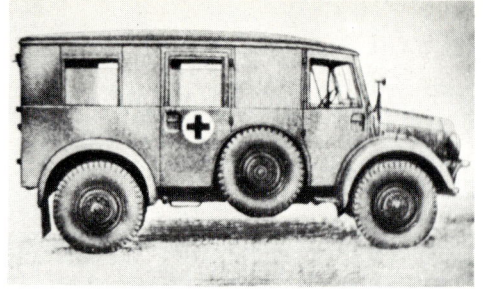

Ambulance, 4×4, Kfz.31 (Auto-Union/Horch) V-8-cyl., 81 bhp, 5F1R×1, wb 3000 mm, 4850×2000×2100 mm, 3050 kg. Standardised *s.E.Pkw.* II chassis. Widely used by *Afrika Korps*. Similar closed bodywork used for *Kfz.24 Verstärker Kw.* (Amplifier vehicle).

Ambulance, 4×2, Kfz.31 (Mercedes-Benz L or LE1100) 4-cyl., approx. 50 bhp, 4F1R. Generally similar to Phänomen Granit 4×2. Produced in 1938 on light truck chassis. Similar body also on Opel Blitz light truck chassis.

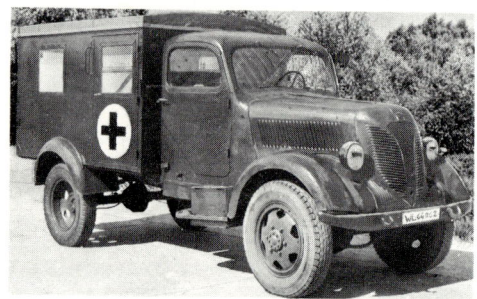

Ambulance, 4×4, Kfz.31 (Phänomen Granit 1500A) 4-cyl., 50 bhp, 4F1R×2, wb 3270 mm, 5200×1980×2085 mm, approx. 3000 kg. Late type with *Einheitskastenaufbau* (standard house type body) on *A-Typ* 1½-ton chassis. Chassis was continued in production after the war (East Germany) as Robur Granit.

GERMANY

TRUCKS, LIGHT
4×2 and 4×4

Types, Makes and Models: *Trucks, Light, 4×2 (l.Lkw. (o), l.Lkw. (S-Typ)):* Adler W61, 60/61 (1- and 1½-ton). Auto-Union AU 1500S (1½-ton). Borgward L1400 (1-ton) and L2300 (1½-ton). Citroën 23, 23R (1½ and 2-ton). Fiat 1100 Mil. (350-kg). Mercedes-Benz L1500S (1½-ton). Opel Blitz 2,0–12 (1-ton), 2,5–32 (1½-ton), 3,5–34–20 (2-ton). Peugeot DK5, DMA (1¼- and 2-ton). Phänomen Granit 25H and 1500S (1½-ton, closed or open cab). Renault AGC, AHS (2-ton), etc.
Trucks, Light, 4×4 (l.Lkw. (A-Typ), l.gl.Lkw.): Auto-Union AU1500A (1½-ton). Fiat/Spa T40 (1-ton) and CL39 (1½-ton). Mercedes-Benz L1500A (1½-ton). Phänomen Granit 1500 (1½-ton). Steyr 1500A/01 and 02 (1½-ton), 2000A (2-ton).
Bodywork: mainly cargo, with or without tilt; some closed.
General Data: The majority of these light trucks were basically commercial types with payload capacities of up to 2 tons. There were many types and to list them all would be prohibitive. In order to reduce the large number of types of vehicles of all kinds which were in production in Germany, General von Schell introduced in 1938 what became known as the *Schell Programm*. This scheme curtailed the multitude of vehicle models to a sensible programme which under the prevailing conditions could be more easily adhered to. Under this programme some of the best trucks were produced by more than one manufacturer and some of these vehicles (in all classes) were made in two versions: the *S-Typ* (*Wirtschaftstyp*, commercial model, 4×2) and the *A-Typ* (*Wehrmachtstyp*, military model, 4×4). In addition to German military and civilian-type light trucks, the German forces used several types which were produced for them in Austria (Steyr) and France (Citroen, Peugeot, Renault) and also large numbers of captured and confiscated vehicles, including the Russian GAZ-AA ('Russki-Ford'). Bodywork was open (*Pritsche*), with tilt (*Spriegel*), or closed (*geschlossen, Kastenaufbau*), ambulance (*Krkw.*), etc. For 6×4 and 6×6 types see next section.

Vehicle shown (typical): Truck, Light, 4×4, Cargo (Steyr 1500A/02) (*leichter geländegängiger Lastkraftwagen, offen; l.gl.Lkw., off., A-Typ*)

Technical Data:
Engine: Steyr 1500A 8-cylinder, V-I-A-F, 3517 cc, 85 bhp @ 3000 rpm.
Transmission: 4F1R×2.
Brakes: hydraulic.
Tyres: 270–16 (oversize for desert use, normally 190–20).
Wheelbase: 3250 mm.
Overall l×w×h: 5230×2035×2460 mm.
Weight: 2485; payload 1675 kg.
Note: also produced in Germany by Auto-Union and with std *Einheits* cab.

Truck, Light, 4×2, Cargo (Adler 60/61) 6-cyl., 58 bhp, 4F1R, wb 3250 mm, chassis 4900×1920 mm, 3250 kg (gross). Tyres 7.00–20 (1½-ton), 6.50–20 (1-ton). Military-pattern open cab. Cargo body with canvas tilt (*Pritschenwagen, mit Spriegel*). Also with ambulance bodywork.

Truck, Light, 4×2, Cargo (Mercedes-Benz L1500S) 6-cyl., 60 bhp, 4F1R, wb 3000 mm, 5320×1960×2340 mm, 2060 kg. Typical *Schell Programm* vehicle, also produced as 4×4 (L1500A) and with various body types. Hinged triangle on cab roof was erected when towing trailer.

Truck, Light, 4×2, Cargo (Borgward B1000) 4-cyl., 33 bhp, 4F1R, wb 2800 mm, 4580×1620×1950 (1780) mm, 1255 kg. Tyres 6.00–16. Borgward 4M 1384-cc (72×85 mm) OHV petrol engine. Payload 1155 kg. Commercial type 1-tonner, also with van bodywork. Classed as *l.Lkw.(o).*

Truck, Light, 4×4, Cargo (Auto-Union AU1500A) Wanderer 6-cyl., 62 bhp, 4F1R×2, wb 2800 mm, 4820×1980×2120 mm, 2210 kg. Prototypes (*S* and *A*), 1942/43. No quantity production because Auto-Union was ordered to produce the Steyr 1500A. 190–20 tyres; 6.00–20 (2DT) on AU1500S.

Truck, Light, 4×2, Gun Tractor (Morris-Commercial CS8) 6-cyl., 60 bhp, 4F1R, wb 2489 mm, 4065×2000 mm approx. Ex-British Army Morris-Commercial 15-cwt infantry truck (see page 156), rebodied by *Wehrmacht*. Main body featured ammunition lockers, crew seats and folding canvas tilt.

Truck, Light, 4×2, Cargo (Citroën 23R) 4-cyl., 48 bhp. 4F1R, wb 3750 mm, 5540×1980×2710 mm, 2200 kg. Over 6000 of these 2-ton Citroëns were producedm, mainly for the German Army. 1½-ton model 23 was discontinued in 1940. Many French vehicles saw service on the *Ostfront* (eastern front).

Truck, Light, 4×2, Cargo (Renault ADK) 4-cyl., 48 bhp, 4F1R, wb 3345 mm, 5700×2000×2610 (2040) mm, 2200 kg. 2383-cc SV engine. Ex-French Army Renault rhd 2-ton truck, used extensively on Eastern (shown) and Western front, also with van-type bodywork. Renault AGC was similar.

Truck, Light, 4×2, Cargo (Renault AHS) 4-cyl., 50 bhp, 4F1R, wb 3125 mm, 5600×1970×2820 (2430) mm, 2520 kg. Renault 603S 2383-cc SV engine. Tyres 6.50–20. Produced during 1941–44, mainly for *Deutsche Wehrmacht*. After the war it was continued as AHS2 and 3.

GERMANY

TRUCKS, LIGHT
6×4 and 6×6

Types, Makes and Models: *Trucks, Light, 6×4 (l.gl.Lkw. (o) and l.gl.Lkw.):* Austro-Daimler ADG, ADGR. Botond. Büssing-NAG G31. Fiat/Spa 'Dovunque' 35. Krupp L2H43 and L2H143. Magirus M206. Mercedes-Benz G3, G3A, B, B/1. Praga RV, RVR. Steyr 640. Tatra T92. *Trucks, Light, 6×6 (l.gl.Lkw. and l.E.Lkw.):* Büssing-NAG HWA 526D (Einh.). Henschel HWA 526D (Einh.). Laffly W15T, S35TL, etc. Magirus (KHD) HWA 526D (Einh.). MAN HWA 526D (Einh.). Mercedes-Benz HWA 526D (Einh.). Tatra T93.

General Data: Bodywork on the above chassis types was of many different types and included open types for transport of cargo and personnel and house types (*Kastenaufbau*) for special technical roles. *Kfz.62*, for example, was a wood-panel type used for seven different roles, incl. printing, meteorological, etc. During the early 1930s these bodies were fitted on 6×4 chassis which were produced to a common basic specification by Büssing-NAG, Daimler-Benz (Mercedes-Benz), and Magirus. In general design they were much like the contemporary British six-wheelers and they were also available to civilian purchasers. The Krupp L2H43 was different in many respects and had an air-cooled 'boxer' engine and all-independent suspension. For most of its roles the Krupp was superseded by special versions of the *s.E.Pkw.*, although complete replacement never took place. In the above listing there are several 'foreign' types which were either produced for the *Wehrmacht* and other Axis forces, or taken over in quantity. The French Laffly W15T was a 6×6 personnel carrier-cum-prime mover and like the S35TL and several other types in this class had exceptional cross-country performance. The *l.E.Lkw.* was a standardised diesel-engined 6×6 chassis, produced by various manufacturers and was commonly known as *Einheitsdiesel*. It was also exported to friendly countries. Some of the early 6×4 chassis were used, in modified form, for the *Sd.Kfz.231* and other armoured cars. The Tatra T93 was a 6×6 version of the firm's T92 and was supplied mainly to the Rumanian Army. It was known as the *Rumänian-Typ*.

Vehicle shown (typical): Truck, Light, 6×4, Light AA, Kfz.81 (Krupp L2H143) (*leichter Flakkraftwagen (Kfz.81) mit Fahrgestell des l.gl.Lkw. (o)*)

Technical Data:
Engine: Krupp M304 4-cylinder, H-I-A-F, 3308 cc, 60 bhp @ 2500 rpm.
Transmission: 4F1R×2.
Brakes: hydraulic.
Tyres: 7.50–17.
Wheelbase: 2905 mm, BC 910 mm.
Overall l×w×h: 4950×1950×2300 mm.
Weight: 2600 kg.
Note: Towed 2-cm Flak AA/AT gun. Superseded L2H43 in 1937. Popularly known as the *Krupp-Schnauzer*.

Truck, Light, 6×4, Prime Mover, Kfz.69 (Krupp L2H43)
4-cyl., 60 bhp, 4F1R×2, wb 2900 (BC 860) mm, 5000 ×1900×2000 mm, 2700 kg. Introduced in 1934, these Krupps were known as *Protzkraftwagen* and used by the *Panzertruppen* to tow AT gun (*Pak*).

Truck, Light, 6×4, Communications, Kfz.19 (Krupp L2H143)
4-cyl., 60 bhp, 4F1R×2, wb 2900 (BC 910) mm, 5100×1920× 2250 mm, 2824 kg. *Fernsprechbetriebs Kw.*, fitted out with telephone switchboard and related equipment. Widely used chassis for variety of body types.

Truck, Light, 6×6, Cargo (l.E.Lkw., various mfrs) 6-cyl. diesel, 85 bhp, 4F1R×2, wb 3650 (BC 1100) mm, 5850×2260 ×2600 mm, 4900 kg. This version also used as *Pionier Kw.III*. Coil-spring independent suspension all round. 210–18 tyres. Also with snow-plough and as recovery vehicle.

Truck, Light, 6×6, Communications, Kfz.61 (l.E.Lkw.)
6-cyl., diesel, 85 bhp, 4F1R×2, wb 3650 (BC 1100) mm, 5850×2200×2530 mm (roof), 5200 kg. Very comprehensively equipped *Verstärker Kw.* Used by *Nachr. Tr.* and *Luftwaffe*. Similar vehicles also for several other roles.

Truck, Light, 6×4, Kfz.61 (Büssing-NAG G31) 4-cyl., 65 bhp, 3F1R×2, wb 3190 (BC 950) mm, 5630×2100×2800 mm, 3900 kg. Typical example of standard 6×4 1½-tonner of the early and mid-1930s. Similar chassis produced by Daimler-Benz and Magirus. Numerous body types, incl. armoured.

Truck, Light, 6×6, Personnel/Prime Mover (Laffly W15T) 4-cyl., 52 bhp, 4F1R×2, wb 2830 mm, 4640×1900×1960 mm, 4500 kg (gross). Low-silhouette variant of Laffly S15T. One of various 4×4 and 6×6 tactical types made by this firm. Many were used by the Germans after the fall of France. Also made by Hotchkiss.

Truck, Light, 6×4, Cargo (Praga RV) 6 cyl., 68 bhp, 4F1R×2, wb 3560 (BC 920) mm, 5690×2000×2500 (2090) mm, 3810 kg. Tyres 6.00–20. 3-ton winch at rear. 2033 produced during 1935–39, first mainly for Czech and Rumanian armies, later used by *Wehrmacht* also.

Truck, Light, 6×4, Radio (Praga RVR) 6-cyl., 68 bhp, 4F1R×2, wb 3560 (BC 920) mm, approx. 5690×2000×2200 mm. 6.00–20 tyres. Used by Czech and later by German Army. Torsion-bar independent front suspension. On similar chassis: AV Command car.

305

Truck, Light, 6×4, Cargo/Personnel (Botond 38M) 4-cyl., 65 bhp, 5F1R, wb 3030 (BC 1120) mm, 5746×2200×2550 mm, 4000 kg. Payload on/off road 2000/1500 kg. Swing axles at rear. Design Winkler. Produced by Rába-Werke in Györ, Hungary. Rába also licence-produced Krupp trucks and buses.

Truck, Light, 6×4, Cargo (Steyr 640) 6-cyl., 55 bhp, 4F1R×2, wb 3030 (BC 1060) mm, 5330×1730×2330 mm, 2300 kg. 7.00–18 tyres. 3780 chassis produced, 1937–41 (majority in 1940–41). Also as ambulance and command car. Superseded 45-bhp model 440 (713 produced, 1935–37).

Truck, Light, 6×4, Cargo (Tatra T92) V-8-cyl., 70 bhp, 4F1R×2, wb 3270 (BC 940) mm, 5495×2000×2610 (2130) mm, 3580 kg. Tyres 6.00–20. Payload 2 tons. Also with ambulance bodywork. Air-cooled 4-litre engine. 500 produced, 1937–40. T93 was 6×6 version (for Rumania).

Truck, Light, 6×4, Personnel (Austro-Daimler ADGR) 6-cyl., 72 bhp, 7F3R, wb 3720 (BC 1200) mm, 6500×2135×2230 mm, 4420 kg. 361 produced by Steyr-Daimler-Puch (1936-40), mainly for export to Rumania. Also with ambulance, tanker and command bodywork. Superseded ADG (1932-35).

GERMANY

TRUCKS, MEDIUM 4×2

Types, Makes and Models: *Trucks, Medium, 4×2 (m.Lkw. (o) and m.Lkw. (S-Typ))* (*Note:* payload 3 tons unless indicated otherwise): Borgward 3 To.GW, B3000S (petrol and diesel). Bussing-NAG Burglöwe 25 (2½-ton). Citroen 45 (3½-ton). Fiat 626BL/NL/N/RNL. Ford BB (2½-ton), B3000G, V3000S, V3000S/III (G198TSST, G398TS, G917TST IIIA, G987T, G997TST IIIB). Hansa-Lloyd Bremen III (2½-ton) and Merkur IIID. Henschel (various). Krupp LD3H62, L3, 5M222/242. Magirus (KHD) L27a, L30a, Klöckner S3000. MAN E2, E3000 (also produced by Fross-Büssing). Matford 198TM, F01W, F19W. Mercedes-Benz L2000, L2500, LGF3000, L3000(S) (2- to 3-ton), L701 (Opel). OM Taurus N/B/C. Opel Blitz 3410, 4000 (2½-ton), 3600, 4200, 3,6-36S. Praga RN. Renault AHN (3½-ton). Spa 38R, 38RA. Tatra T27b, etc. *Bodywork:* various, open and closed; also tankers.

General Data: Most of the above models were commercial types. The 3000S or *S-Typ* models (Borgward, Magirus, Mercedes-Benz, Opel) were the result of the *Schell Programm* and were also produced as 4×4 (3000A or *A-Typ*). The French Citroen, (Mat)Ford and Renault were built for the German Army. In addition to the above the Germans (especially the *NSKK*) used large quantities of confiscated civilian and captured military medium trucks. Particularly 'popular' were 1937-40 models of the US Chevrolet, Ford, and Oldsmobile. Many vehicles were converted for producer gas (chiefly Imbert). In the early stages several of the above vehicle types were fitted with a standard soft-top *Wehrmacht* cab, instead of the civilian all-steel type, but in 1944 a standardised substitute was introduced for practically all types of trucks then in production. This *Ersatz* cab was known as *Einheitsfahrerhaus* and was made of wood and pressed cardboard. It continued in production on several German trucks of the immediate post-war period (1945-47). Some of the above 3-ton trucks were fitted with track bogies, replacing the rear wheels for improved off-road performance. These semi-track vehicles were known as *Maultier* (see also relevant section).

Vehicle shown (typical): Truck, Medium, 4×2, Cargo (Borgward 3 To.GW) (*mittlerer Lastkraftwagen, offen; m.Lkw. off. (o)*)

Technical Data:
Engine: Borgward L3500R 6-cylinder, I-I-W-F-, 3485 cc, 63 bhp @ 3000 rpm.
Transmission: 5F1R.
Brakes: mechanical.
Tyres: 7.50–20 (2DT).
Wheelbase: 3650 mm.
Overall l×w×h: approx. 6200×2280 ×2780 mm.
Weight: approx. 3000 kg. *Payload:* 3 tons.
Note: early type military pattern soft-top cab (also used on other trucks, incl. Büssing, Ford and Opel).

Truck, Medium, 4×2, Signals (Bussing-NAG Burglowe 25) 4-cyl., 65 bhp, 4F1R, wb 4350 mm, approx. 6600×2150×2600 mm. Commercial-type 2½-ton chassis, produced in 1938. Cab integral with house-type radio-transmitter body. IWM photo BU 5022.

Truck, Medium, 4×2, Cargo (Ford V3000S/G198TS) V-8-cyl., 95 bhp, 5F1R, wb 4013 mm, 6385×2250×2175 mm, 2540 kg. Shown with *Einheits* cab. Most had Ford steel cab. Some had flat-section front wings. 3·9-litre engine. 1941–45. Preceding G917T had 3·6 or 3·9-litre engine with 4F1R gearbox.

Truck, Medium, 4×2, House Type, Kfz.305 (Mercedes-Benz L3000S/O66) 4-cyl. diesel, 75 bhp, 5F1R, wb 4250 mm, 6750×2250×2860 mm, 6690 kg (gross). Std house-type body used for wide range of roles (*Kfz.* Nos. between 305/1 and 305/137) on various 3-ton 4×2 and 4×4 chassis, notably Opel.

Truck, Medium, 4×2, Fuel Tanker Kfz.385 (Opel Blitz 3,6-36S) 6-cyl., 68 bhp, 5F1R, wb 3600 mm, approx. 6700×2265 ×2300 mm, 5800 kg (gross). Military chassis, from 1944 also produced by Daimler-Benz AG (Model L701). Various body types, incl. *Kfz.305*, usually with closed cab.

Truck, Medium, 4×2, Aircraft Maintenance (Borgward B3000S/O) 6-cyl., 78 bhp, 5F1R, wb 3700 mm. Also with diesel engine (6-cyl., 75 bhp; Model B3000S/D). Produced during 1939–44. Truck shown was used by *Luftwaffe* and fitted with crane for replacement of aircraft engines, etc.

Truck, Medium, 4×2, Fire Fighting (Klöckner-Deutz-Magirus S3000) Deutz F4M513 4-cyl. diesel, 80 bhp, 5F1R (OD top), wb 4200 mm. Standardised Type 15 *Tanklöschfahrzeug (TLF 15/43)* produced during 1942–45. Magirus 1500 l/min. pump at rear. Chassis introduced in 1941 as S330.

Truck, Medium, 4×2, Service Tower (Citroën 45) 6-cyl., 73 bhp, 4F1R, wb 4600 mm. Over 15,000 chassis produced during 1941–44, mainly for Germany. Payload 3500 kg. Also with Imbert gas producer. Usually cargo body but illustrated with telescoping platform for rocket servicing.

Truck, Medium, 4×2, Cargo (Renault AHN) 6-cyl., 75 bhp, 4F1R, wb 3730 mm, 6420×2350×2990 (2500) mm, 3540 kg. Renault 622B 4086-cc SV engine. Tyres 210–20 or 34×7. Payload 3500 kg. Produced during 1941–44, mainly for Germany. In 1945 continued as AHN2.

GERMANY

TRUCKS, MEDIUM
4×4

Types, Makes and Models: *Trucks, Medium, 4×4 (m.gl.Lkw. and m.Lkw. (A-Typ) (Note:* payload 3 tons unless indicated otherwise*):* Borgward B3000A (petrol and diesel). Fiat/Spa T40 (2½-ton, diesel). Magirus (KHD) Klöckner-Deutz A3000 (diesel). Mercedes-Benz LG65/2, LG2000 (2·6-ton), L3000A/066 (diesel). Opel Blitz 3,6-6700A (petrol). *Bodywork:* various, open and closed; also tankers and wreckers.

General Data: Like the British, the Germans seemed uninterested in the medium 4×4 truck during the 1930s. Both countries concentrated on the development of 6×4 six-wheelers. The advantage of the 6×4 were that it was more suitable for the mounting of specialist bodies and that it did not require the intricate front-wheel drive. Among the disadvantages were the number of wheels and tyres required. Once development of the 4×4 got under way, however, it was found that vehicles of this configuration were well suited for most transport needs. Opel became the main producer of this type in Germany and their Brandenburg/Havel plant, under the direction of the late Heinz Nordhoff (later boss of Volkswagen), turned out about 25,000 of the successful Opel *Allrad*. This model was very similar in design to the American Chevrolet 1½-ton 4×4. Like the Americans, the Germans seldom fitted large single tyres, although the British and Canadians proved without any doubt, first in North Africa, the value of such tyre equipment. In fact, Rommel once put out a directive that for reconnaissance work in the desert only (captured) Canadian and British trucks were to be used 'because ours stick in the sand too often'. In the German Army the *A-Typ* 3-ton 4×4 replaced various types of medium 4×2 and *I.E.Lkw.* (6×6) trucks. After the war Borgward, Klöckner-Humboldt-Deutz (Magirus-Deutz) and Daimler-Benz (Mercedes-Benz) re-introduced medium-type 4×4 trucks which were very similar to the war-time models. The Magirus-Deutz A3000 of the 1950s, for example, was almost identical to the *Wehrmacht*'s Klöckner-Deutz A3000, the main difference being its air-cooled engine.

Vehicle shown (typical): Truck, Medium, 4×4, Cargo (Opel Blitz 3,6-6700A) (*mittlerer (geländegangiger) Lastkraftwagen, A-Typ*) (*Luftwaffe*)

Technical Data:
Engine: Opel 3·6-litre 6-cylinder, I-I-W-F, 3626 cc, 68 bhp @ 3000 rpm.
Transmission: 5F1R×2.
Brakes: hydraulic.
Tyres: 190–20, 7.50–20 (2DT).
Wheelbase: 3450 mm.
Overall l×w×h: 6020×2265×2175 (cab) mm.
Weight: 2810 kg. *Payload:* 3290 kg.
Note: fitted out for carrying Air Force equipment. Also with house-type body (*Wehrmachts-Einheitskofferaufbau*).
A-Typ swb version of Opel 3,6-36S.

Truck, Medium, 4×4, Cargo (Borgward B3000A/D) 6-cyl. diesel, 75 bhp, 5F1R×1, wb 3700 mm, 6450×2300×2930 (2220) mm, 3375 kg.   190–20 (2DT) tyres.   Also with 78-bhp petrol engine (B3000A/O) and as 4×2 'S-Typ' (B3000S).   Also with flat-section front wings.

Truck, Medium, 4×4, Cargo (Klöckner-Deutz A330) Deutz F4M513 4-cyl. diesel, 80 bhp, 5F1R×2 (OD top), wb 3700 mm, 6100×2220×2900 (2330) mm, 3660 kg.   Payload on/off roads 3100/2700 kg.   Tyres 190–20.   Later redesignated A3000.   4941-cc water-cooled engine.

Truck, Medium, 4×4, Cargo (Mercedes-Benz L3000A/066) 4-cyl. diesel, 75 bhp, 5F1R×2, wb 3800 mm, 6255×2350 ×2600 mm (w/tilt), 4040 kg.   190–20 (2DT) tyres (= 7.50–20). Also produced as 4×2 (L3000S).   Soft-top (shown) or all-steel cab.   Built in Gaggenau and Mannheim.

Truck, Medium, 4×4, House Type, Kfz.305 (Opel Blitz 3,6-6700A) 6-cyl., 68 bhp, 5F1R×2, wb 3450 mm, 6100 ×2200×2900 mm approx.   Standard multi-purpose house type van body (3500×2010×1775 mm, 845 kg (empty)). Designated LC-Koffer (mot.) Kfz.305/.. and Kfz.305K, Ausf.B. 1941.

GERMANY

TRUCKS, MEDIUM 6×4 and 6×6

Types, Makes and Models: *Trucks, Medium, 6×4 or 6×6 (m.gl.Lkw. (o)) (Note:* 6×4 unless indicated otherwise*):* Büssing-NAG G31K (V8), KD, III GL6 (or: 3 GL6). Henschel 33D, 33D1, 33D2, 33G1. Krupp L3H 163. Magirus (KHD) 33G1 (Henschel licence). Mercedes-Benz LG65/3 (6×6), LG3000/LG63, LG4000 (6×6). *Bodywork:* various, open and closed; also tankers, fire appliances, searchlight, etc.

General Data: Most of the above chassis were conventional 6×4 types, built to a common general specification and rather similar to contemporary British medium six-wheelers. Like the British types they were also available to commercial users (with subsidy) and for export (Spain, Turkey, etc.), mainly in order to boost production and to reduce unit costs. In Germany these trucks were also used by other government departments including the *Deutsche Reichsbahn* (railways). Relatively large numbers were produced and a wide variety of standard and special bodies was used, incl. house types for signals and other specialist roles. The 6×6 models of Daimler-Benz were rather unusual, featuring coil-spring independent suspension, etc. Their LG65/3 was in fact a 6×6 version of the LG65/2 4×4 with which it had many components in common. It had, however, a 6-cyl. engine, like the larger LG65/4 which was the 8×8 version of this interesting range of diesel-engined 'pseudo-military' chassis. 200 of the LG65/3 were supplied to the Greek Army. The Büssing-NAG G31K was interesting in that it had a twin-carburettor 150-bhp V8 engine (BN model L8V). This chassis was produced during 1935–36, mainly for fire-fighting trucks. The same engine was also used for the Büssing-NAG model GS 8×8 armoured-car chassis, although during the later war years they were uprated to 180 bhp (larger cyl. bore, different carburation). In addition, 200 of these V8 power units were supplied to Hungary, for unknown purposes.

Vehicle shown (typical): Truck, Medium, 6×4, Printing, Kfz.72 (Büssing-NAG III GL6) (*Druckereikraftwagen (Kfz.72) mit Fahrgestell des m.gl.Lkw. (o)*)

Technical Data:
Engine: BN C4 6-cylinder, I-I-W-F, 9348 cc, 90 bhp @ 1250 rpm.
Transmission: 5F1R.
Brakes: air.
Tyres: 7.50–20 (4DT).
Wheelbase: 4525 mm, BC 1250 mm.
Overall l×w×h: 7100×2300×2800 mm.
Weight: 9500 kg (gross).
Note: approx. 300 chassis produced, 1933.

Truck, Medium, 6×4, Engineers I (Henschel 33G1) 6-cyl. diesel, 100 bhp, 5F1R, wb 4300 (BC 1100) mm, approx. 7400×2250×3200 mm, 6800 kg. Produced by Henschel, 1934–43, and Magirus (KHD), 1938–40. 7.25–20 (4DT) tyres. Knorr air brakes. Model 33D1 had petrol engine.

Truck, Medium, 6×4, Cargo (Krupp L3H 163) 6-cyl., 110 bhp, 4F1R×2, wb 4200 (BC 1100) mm, approx. 7400×2500 ×3200, 5695 kg. Tyres 7.25–20 (4DT). Produced until 1938. Also used as searchlight truck, telephone truck, etc. Licence-produced in Hungary (Rába).

Truck, Medium, 6×4, Radio Antenna, Kfz.301 (Mercedes-Benz LG3000/LG63) 6-cyl. diesel, 95 bhp, 5F1R×2, wb 4425 (BC 1050) mm, approx. 6500×2500×2800 mm, 8450 kg (gross). 7434 chassis produced, 1934–38. Fitted with winch. Also with other body types.

Truck, Medium, 6×6, Cargo (Mercedes-Benz LG65/3) 6-cyl. diesel, 80 bhp, 4F1R×2, wb 3140 (BC 1100) mm. Chassis weight 3500 kg. Coil-spring ifs and irs. Four lockable differentials (one in transfer case). Also as 4×4 (4-cyl., LG65/2) and 8×8 (6-cyl., LG65/4).

GERMANY

TRUCKS, HEAVY
4×2 and 4×4
and BUSES, 4×2

Types, Makes and Models: *Trucks, Heavy, 4×2 (s.Lkw. (o), s.Lkw. (S-Typ)):* Alfa-Romeo 800RE (6½-ton). Berliet GDR28F (6-ton). Büssing-NAG 500S, 4500S (4½-ton), 654 (6½-ton). Fiat 666N, 666NM (6-ton). Henschel 6J (6½-ton). Klöckner-Deutz S4500, GS145 (4½-ton). Krupp LD6, 5N242 (6½-ton). Lancia 3RO (6½-ton). Latil (ex-French Army). MAN F4 (6½-ton), ML4500S (4½-ton). Matford F917WS, F997WS. Mercedes-Benz L3750 (4-ton), L4500S (4½-ton), L5000 (5-ton), L6500 (6½-ton). ÖAF ML4500S (4½-ton). Renault AGR (4½-ton), AHR (5-ton), AGK (6-ton). Saurer 4 BTDVS, BT4500 (4½-ton), 4BT (5-ton. Skoda 706 (7-ton). Vomag 6LR652, 6LR647 (6½-ton), etc. *Bodywork:* various, open and closed; also tankers, fire appliances, wreckers, etc.
Trucks, Heavy, 4×4 (s.Lkw. (A-Typ)): Büssing-NAG 454, 500A, 4500A (4½-ton), 454 (6½-ton). MAN ML4500A (4½-ton). Mercedes-Benz L4500A (4½-ton). ÖAF ML4500A (4½-ton).
Bus, Light, 4×2 (up to 15 seats) (l.Kom. (o)), Bus, Medium, 4×2 (up to 30 seats) (m.Kom. (o)), Bus, Heavy, 4×2 (over 30 seats) (s.Kom. (o)): Ford, Henschel, Magirus, MAN, Mercedes-Benz, Opel, Vomag, etc.

General Data: Most of the 4×2 trucks listed above were basically commercial types or *S-Typ* versions of the *Schell Programm*. The others (and many more which are not listed) were captured or taken over from other armies. The ÖAF models were Austrian-built replicas of the corresponding MAN models. Buses were also basically commercial types and large numbers of captured/confiscated buses and coaches were also used, mainly for transporting troops. Some early-type buses were employed for Signals roles. The designation *Kom* was the standardised abbreviation of *Kraftomnibus*. The Opel Blitz 3,6-47S, known as *Wehrmachts-Omnibus*, was used as 26-str bus but could be easily converted into an ambulance or command vehicle (see also General Data on page 297). The Italian types listed were taken over from the Italian forces.

Vehicle shown (typical): Truck, Heavy, 4×4, Cargo (Mercedes-Benz L4500A) (*schwerer geländegängiger Lastkraftwagen, offen; s.Lkw. (A-Typ)*)

Technical Data:
Engine: Daimler-Benz OM67/4 diesel 6-cylinder, l-l-W-F, 7274 cc, 112 bhp @ 2250 rpm.
Transmission: 5F1R×2.
Brakes: hydraulic front, air rear.
Tyres: 270–20 (2DT).
Wheelbase: 4600 mm.
Overall l×w×h: 7860×2350×3345 mm.
Weight: 5715 kg.
Note: S-Typ same but 4×2, without two-speed transfer; height 3215 mm, wt 4930 kg.

314

Truck, Heavy, 4×4, Cargo (Bussing-NAG 4500A-1) 6-cyl. diesel, 105 bhp, 5F1R×1, wb 4800 mm, 8165×2350×3100 mm, 5900 kg. Also as 4×2 (4500S-1) and with other body types. Produced 1941–45, superseding 1940–41 500A (and 500S) which differed in detail.

Truck, Heavy, 4×4, Cargo (MAN ML4500S) 6-cyl. diesel, 110 bhp, 5F1R (OD top), wb 4600 mm, 7500×2350×3350 (2610) mm, 5050 kg. Payload 4500 kg. Produced 1940–46. Also as A-Typ (ML4500A, 4×4). During 1941–44 both were co-produced by ÖAF in Vienna, Austria.

Truck, Heavy, 4×2, Cargo (Henschel 6J2) 6-cyl. diesel, 125 bhp, 5F1R, wb 5000 mm, 8150×2500×2600 (cab) mm, 7100 kg. Dim. and wt apply to conventional truck. Picture shows conversion for use on railroads (twin-unit *Schienenzepp*). Similar conversion: Mercedes-Benz L4500A.

Truck, Heavy, 4×2, Cargo (Saurer BT 4500) 6-cyl. diesel, 105 bhp, 5F1R, wb 5000 m, 8340×2350×3470 mm, 5435 kg. Payload 4500 kg. Some 6000 produced (Models 4BT, 4BTDVS, BT4500) by the Austrian Saurerwerke, which also built Mercedes-Benz L4500 trucks.

Truck, Heavy, 4×2, Fire Fighting (Klöckner-Deutz S4500) 6-cyl. diesel, 120 bhp, 5F1R.   Std Type 25 *Löschfahrzeug (LF25).* Metz 2500 l/min. pump at rear.   1500-litre water tank.   Also on Mercedes-Benz L4500F chassis.   TFL25 (*Tanklöschfahrzeug*) had same pump but 4000-litre tank.

Truck, Heavy, 4×2, Cargo (Klöckner-Deutz GS145) Standardised 4½-ton chassis (*Gemeinschaftsfahrgestell GS145*) with ditto engine (*Gemeinschaftsdieselmotor GM145*) designed jointly by Henschel, Klöckner-Deutz (Magirus) and Saurer. Relatively few were produced.

Truck, Heavy, 4×2, Cargo (Renault AGK) 6-cyl., 85 bhp, 4F1R, wb 4000 mm, 7330×2400×2650 mm.   Payload 6000 kg. Introduced in 1938, produced until 1942.   Renault 441 5·9-litre engine. Basically commercial truck.   4½-ton AGR (with 4-litre engine) similar in general appearance.

Truck, Heavy, 4×2, Cargo (Renault AHR) 6-cyl., 75 bhp, 6F1R, wb 4440 mm, 7150×2250×3070 mm, 3700 kg.   Payload 5000 kg.   Produced during 1941–44, mainly for *Wehrmacht.*   Tyres 230–20.   Mechanical brakes with servo assistance.   Max. speed 60 km/h.

GERMANY

TRUCKS, HEAVY
6×4 and 6×6

Types, Makes and Models: *Trucks, Heavy, 6×4 (s.Lkw.(o), s.gl.Lkw.(o)):* Büssing-NAG 900. Faun L900D567 (9-ton). Skoda S (pre-1939: 6 ST6-L), SD, H (pre-1939: 6ST6-T), HD (4-ton). Tatra T81 (6½-ton), etc. *Trucks, Heavy, 6×6 (s.gl.Lkw. (o) and s.gl.Lkw.):* Büssing-NAG 900A (9-ton). Laffly S45TL (7-ton). Mercedes-Benz LG4000, LG68 (4- to 5-ton). Skoda 6S, 6SD (pre-1939: 6 STP 6-LD) (4-ton), 6V, 6VD (5-ton), 6K (pre-1939: 6VTP6-T) (11-ton tank transporter). Tatra T81, 6500/111 (6½-ton), 8000/111 (8-ton).
Bodywork: various, open, incl. tank transporters and other specialist bodies.

General Data: German-made heavy six-wheelers (Büssing-NAG, Faun) were used chiefly as transporters for light tanks. They were not numerous. Tanks were usually transported by rail, but for road transport of the heavier types there was a long low-loading trailer with twin-axle bogies at either end, towed by the heavy semi-track tractor (18-ton *Sd.Kfz.9*). Several types of Czech heavy six-wheeler trucks were taken over by the German Army and used for transport of engineer's and other heavy equipment. The Skoda S and 6V, built prior to 1940, were used by German *Pionier* units to haul bridging equipment. The latter had a forward-control tilt cab and was also used by the Swedish forces (with extra crew cab, later to be replaced by the Volvo-built TVC m/42B and C). The Tatra 6500/111 and 8000/111 superseded the T81 in 1942 and after the war continued in production for many years as the T111. They had a V-12-cylinder air-cooled diesel engine of almost 15-litre cubic capacity, a tubular backbone-type chassis, all-enclosed drive shafts, all-independent suspension and many other interesting features. One of the post-war developments of the T111 was the T141 heavy prime mover which had additional reduction gearing in the wheel hubs. Some French heavy 6×4 trucks were also used by the Germans, such as Bernard and Willème.

Vehicle shown: Truck, Heavy, 6×4, Tank Transporter (Faun L900D567) (*schwerer Lastkraftwagen (o)*).

Technical Data:
Engine: Deutz F6M517 diesel, I-I-W-F, 13540 cc, 150 bhp @ 1600 rpm.
Transmission: 5F1R.
Brakes: air.
Tyres: 13.50 or 14.00–20.
Wheelbase: 6175 mm, BC 1400 mm.
Overall l×w×h: approx. 10,400×2500 ×2600 mm.
Weight: 9200 kg.
Note: shown carrying *Pz.Kpfw.II* light tank. Payload 8800 kg.

Truck, Heavy, 6×4, Cargo (Büssing-NAG 900) 6-cyl. diesel, 150 bhp, 5F1R, wb 6425 (BC 1450) mm, 10400×2500×2600 mm, 8900 kg approx. This 9-ton *s.Lkw.(o)*, introduced in 1937, was used in small numbers mainly for carrying light tanks. 6×6 version experimental only.

Truck, Heavy, 6×4, Cargo (Skoda 6 ST6-T) 6-cyl., 100 bhp, 4F1R×2, wb 4125 (BC 1250) mm, 7100×2160×2650 mm, 7800 kg. Payload 4 tons, trailer wt 8 tons. Tyres 10.50-20. Fitted with 7-ton winch. Independent suspension. Max. speed 50 km/h. 1939 designations: H (petrol), HD (diesel).

Truck, Heavy, 6×4, Cargo (Tatra T81) V-8-cyl., 150 bhp, 4F1R×2, wb 4560 (BC 1220) mm, 9030×2500×2750 mm, 7800 kg. Payload 6½ tons. Tyres 9.75-20. Produced 1939-42, also as 6×6. engine was 14,726-cc (125×150 mm) ohv water-cooled Tatra T81H. Speed 65 km/h.

Truck, Heavy, 6×6, Cargo (Tatra 6500/111, 8000/111) V-12-cyl. diesel, 210 bhp, 4F1R×2, wb 4785 (BC 1220) mm, 8550×2500×3100 mm, 8350 kg. Also known as 6500A/III and 8000A/III. Difference mainly in rear springs. Ifs, irs. Air-cooled 14,825-cc engine. Speed 75 km/h. *Einheits* cab.

GERMANY

TRACTORS
WHEELED

Principal Types: Tractor, Wheeled, Light (*l.Rd.Schlp. (o)*), Tractor, Wheeled, Medium (*m.Rd.Schlp. (o)*), Tractor, Wheeled, Heavy (*s.Rd.Schlp. (o)*), Tractor, Wheeled, East (*OstRd.Schlp.*).

Makes and Models: Austro-Daimler ADZK (4×4), ADAZ (6×6). Deuliwag DA32 (4×2). Famo XL, XS (4×2). Faun ZR and ZRS (4×2). Fiat/Spa TM40 and Pavesi P4-100 (4×4). Hanomag RL20, SS20, SS55, SS60, HDS, R36, R38, SR38, SS100N, SS100LN (4×2). Kaelble (4×2, 6×4, 6×6). Laffly V15T (4×4), W15T, S35T, S45T (6×6). Lanz-Bulldog D7506, D8506, D9506 (4×2). Latil M2TL7, M2TL7 RR (Rail/Route), M7TLPS3, FTARH (4×4), M2TZ (6×6). Normag Miag LD20 (4×2). Skoda 175 *Radschlepper Ost* (4×4).

General Data: Wheeled tractors or prime movers were used by the Germans for a large variety of roles such as the towing of refueller trailers on airfields, low-loader trailers, rocket equipment, etc. Hanomag was the main supplier of heavy road tractors. This type was widely used in Germany for civilian purposes, mainly to haul heavy trailer combinations on motorways (*Autobahnen*). Many were fitted with gas-producers, usually Imbert, or compressed gas, due to fuel-oil shortage. The *Ostradschlepper* was designed by the Dr.Ing.h.c. F. Porsche KG specially for use on the Eastern front and was produced by Skoda in Czechoslovakia. Only some 200 were made and many of these were employed on the Western front. Their performance on metalled roads, especially under winter conditions, was substandard, if not dangerous, owing to the lack of traction of their large steel wheels. These wheels were reminiscent of German and Austrian heavy artillery tractors of World War I, two of which, incidentally, were still shown in official German military vehicle literature (D600) as late as 1936, namely the Krupp-Daimler Kw.19 gun mount or heavy tractor as *Sd.Kfz.1* and the Krupp-Daimler K.D.I heavy tractor of 1917 as *Sd.Kfz.2*. It is not likely, however, that these elderly machines saw active service in World War II, unlike some French World War I Latil tractors.

Vehicle shown (typical): Tractor, Wheeled, Heavy, 4×2 (Hanomag SS100) (*schwerer Radschlepper, 5 to (o)*)

Technical Data:
Engine: Hanomag D85 diesel, 6-cylinder, I-I-W-F, 8553 cc, 100 bhp @ 1500 rpm.
Transmission: 4F1R.
Brakes: air/hydraulic or hydraulic.
Tyres: 270–20 or 10.50–20 (2DT).
Wheelbase: 3000 mm.
Overall l×w×h: 5545×2460×2420 mm.
Weight: 6540 kg.
Note: produced 1936–45; continued after 1945 as ST100 (for French army).

Tractor, Wheeled, Heavy, 4×2 (Faun ZR) Deutz diesel 6-cyl., 150 bhp, 4F1R×2, wb 3600 mm, 6450×2440×2550 mm, 10,000 kg. 13,540-cc engine. Overdrive gear between engine and gearbox. 9.75–20 (2DT) tyres. Also used on railroads as Faun ZRS *schg.Zgm.* (*Schienenzepp*).

Tractor, Wheeled, East, 4×4 (Skoda 175) Porsche diesel 4-cyl., 80 bhp, 5F1R×1, wb 3000 mm, 5475×2300×2780 mm, 10,000 kg (gross). Also with petrol engine. Porsche design. 200 produced by Skoda in Pilsen, 1942, specially for Eastern front. Wheels 1500×300 mm front, 1500×400 mm rear.

Tractor, Wheeled, Medium, 4×2 (Hanomag SS55N) 4-cyl. diesel, 55 bhp, 4F1R, wb 2680 mm, 4450×2200×2350 mm, 4800 kg (incl. ballast). Tyres 9.00–20. Max. speed 33 km/h. Produced 1938–43. Special rear body for *Luftwaffe* (similar body on Kaelble Z6GN125).

Tractor, Wheeled, Heavy, 4×4 (Latil FTARH) 4-cyl., 68 bhp, 5F1R, wb 2900 mm, 5800×2400 mm approx. 6500 kg approx. Special *Ostfront* version of std TARH tractor, 1943. Towed load on roads up to 40 tons. During the war, Latil (of Suresnes, Seine, France) was controlled by Daimler-Benz AG.

GERMANY

SEMI-TRACKS
LIGHT

Basic Types, Makes and Models: *Motorcycle Tractor, Sd.Kfz.2 (kl.K.Krad):* NSU HK101 (1940–44). Derivatives: *Sd.Kfz.2/1* and *2/2*, Cable Layers.
1-ton Series, Sd.Kfz.10 (l.Zgkw. 1 to): Demag D6 (1937–38), D7 (1939–44). Also produced by: Adler, Büssing-NAG, M.W. Cottbus, MIAG, MRH, Phänomen, Saurer. Derivatives included SP (*2-cm Flak*) *Sd.Kfz.10/4* and the *Sd.Kfz.250/252/253* armoured series (some 12 versions on D7p chassis).
HK300 Series and LWS: Adler A1 (1938–39), A2, A3 and A3F (1939–40), HK301 (1941), LWS (1942–44). No quantity production.
3-ton Series, Sd.Kfz.11 (l.Zgkw. 3 to): Hansa-Lloyd (later Borgward) HL kl 5 (1937–38), HL kl 6 (1939–44). Also produced by: Adler, Auto-Union/Wanderer, Hanomag (H kl 6), Skoda. Derivatives included decontamination vehicles, an ambulance and the armoured *Sd.Kfz.251* series (some 25 versions on H kl 6p chassis).
HK600 Series: Hanomag HK601 (1939–40) and Demag HK605 (1941–42) (also projected armoured versions: HKp 602, HKp 603, HKp 606, HKp 607, by Hanomag and Demag). No quantity production. Projected replacements for 1- and 3-ton series.
Miscellaneous types: Mercedes-Benz LR75 (commercial, 1937–38). Citroën P14P (ex-French). Somua MCG and MCL6 (ex-French). Unic TU1 and P107 (ex-French). *Maultier* types (see separate section).

General Data: The 1-ton Series was initiated with the Demag DII 1 (with BMW engine) of 1934. The HK300 Series (Adler) was developed concurrently with the 1-ton Series but discontinued in favour of the Demag design. The LWS (*leichter Wehrmachtsschlepper*) was intended to replace the 1-ton series but this project was discontinued in 1944. Hanomag became chief producer of 3-ton Series and turned out 6170 units (incl. armoured versions). *Sd.Kfz.251* was produced after the war in substantially the same form in Czechoslovakia as OT810 (Tatra).

Vehicle shown (typical): Motorcycle Tractor, Semi-Track, Sd.Kfz.2 (*NSU HK101*) (*kleines Kettenkraftrad*)

Technical Data:
Engine: Opel Olympia 4-cylinder, I-I-W-C, 1478 cc, 36 bhp @ 3400 rpm.
Transmission: 3F1R $\times$ 2.
Brakes: mech. (controlled diff.).
Tyres: 3.50–19 (front only).
Tracks: 40 track links, 170-mm wide.
Overall l $\times$ w $\times$ h: 3000 $\times$ 1000 $\times$ 1200 mm.
Weight: 1235 kg. *Crew:* three.
Note: Sd.Kfz.2/1: kl.K.Krad für Feldfernkabel, 2/2: für schw. Fernkabel. 8345 built by NSU and Stoewer. Larger HK102 (*grosses Kettenkraftrad*) (2-litre Stump K20 engine, crew five) only as prototype.

Semi-Track, Light, 1-ton (Demag DII 3) BMW 319 1971-cc 6-cyl., 42 bhp, approx. 4750×1900×1750 mm. Produced in 1936 as one of the predecessors of the *Sd.Kfz.10*. Models DII 1 and 2 (1934/35) had 28-bhp 1479-cc BMW 315 engine.

Semi-Track, Light, 1-ton, Sd.Kfz.10 (Demag D7) Maybach HL42TRKM 6-cyl., 100 bhp, 7F3R, 4750×1840×1620 mm, 3400 kg. Used by *Panzer* troops. Crew eight. Model D6 had smaller 90-bhp NL38 engine (1937–38). Most numerous *Zgkw.* of World War II. Similar vehicle built by Volvo (Sweden).

Semi-Track, 2-cm AA, Sd.Kfz.10/4 (Demag D7) Maybach HL42TRKM 6-cyl., 100 bhp, 7F3R, 4750×2155×2000 mm, 4350 kg. *Selbstfahrlafette* (SP Mount), used by *Flak* units and *Luftwaffe*. Lightly armoured cab but no protection for crew. Gun: *2-cm Flak 30*.

Semi-Track, APC, Light, Sd.Kfz.250/1 (Demag D7p) Chassis as D7 except shorter track bogies. 4560×1945×1980 mm, 4600 kg. Crew six. Armament (not shown) incl. two mgs and two machine carbines. Produced by Demag and Evens & Pistor of Helsa. Chassis also by Büssing-NAG.

Semi-Track, Light, Personnel (Adler HK300/A1) Maybach HL25 4-cyl., 65 bhp, 6F1R. Cletrac type controlled diff. with final reduction gears. Engine capacity 2543 cc. Max. road speed 65 km/h. Produced as prototype in 1938/39. Front susp. with transversal leaf spring.

Semi-Track, Light, Personnel (Adler HK300/A2) Maybach HL28 4-cyl., 78 bhp, 6F1R. Engine capacity 2800 cc. Further prototype in HK300 series, produced in 1939. A3 generally similar but with HL25 engine (as A1). All had max. speed of 65 km/h.

Semi-Track, Light, Command (Adler HK300/A3F) Maybach HL28 4-cyl., 78 bhp, 6F1R. Produced in 1940 as semi-track command car in HK300 series of prototypes. Convertible body (*Cabrio-Limousine*). Max. speed of this model was almost 75 km/h.

Semi-Track, Light, Personnel and Prime Mover (Adler HK301) Maybach HL30 4-cyl., 95 bhp, 6F1R. Five produced in 1941. Further order for 50 was later cancelled and HK300 series project was discontinued. Engine had 3119-cc capacity. Max. speed 75–80 km/h.

Semi-Track, Light, 3-ton, Sd.Kfz.11 (Hanomag H kl 6)
Maybach HL42 TUKRM 6-cyl., 100 bhp, 4F1R×2, 5500×2000 ×2200 mm, 5550 kg. Std 3-ton prime mover, introduced in 1937. Models HL kl 2 to 5 (1934–36) had HL/Borgward engine. Later had wooden truck-type body.

Semi-Track, Armoured, Flamethrower, Sd.Kfz.251/16 (Hanomag H kl 6p) Chassis essentially as H kl 6. 5800×2100 ×2100 mm, 7000 kg. Equipped with three flamethrowers, one mg (mounted), etc. Crew five. Basically APC, used in over 20 different versions. Nearly 15,000 produced (all types).

Semi-Track, Prime Mover, Light (Somua MCG) 4-cyl., 60 bhp, 5F1R, 5300×1800×2600 mm, 6250 kg. German designation for this French vehicle: *'le. Zgkw. S(f) Typ MCG'* or (D50/12) *'Zgkw. S 307(f)'*. Somua MCL6 was *'Zgkw. S 303(f)'*. Both used by the *Panzer Tr.* Note hoisting gear.

Semi-Track, Prime Mover, Light (Unic P107) 4-cyl., 55 bhp, 5F1R, 4850×1800×2280 mm, 4000 kg. Another French production. German nomenclature: *'le. Zgkw. U(f) Typ P107'* or (D50/12) *'Zgkw. U304(f)'*. Unic TU1 (*Zgkw. U305(f)*) was smaller and very low (4200×1500×1310 mm, 2435 kg).

GERMANY
SEMI-TRACKS
MEDIUM and HEAVY

Basic Types, Makes and Models: *5-ton Series, Sd.Kfz.6 (m.Zgkw. 5 to):* Büssing-NAG BN I 4 (1934), BN I 5 (1935–36), BN I 7 (1936–37), BN I 8 (1938), BN 9 and 9b (1939–43). Also produced by Krauss-Maffei (KM I 4), Daimler-Benz (DB I 5, DB I 7, DB I 8), Praga. Derivatives included: artillery tractor *Sd.Kfz.6/1*, SP (37-mm AA) *Sd.Kfz.6/2*, snow-plough.
SWS (schwerer Wehrmachtsschlepper): Büssing-NAG and Tatra. Load-carrying and armoured versions (1943–45). Derivatives: SP mount for 37-mm Flak 43 and 10-barrel rocket launcher 42. About 400 built.
8-ton Series, Sd.Kfz.7 (m.Zgkw. 8 to): Krauss-Maffei KM m 7 (1933), KM m 8 (1934–35), KM m 9 (1936), KM m 10 (1936), KM m 11 (1937–45), Km m 12 (1939). Also produced by Büssing-NAG (BN m 8), Daimler-Benz (DB m 8), Hansa-Lloyd/Borgward (HL m 10, HL m 11), Saurer. Copied by Vauxhall (GB) and Breda (I). Derivatives included: SP mounts *Sd.Kfz.7/1* and *7/2*.
HK900 Series: Krauss-Maffei: HK901 (1940), HK904 (1941), HK905 (1941). Büssing-NAG: HKp 901, 902, 903. No quantity production.
12-ton Series, Sd.Kfz.8 (s.Zgkw. 12 to): Daimler-Benz DB s 7 (1934), DB s 8 (1936), DB 9 (1938–39), DB 10 (1939–44), DB 11 (1939). Also produced by: Krauss-Maffei, Krupp, Skoda.
HK 1600 Series: Daimler-Benz HK1601 (1941), HK1604. No quantity production.
18-ton Series, Sd.Kfz.9 (s.Zgkw. 18 to): Famo FM gr 1 (1936), F2 (1938), F3 (1939–44), F4 (1940). Also produced by Tatra. Derivatives: mobile cranes *Sd.Kfz.9/1* and *9/2*, and SP gun mounts.

General Data: The most important types (in terms of production and use) were the 5-ton BN 9 and 9b, the 8-ton KM m 11 and HL m 11, the 12-ton DB 9 and 10, and the 18-ton F3. In these designations, the capital letters indicate the mfr (BN—Büssing-NAG; KM—Krauss-Maffei, etc.), the small letters (until 1938/39) the size (kl—*klein*/small; l—*leicht*/light; m—*mittler*/medium; s—*schwer*/heavy; gr—*gross*/large), the figure indicates the 'mark'.

Vehicle shown (typical): Semi-Track, Medium, 5-ton, Sd.Kfz.6 (Büssing-NAG BN I 7) *(mittlerer Zugkraftwagen 5 to, mit Pi-Aufbau)*.

Technical Data:
Engine: Maybach NL38 TUK 6-cylinder, I-I-W-F, 3790 cc, 100 bhp @ 3000 rpm.
Transmission: 8F2R.
Brakes: air; hydraulic track brakes.
Tyres: 7.00 or 7.50–20 (front).
Tracks: bogie length 1270 mm.
Overall l×w×h: 6325×2260×2270 mm.
Weight: 7500 kg. *Crew:* 15.
Note: Sd.Kfz.6/1 was art. tractor with 11 seats and rear amn compt. (*A-Aufbau*) BN I 5, DB I 5 and DB I 7 similar in appearance.

Semi-Track, Medium, 5-ton, Sd.Kfz.6 (Praga) Maybach HL54 TUKRM 6-cyl., 115 bhp, 4F1R×2, 6325×2260×2500 mm, approx. 8500 kg. Czech-built model BN 9, featuring longer track bogies (2025 mm) than BN I 7 and different wings. BN I 8 and DB I 8 were similar in appearance.

Semi-Track, Medium, 8-ton, Sd.Kfz.7 (Krauss-Maffei KM m 8) Maybach HL 52 TU 6-cyl., 115 bhp, 4F1R×2, 6690×2360 ×2760 mm, 9500 kg. Early model, 1934–35. Used by Artillery to tow 10-cm gun and 15-cm how. Also produced by Bussing-NAG (BN m 8) and Daimler-Benz (DB m 8).

Semi-Track, Medium, 8-ton, Sd.Kfz.7 (Krauss-Maffei KM m 10) Maybach HL62 TU 6-cyl., 140 bhp, 4F1R×2, approx. 7175×2350×2613(2475) mm, 8940 kg. Front wings like BN I 7 (but spoke wheels). KM m 9 had similar appearance but HL57 TU 130-bhp engine. Vehicle tows 88-mm *Flak.*

Semi-Track, Medium, 8-ton, Sd.Kfz.7 (Krauss-Maffei KM m 11) Maybach HL62 TUK 6-cyl., 140 bhp, 4F1R×2, 6850 ×2400×2760 mm, 9750 kg. Final model in 8-ton series. Also by Borgward (HL m 11). Produced 1937–45. Bogies with three 'outer wheels'. Later also with truck-type bodywork (like SWS).

Semi-Track, Quadruple 2-cm AA, Sd.Kfz.7/1 (Krauss-Maffei KM m 11) Chassis as *Sd.Kfz.7* (KM m 11, HL m 11). 6800×2440×2950 mm, 11,020 kg. Known officially as *Sfl. (Sd.Kfz.7/1) mit 2 cm Flakvierling*, used by *Flak* units and *Luftwaffe*. Combat weight 11,430 kg.

Semi-Track, 37-mm AA, Sd.Kfz.7/2 (Krauss-Maffei KM m 11) Chassis as *Sd.Kfz.7/1*. 6800×2440×2950 mm, 11,550 kg. Officially: *Sfl. (Sd.Kfz.7/2) mit 3,7 cm Flak 36*. (sfl. = SP). Combat weight 11,870 kg. Also produced without cab and front armour.

Tractor, Semi-Track, Heavy (Bussing-NAG and Tatra SWS) Maybach HL42 TRKMS 6-cyl., 100 bhp, 4F1R×2, 6675×2500×2830 mm, 9500 kg. *Schwerer Wehrmachtsschlepper* superseded the 5-ton *Sd.Kfz.6*. Tatra later built modified version, T809, with 140-bhp V-12 engine.

Tractor, Semi-Track, Heavy, Gun Mount (Bussing-NAG SWS) Chassis as model shown on left. Armoured cab and front end. Used as mount for *37-mm Flak 43*. Also fully armoured version as *15-cm Panzerwerfer 42* (10-barrel rocket launcher).

Semi-Track, Heavy, 12-ton, Sd.Kfz.8 (Mercedes-Benz DB 9) Maybach HL85 TUKRM V-12-cyl., 185 bhp, 4F1R, 7100×2400×2800 mm, 12700 kg. DB s 8 similar in appearance but with Maybach DSO8 150-bhp V-12 engine. 11.25–20 front tyres. Max. speed 50 km/h. Crew 13.

Semi-Track, Heavy, 12-ton, Sd.Kfz.8 (Mercedes-Benz DB 10) Maybach HL85 TUKRM V-12-cyl., 185 bhp, 4F1R, 7100×2500×2800 mm, 13,100 kg. Most widely used version in 12-ton series. Differs from DB 9 chiefly in having disc-type front wheels. Towed 15-cm gun and 21-cm mortar.

Semi-Track, Heavy, 18-ton, Sd.Kfz.9 (Famo F2) Maybach HL98 TUK V-12-cyl., 230 bhp, 4F1R×2, 8325×2600×2850 mm, 15,470 kg. Heaviest of German semi-tracks. F3 was similar in appearance. Used mainly as tank retrievers (often in tandem). Few were built with art. prime-mover bodywork.

Semi-Track, Heavy, 6-ton Crane, Sd.Kfz.9/1 (Famo F3) Maybach HL108 TUKRM V-12-cyl., 250 bhp, 4F1R. Manually operated revolving crane by Bilstein. *Sd.Kfz.9/2* had 10-ton petrol-electric crane with all-round power traverse, counterweight-ballast boxes, etc. AFV is *Wespe* SP.

GERMANY

TRUCKS, SEMI-TRACK
'MAULTIER'

Types, Makes and Models: *Trucks, 2-ton, Semi-Track, Cargo (Gleisketten-Lkw., 2 t, offen (Maultier), Sd.Kfz.3 & 4)*: Ford V3000S/SSM. Klöckner (KHD/Magirus) S3000/SSM. Opel Blitz 3,6-36S/SSM, also armoured version (*Sd.Kfz.4/1*).
Truck, 4½-ton, Semi-Track, Cargo (Gleisketten-Lkw., 4½ t, offen (Maultier)): Mercedes-Benz L4500R.
Bodywork: usually open cargo body; few with house type. 4½-ton also with crane.

General Data: The *Maultier* (Mule) reportedly originated from a field modification by the SS division *Das Reich*. During the Russian campaign of 1941/42 a Ford 3-ton truck was converted into a semi-track by replacing the rear wheels with modified British Carden-Loyd track bogies. The Ford production version was a modification of the standard 3-ton Ford model V3000S and was supplied by the Ford factories in Cologne, Amsterdam and Asnières (France). The latter plant turned out 1000 units. The conversion comprised the moving forward of the rear axle and shortening the drive line, reinforcing the chassis and fitting twin Carden-Loyd type two-wheel bogies on either side with *Pz.* I or II tracks. The rear drum brakes were utilised for track steering, operated by two additional hand-brake levers in the cab. The Opel Blitz version was produced at Opel's Brandenburg/Havel plant. According to an engineer who worked there at the time the output was 'up to 120 daily if the material was at hand'. 300 Opels were fitted with an armoured body, *Sd.Kfz.4/1*, mounting the 10-barrel rocket launcher: *15-cm Panzerwerfer 42 (Sf)*, or *15-cm Zehnling Nb.W.42*. The Klockner with its low-speed diesel engine was the most successful of the 2-ton types. The Mercedes-Benz 4½-ton *Maultier* was introduced in 1943. Altogether well over 5000 2-ton *Maultier* trucks were produced until it was decided that the new SWS semi-track should take over the functions of both the *Maultier* and the 5-ton *Zgkw. Sd.Kfz.6*. The British and Australians also produced some very similar semi-tracks (see relevant sections).

Vehicle shown (typical): Truck, 2-ton, Semi-Track, Cargo, Sd.Kfz.3 (Opel Blitz 3,6-36S/SSM) (*Gleisketten-Lkw., 2 t, offen (Maultier)*).

Technical Data:
Engine: Opel 3·6-litre 6-cylinder, I-I-W-F, 3626 cc, 68 bhp @ 3120 rpm.
Transmission: 5F1R.
Brakes: hyd., also mech. track brakes.
Tyres: 190–20 (front).
Overall l × w × h: 6000 × 2280 × 2710 mm.
Weight: 3930 kg.
Note: also known as *Lkw. 2 to auf Gleisketten* or *Lkw. 3 t mit Gleiskette*. Experimental model shown.

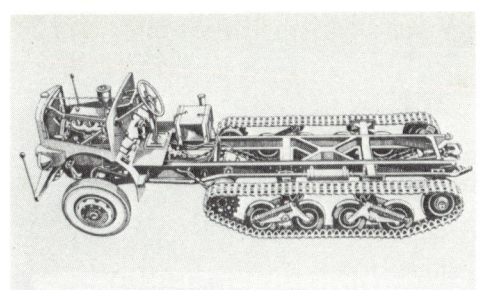

Truck, 2-ton, Semi-Track, Chassis (Klöckner-Humboldt-Deutz S3000/SSM) Deutz diesel 4-cyl., 70 bhp, 5F1R, complete vehicle 6120×2220×2800 mm, 4650 kg. Minimum speed 3 km/h. Based on KHD S3000 3-ton 4×2. Track bogies similar to Ford and Opel versions. 'SSM' stood for *SS Maultier*.

Truck, 2-ton, Semi-Track, Cargo (Opel Blitz 3,6-4200) 6-cyl., 73.5 bhp, 5F1R, wb 4200 mm, 6970×2265×2900 mm, approx. 3500 kg. Experimental conversion whereby rear axle remained in original position. Sprockets replaced the rear wheels and drove the tracks. Bogies on pivoting subframe.

Truck, 4½-ton, Semi-Track, Cargo (Mercedes-Benz L4500R) 6-cyl. diesel, 112 bhp, 5F1R, approx. 7900×2360×3230 mm. Conversion of standard Mercedes-Benz L4500S 4½-ton 4×2 (*S-Typ*) truck. *Panzer II* (light tank) track bogies. This unit fitted with the wooden *Einheits* cab. Produced 1943–44

Truck, 4½-ton, Semi-Track, Chassis (Mercedes-Benz L4500R) Chassis similar to that shown on left but all-steel cab and different track bogies with subframe suspended on four semi-elliptic springs. Track brakes were provided (for sharp turns) which operated in conjunction with conventional differential.

GERMANY

VEHICLES
FULL-TRACK and WHEEL-CUM-TRACK

Types, Makes and Models: *Prime Movers, Full-Track:* Hanomag. Kaelble. KHD/Magirus RSO/03 (*Raupenschlepper-Ost*). Praga T3 (from 1938), T5, T6R-P (1942), T6-SS (1944), T9 (from 1938). Sachsenberg (and others) *Land-Wasser-Schlepper.* Steyr RSO/01 (*Raupenschlepper-Ost*; also produced by Auto-Union, Gräf & Stift, KHD/Magirus).
Wheel-cum-Track Vehicles: Austro-Daimler ADMK, 1935–1938. Saurer RK7, RR7, RR7/2 (armoured, *Sd.Kfz.254*).

General Data: The *Raupenschlepper-Ost* or *RSO* was designed specifically for use on the Eastern front, but during the later part of the war they were also used in the West. The latest edition (RSO/03) was produced by KHD/Magirus and fitted with a 4-cylinder Deutz diesel engine and a soft-top cab. This model was continued in production for a while after the War as a half-track vehicle (shortened track bogies, conventional front axle under the cab) as *Waldschlepper RS1500* (forest tractor), presumably using left-over parts. The *Land-Wasser-Schlepper* was an amphibious tractor-cum-tug, designed principally for working in water. It was first ordered in 1935 by the Engineer Branch of the *Heereswaffenamt* and designed by Rheinmetall-Borsig in Düsseldorf, but was not put into service until about 1940. Co-operating in its production were the firms of Alkett GmbH in Spandau, Gebr. Sachsenberg AG in Dessau, and Hüttenwerke in Southofen. The vehicle could be discharged from a landing-craft and some were proof against small-arms fire. Towards the end of the War they were also used as ferries with buoyant decks slung between two tugs. The Maybach engine was fitted with an inertia-type starter (like the heavier Semi-Track *Zgkw.*) driven by a handle from the rear and clutched in by the driver through a control lever. The Austrian wheel-cum-track machines were not numerous. Numbers of captured artillery tractors were also used by the *Wehrmacht*, amongst them the French Renault R-602(f), and the Soviet Stalinets 65 (St(r)), CT3-601(r), and Stalin 607(r).

Vehicle shown (typical): Tractor, 1½-ton, Full-Track (Steyr RSO/01) (*Raupenschlepper-Ost*—Tracked Tractor, East)

Technical Data:
Engine: Steyr 1500A V-8-cylinder, V-I-A-F, 3517 cc, 70 bhp @ 2500 rpm (85 @ 3000).
Transmission: 4F1R.
Brakes: hyd., disc (Argus).
Track width: 340 mm (snowtracks 600 mm).
Overall l×w×h: 4425×1990 (w/snowtracks 2050)×2530 mm.
Weight: 3500 kg. *Payload:* 1500 kg.
Note: RSO/03 (KHD) had open cab. Chassis also as SP gun mount.

Tractor, Light, Full-Track, Artillery (Praga T3) 6-cyl., 77 bhp, 4F1R×2, 4100×1730×2300 mm, 4650 kg. 3-ton winch. Drawbar pull 3-ton. Crew six. Payload 600 kg. Max. speed 51 km/h. Used by *Panzer Tr.* as *leichter Raupenschlepper (Praga T3)*. Praga Art. Tractors were employed by the armies of many countries.

Tractor, Heavy, Full-Track, Artillery (Praga T9) V-8-cyl., 140 bhp, 4F1R×2, 5595×2450×2540 mm, 10,100 kg, 7½-ton winch. T6R-P and T6-SS similar in appearance but somewhat smaller and lighter (7500 kg), with 6-cyl. 110-bhp engine, and 3 track return rollers.

Tractor, Full-Track, Amphibious (Sachsenberg L-W-S) Maybach V-12-cyl. 300-bhp engine, mounted amidship. Twin propellers; rudder in slipstream of each. Weight 13 tons. Crew 20. Used as tractor on land and tug in (sheltered) water with pontoon bridging equipment. Bollard winch at rear.

Carriage, Wheel-cum-Track, Machine Gun (Austro-Daimler ADMK) 4-cyl., 18 bhp, 4F1R, wb (on wheels) 1770 mm, 3570×1500×1300 mm, on tracks 2775×1060×1185 mm, 1450 kg. Max. speed 45 km/h, on tracks 15 km/h. Later production had four-door open body. 334 produced.

GERMANY

COMBAT VEHICLES
WHEELED

Principal Types, Makes and Models: *Armoured Cars, 4×4:* Auto-Union/Horch *Sd.Kfz.* 221, 222, 223, 260, 261 (models 801, EG, v, rear-engined chassis I for std heavy car). Auto-Union/Horch *Sd.Kfz.247/I and II* (model 1c, front-engined chassis II for std heavy car).
Armoured Cars, 6×4: Büssing-NAG *Sd.Kfz.* 231, 232, 263 (chassis model A5P). Krupp *Sd.Kfz.* 247 (chassis model L2H43, L2H143). Magirus *Sd.Kfz.* 231, 232, 263 (chassis model M206p). Mercedes-Benz *Sd.Kfz.* 231, 232, 263 (chassis model G3a/P).
Armoured Cars, 8×8: Büssing-NAG *Sd.Kfz.* 231, 232, 233, 263 (chassis model GS). Büssing-NAG/Tatra *Sd.Kfz.234/1, 2, 3, 4.*

General Data: Development of the armoured cars which were used before and during World War II was started in 1931/32. The principal type then was the *schw.Pz.Spähwg. Sd.Kfz.231*. This was a heavy armoured reconnaissance car utilising a modified version of the contemporary 6×4 military truck chassis as produced by Büssing-NAG, Magirus and Daimler-Benz. Modifications of this basic type were the *Sd.Kfz.232* (fitted with radio) and *263* (command vehicle). These vehicles were intended as an interim measure until they could be replaced by a new 8×8 type, development of which began in 1934 by Büssing-NAG. This was a completely new model, designed entirely as an armoured combat vehicle. The chassis featured eight independently sprung wheels and a rear-mounted 180-bhp V-8 engine. With a crew of four and capable of moving in either direction at the same speed these vehicles, designated *Sd.Kfz.231*, formed the backbone of most armoured reconnaissance units. The radio version was the *Sd.Kfz.232*. The *Sd.Kfz.233* had a 7·5-cm gun, while the *263*, a command vehicle, had only one mg. From July, 1943, an improved 8×8, the *Sd.Kfz.234*, went into production, featuring a rear-mounted Tatra V-12 air-cooled 220-bhp diesel engine. There were four versions. As with all other types, several manufacturers were involved in their production. The 4×4 types were produced from 1935 and were based on the heavy Std car chassis I and II (*s.E.Pkw.*).

Vehicle shown (typical): Armoured Car, Light, 4×4, Radio, Sd.Kfz.223 (Auto-Union/Horch EGI) *(leichter Panzerspahwagen (Funk)* or *le.Pz.Spahwg (Fu) (Sd.Kfz.223) (Ausf.A))*

Technical Data:
Engine: AU/Horch 3·5-litre, V-8-cylinder, V-L-W-R, 3517 cc, 75 bhp @ 3600 rpm.
Transmission: 5F1R.
Brakes: mechanical.
Tyres: 210-18.
Wheelbase: 2800 mm.
Overall l×w×h: 4800×1950×1750 mm.
Combat weight: 4200 kg. *Crew:* three.
Note: vehicle shown in Chinese Army service. *Ausf.B* (1941–43) had 3·8-litre 90-bhp engine and hyd. brakes.

Armoured Car, Light, 4×4 (Manfred Weiss/Straussler 39M Csaba) 4-cyl. twin-ohc rear-mounted engine with dual ignition. 4520×2100×2270 mm, 5900 kg. Two- or four-wheel steering with steering box at each wheel. Ifs, irs. Built originally for Hungarian Army to Straussler design.

Armoured Car, Heavy, 6×4, Sd.Kfz.231 (Mercedes-Benz G3a/P) 6-cyl., 68 bhp, 4F1R×2, wb 2525 (BC 950) mm, 5600×1850×2250 mm, 5600 kg. Produced 1930–34. Sd.Kfz.232 6×4 similar but with radio and large overhead antenna (vehicle height 2870 mm). Superseded by 8×8.

Armoured Car, Heavy, 8×8, Sd.Kfz.232 (Büssing-NAG GS) V-8-cyl., 150 (later 180) bhp, 3F3R×2, wb 2750 (BC 1350, f. and r.) mm, 5850×2200×2900 mm, 7900 kg. 8-wh. steering. 1235 produced, 1935–42. Hull and final assembly by Deutsche Werke, Kiel, and F. Schichau, Elbing.

Armoured Car, Heavy, 8×8 (Austro-Daimler ADGZ) 6-cyl., 150 bhp, Voith fluid drive transm.+3-speed aux., wb overall 4750 mm, centre bogie 1050 mm, 6260×2160×2564 mm, 7950 kg. 27 produced 1935–37, another 25 in 1942. Chassis had much in common with 6×6 art. tractor ADAZ.

GERMANY

COMBAT VEHICLES
TRACKED

Principal types of tanks used by the *Deutsche Wehrmacht*, with tank-hunter and assault-gun (SP) variants, where applicable, in brackets, were: *Pz.I*, 1935–40; *Pz.II 'Luchs'*, 1935–42; *Pz.III*, 1938–45 (*SturmG III*, 1942–44); *Pz.IV*, 1938–45 (*JagdPz.IV*, 1942–45); *Panther*, 1943–45 (*Jagdpanther*, 1944–45); *Tiger I*, 1942–45 (*Elefant*, 1943); *Tiger II*, 1944–45 (*Jagdtiger*, 1944). Notable among foreign types in German service was the Czech *Pz.38(t)*, 1937–42 (*Hetzer*, 1944). In addition there were small demolition vehicles, SPs on captured French tank chassis, etc. Only a few typical examples are shown here, including the experimental 'Maus' which could be termed a mobile fort. Some German AFVs of WWII were kept in service long after the war, notably in Finland, Norway and Syria.

Vehicle, Demolition, Sd.Kfz.302 (Zündapp 'Goliath') Powered by electric motor, radio-controlled (or, as *Sd.Kfz.303*, with motorcycle engine and wire-guided). Carried 75 kg of explosives. Shown with 2-wh. transporter. Several larger types existed, produced by Borgward and NSU (*'Springer'*).

Tank, Light, Pz.Kpfw.38(t) (CKD/Praga TNHP) Praga EPA 6-cyl., 125 bhp, Praga-Wilson pre-selective 5F1R, 4900×2060 ×2370 mm, 10·5 tons. Czech tank, widely used by *Wehrmacht*. Chassis also employed for various SP guns and, with air-cooled Tatra T103 engine, as prime mover.

Tank, Medium, Sd.Kfz.141, Pz.Kpfw.III (various mfrs) Maybach HL 120 TRM V-12-cyl., 300 bhp, Variorex pre-selective 10F4R, 5400×2910×2435 mm, 20·3 tons. *Ausf. G* (with tropical equipment) shown. 5-cm gun. 5644 *Pz.III*s were produced by Alkett, Daimler-Benz and five other firms.

Tank, Medium, Sd.Kfz.161/2, Pz.Kpfw.IV (various mfrs)
Maybach HL120 TRM V-12-cyl., 300 bhp, ZF 6F1R, 7020 (w/o gun 5930)×2880×2680 mm, 25 tons. *Ausf.H* with long 7·5-cm gun shown, minus 'skirt armour'. 10 other versions (*Ausf.A-J*). Just over 8200 produced by Krupp, Vomag and other firms.

Tank, Heavy, Sd.Kfz.171, Pz.Kpfw.V, 'Panther' (various mfrs)
Maybach HL210P30 V-12-cyl., 650 bhp, ZF 7F1R, 8860 (w/o gun 6880)×3430×2950 mm, 44 tons. *Ausf.D* shown. Over 5500 *'Panthers'* (3 versions) were built by MAN and others, all with 7·5-cm gun. There was also an ARV variant ('*Bergepanther*').

Tank, Heavy, Sd.Kfz.181, Pz.Kpfw.VI, 'Tiger I' (Henschel)
Maybach HL210P45 V-12-cyl., 650/700 bhp, Olvar preselective 8F4R, 8240 (w/o gun 6200)×3730(3150)×2880 mm, 56 tons. 8·8-cm gun. 1348 built, from 1942, as well as 485 of the improved *'Tiger II'* (*'Konigstiger'*).

Tank, Ultra-Heavy, Pz.Kpfw. 'Maus' (Porsche 205) Daimler-Benz MB509 V-12-cyl., 1200 bhp, transmission/final drive by electric motors, 10,080 (w/o gun 9030)×3670×3660 mm, 188 tons. Largest German tank. Two prototypes only, one with MB517 diesel engine. Submersible for river crossing. 1944/45.

ITALY

Like Germany, Italy had started building up its war machine long before the outbreak of World War II. In fact, for the Italians war had started earlier than for the other main powers involved. In October, 1935, Italy had invaded Ethiopia, from 1936–39 Italian (as well as German) troops were involved in the Spanish Civil War, in April, 1939, they occupied Albania, and in the following year they entered the war in Libya.

Italy's Fascist government collapsed in 1943 and from then until 1945 a large part of Italy was under German control. During this time the Italian automotive industry was engaged on production for Germany and the existing military equipment was taken over by the *Wehrmacht*. This explains why numerous Italian motor vehicles, chiefly trucks, gun tractors and AFVs, officially entered service with the *Wehrmacht*. In some cases their quantities justified the translating, printing and distribution of so-called *D-Vorschrifte* (manuals, parts list, etc.) for use by German personnel. This had applied, earlier, to certain other equipment of foreign origin, which had been captured in sufficient quantity and taken into German service.

By far the largest supplier of automotive equipment and aircraft in Italy was, and still is, the Fiat concern, which included several subsidiaries, wholly or partly owned, such as OM, Spa and others. Not that it helped Germany much. It is on record that Fiat did everything it could on the one hand to produce as little as possible and on the other to keep its labour force from deportation to the *Reich*. From September, 1943 when Hitler's troops occupied Turin, until April, 1945, the Germans tried everything to make full use of Fiat's production facilities, only to be met with sabotage, strikes and other forms of collective resistance. Instead of the 180 aircraft which Fiat were supposed to deliver every month, the maximum during the first months of occupation was 18. In 1945 no aircraft were delivered at all. Aircraft engines—300 (later 90 or fewer) instead of the 1500 required each month. Trucks— during the first three months of 1945 the daily average was only ten, five of which disappeared 'in a mysterious way'. As an early result of all this organised trouble and corruption, Berlin had ordered in the beginning of 1944 that all Fiat tooling and machinery be transferred to Germany. The reaction at Fiat was a general strike which soon spread over a large part of Northern Italy. Fiat stayed put.

Under Mussolini, incidentally, Fiat's war production efforts had not been outstanding either. Possibly this explains, in part, why Italy's fighting vehicle force was comparatively weak, especially the armoured variety, although in many cases their design was excellent.

Vehicle Types

The following main types of motor vehicles were used:

Motorcycles: light and heavy (solo and w/sidecar) and motor-
 tricycles.
Cars: standard commercial types and field cars (open, mil. pattern).
Trucks: light, medium and heavy.
Std. Trucks: light (*Autocarretta*), medium (*Autocarro Unificato
 Medio*) and heavy (*Autocarro Unificato Pesante*).
Tractors/Prime Movers: light, medium and heavy.
AFVs: armoured cars, tanks, and SPs.

The following general specifications were laid down for medium and heavy trucks to be standardised for military use:

Autocarro Unificato (Truck, standardised)	*Medio* (Medium)		*Pesante* (Heavy)
Gross vehicle weight	6500	kg	12,000
Payload, not less than	3000	kg	6000
Engine type	Petrol or Diesel		Diesel
Overall vehicle width, maximum	2340	mm	2350
Body, internal dimensions	4000×2000	mm	4750×2200
Ground clearance, minimum	200	mm	200
Turning radius, minimum	7000	mm	7000
Max. road speed not less than	60	km/h	45*

*38 km/h with 12-ton trailer.

Trucks of the standard types were produced by Alfa-Romeo, Bianchi, Fiat, Isotta-Fraschini, Lancia and OM.

OM Autocarretta 36DMs on parade

Many of the German *Wehrmacht*'s vehicles were of Italian origin
(IWM photo NA25133)

Italian Vehicle Nomenclature

The following Italian designations are relevant:

Motorcycle	*Motociclo*
Passenger Car	*Autovettura* (mil. type Tourer: *Torpedo Militare*)
Ambulance	*Autoambulanza*
Light Truck	*Autocarretta, Autocarro Leggero (CL)*
Medium Truck	*Autocarro Medio (CM)*
Heavy Truck	*Autocarro Pesante (CP)*
Light Tractor	*Trattore* Leggero (TL)*
Medium Tractor	*Trattore* Medio (TM)*
Heavy Tractor	*Trattore* Pesante (TP)*
Semi-Track	*Semicingolato*
Armoured Car	*Autoblinda (AB)*
Tank	*Carro Armato*
SP Gun, on truck chassis	*Autocannone*
on tank chassis	*Carro Semovente*
Chassis, complete	*Autotelaio*
Trailer	*Rimorchio*

* Also *trattrice*.

Military vehicle model designation symbols, where used, were simple and straightforward. Examples:

Light Tractor, 1937 model	TL37
Armoured Car, 1940 model	AB40
Tank, medium, 13-ton, 1940 model	M13/40

Colours

Italian military vehicles in their homeland were painted a shade of greyish green. For use overseas, vehicles were painted desert-sand yellow. Various camouflage patterns were used additionally.

Registration Numbers

Military vehicle registration numbers were generally carried in conventional fashion, i.e. on number-plates front and rear. The number was prefixed by two letters indicating the using arm, i.e. *RE* for Army, *RA* for Air Force.

Who's Who in the Italian Automotive Industry

The following manufacturers were the main suppliers of automotive equipment at the time:

Alfa Romeo	Alfa Romeo SpA, Milano.
Ansaldo-Fossati	Ansaldo-Fossati, Genova-Sestri.
Bianchi	Edoardo Bianchi-Moto Meccanica SpA, Milano.
Breda	Soc. Ernesto Breda, Milano.
Ceirano*	SA Giovanni Ceirano, Torino.
Fiat	Fiat SpA, Torino (main plant).
Isotta-Fraschini	Fabbrica Automobili Isotta-Fraschini, Milano.
Lancia	Lancia & Co., Fabbrica Automobili-Torino-SpA.
OM*	OM (Officine Meccaniche) SpA, Brescia & Milano.
Pavesi	Pavesi-Tolotti SA, Milano.
Spa*	Soc. Ligure Piemontese Automobili, Torino.

* Part of or absorbed by the Fiat combine.

ITALY

CARS and MOTORCYCLES

Makes and Models: *Cars, 4×2:* Alfa Romeo, Bianchi, Fiat, Lancia, etc.
Field Cars, 4×2: Alfa Romeo 6C2500. Bianchi S4, S6. Fiat 508 Mil., 508C Col., 508C Mil./1100, 518C 'Ardita 2000', 2800C Mil. Col.
Motorcycles, Solo: Benelli 250 cc, 493 cc. Bianchi 500 cc. Gilera 500 cc. Guzzi 500 cc and 'Alce' 500 cc. Sertum 250 cc, etc.
Motorcycles with Sidecar, 3×2: Gilera 'Marte' 500 cc. Guzzi 'Alce' 500 cc.
Motortricycles, 3×2: Benelli 500 cc. Guzzi 'Trialce' 500 cc.

General Data: The majority of cars used by the Italian forces were of Fiat manufacture, either in standard form or as *Torpedo Militare*. The latter were rather similar in general design to the German *Kübelsitzwagen*. Like the Germans, the Italians had designed and produced many pseudo-military type vehicles before the war. In Italy such vehicles were known as 'Colonial' vehicles. They were specially designed or adapted for rough use off the beaten track in overseas territories, notably Africa, and very little further modification was necessary to turn such types into military vehicles. The Alfa Romeo 6C2500 and most of the Fiat field cars were in this category, together with many types of trucks. Of the Italian motorcycles (*motociclo*) listed above, the Bianchi, Guzzi, Sertum and Gilera were specially designed for military use and the latter type was probably also produced by other firms, under licence. The Sertum was a 250-cc 7-bhp four-stroke with a good ground clearance and a max. speed of 75 km/h. Like most Italian machines it had swinging-fork rear suspension. The Guzzis were extensively used in North Africa and captured machines were popular with British despatch riders. The low-compression 500-cc engine was flexible, the ground clearance good and the spring frame effective. The engine had a single horizontal cylinder with overhead inlet and side exhaust valves and was integral with the four-speed gearbox. Motortricycles were used for a variety of purposes and were known as *mototriciclo*.

Vehicle shown (typical): Field Car, Light, 4-seater, 4×2 (Fiat 508C Mil./1100) (*Torpedo Militare*)

Technical Data:
Engine: Fiat 108C 4-cylinder, I-I-W-F, 1089 cc, 32 bhp @ 4400 rpm.
Transmission: 4F1R.
Brakes: hydraulic.
Tyres: 5.00–18.
Wheelbase: 2427 mm.
Overall l×w×h: 3615×1480×1700 mm.
Weight: 1152 kg (laden).
Note: produced 1939–45. Superseded 508C Col. Based on 'Balilla' car.

Field Car, Light, 4-seater, 4×2 (Bianchi S4) 100 cars of this model were supplied by Bianchi to the Italian War Ministry during 1938. Picture shows Mussolini touring battlefields of SE France during the early part of the war. Models S6 and S9 *Coloniale* were similar in appearance.

Field Car, Medium, 5-seater, 4×2 (Alfa Romeo 6C2500 Col.) 6-cyl., 87 bhp, 4F1R, wb 3100 mm, 1650 kg. 2443-cc (72 ×100 mm) twin-OHC engine with 7:1 CR. Front suspension with coil springs, rear suspension with torsion bars. Differential lock. Ground clearance 260 mm. Introduced in 1939.

Field Car, Heavy, 5-seater, 4×2 (Fiat 2800C Mil. Col.) 6-cyl., 85 bhp, 4F1R, wb 3000 mm, 4795×1884×1768 mm, 2420 kg (laden). Load capacity five passengers plus 70 kg of luggage. Produced 1939–41. Model 2800MC 2852-cc engine. 7.50–18 tyres. Max. speed 115 km/h.

Scooter, Motor, Airborne (Volugrafo) Single-cyl. two-stroke engine, built *en bloc* with clutch and gearbox. Exhaust through frame tubes. Folding handle-bars. Fuel tank under saddle. Dual tyres front and rear. Small pintle hook at rear. Made by Aermoto Volugrafo, Torino.

Motorcycle, Solo, Machine Gun (Guzzi 'Alce') 1-cyl., 13·2 bhp, 4F, wb 1455 mm, 2220×790×1065 mm, 180 kg (w/o machine gun, tripod, etc.). Tyres 3.50–19. 500-cc horizontal engine. Sprung frame. Ground clearance 210 mm. Max. speed 90 km/h. Also with pillion. 1939.

Motorcycle with Sidecar, Heavy, 3×2 (Gilera 'Marte') 1-cyl., 12 bhp, 4F, approx. 2300×1600×1000 mm, 295 kg. 498-cc (84×90 mm) side-valve engine. All wheels sprung. No differential. Could also be used as solo machine. Max. power 12 bhp @ 3800 rpm.

Motorcycle with Sidecar, Heavy, 3×1 (Guzzi 'Alce') 1-cyl., 13·2 bhp, 4F, wb 1455 mm, 2220×1575×1065 (min.) mm, 276 kg. Known as *Motocarrozzetta*. Standard military sidecar. Track width 1090 mm. Tyres 3.50–19. Note coil-spring sidecar wheel suspension. 1941.

Motortricycle, Machine Gun, 3×2 (Guzzi 'Trialce') 1-cyl., 13·2 bhp, 4F (w/reduction gearbox), wb 1880 mm, 2825×1240×1050 mm, 354 kg. Known as *Portamitraglia Pesante* (heavy machine-gun carrier). Track 1120 mm. Tyres 3.50–19. Max. speed about 73 km/h. 1941.

ITALY

TRUCKS, LIGHT
3×2, 4×2 and 4×4

Makes and Models: *3×2 types:* Benelli 500, Guzzi 'Trialce' and 500U. *4×2 types:* Fiat 508 Camioncino (0·35-ton, 1932–37). Fiat 508L Col. (0·6-ton, 1939–40). Fiat 508C Mil. (0·35-ton, 1939–45). Fiat 618 Mil. Col. (1·3-ton, 1934–37).
4×4 types: Fiat/Spa Autocarro AS37 and TL37 (1-ton, 1937–48). Fiat/Spa Autocarro Leggero CL39 (1·1-ton, 1940–45), CL39 Col. (1-ton, 1940–45). OM Autocarretta 32, 36P, 36M, 36DM, 37 (from 1932).

General Data: One of the most interesting Italian vehicles of this period was the OM *Autocarretta da Montagna*. It was first introduced in 1932 (Tipo 32) as a 4×4 light multi-purpose truck intended chiefly for operation in mountainous regions. The prototype had been produced by Ansaldo to the design of Ir. Cappa. The air-cooled diesel power unit proved very reliable and would stand up to hard service in extreme temperatures. These vehicles were built in Brescia and later also in Eritrea. British army personnel who used captured specimens spoke highly of them and one was taken to England for a thorough examination by the WVEE at Farnborough. Suspension was independent, front and rear, with transversal leaf springs above and below the differentials. The gearbox was mounted amidship and drove front and rear final drives directly. Cross-country performance was extremely good. It was superseded in 1937 by the slightly modified Tipo 37. In 1936 a pneumatic-tyred version was introduced (Tipo 36M), as well as two personnel carriers, viz. the Tipo 36P seating ten plus driver, and the Tipo 36DM seating seven plus driver and equipped with a pedestal-mounted AA mg. The Spa CL39 was in some respects a scaled-up version of the OM *Autocarretta*. The little Fiat 508 types (*Camioncino*) were based on the popular Fiat 1100 cars of the period. The front end of the truck versions was the same as that of the contemporary 1100 'Balilla' cars (rounded radiator grille on early models, V-shaped later). The 508C *Militare* was also used by the German Army, designated *1100 Mil*. It had wooden half-doors. All Italian military vehicles had right-hand drive.

Vehicle shown (typical): Truck, Light, 4×4, Cargo (OM Autocarretta 32)

Technical Data:
Engine: OM Autocarretta 32 4-cylinder, I-I-A-F, 1616 cc, 21 bhp @ 2400 rpm.
Transmission: 4F1R. Lockable diffs.
Brakes: mechanical.
Tyres: 670×120 semi-pneumatic.
Wheelbase: 2000 mm.
Overall l×w×h: 3800×1300×2150 mm.
Weight: 1615 kg.
Note: Tipo 37 was improved version; 36M had pneumatic tyres. All had four-wheel steering.

Truck, Light, 3×2, Cargo (Guzzi 500U) 1-cyl., 17·8 bhp, 4F, wb 2300 mm, 3580×1498×2025 (1165) mm, 620 kg. Payload 1000 kg. Standard *Motocarro 500 Unificato*, 1942. Tyres, rear 6.00–16, front 3.50–19 (shown) or 6.00–16. Similar type by Benelli. Several variants, incl. SP AA gun.

Truck, Light, 4×2, Cargo (Fiat 508C Mil.) 4-cyl., 28 bhp, 4F1R, wb 2427 mm, 4005×1510×1922 mm, 1170 kg. 6.00–16 tyres. Payload 350 kg. Later production had V-shaped radiator grille. Some had twin AA mg on fixed mount in rear body. Based on 1100 'Balilla' car.

Truck, Light, 4×4, Cargo (Fiat/Spa TL37) 4-cyl., 57 bhp, 5F1R×1, wb 2500 mm, 4700×2100×2120 mm, 3500 kg. 9.00–24 tyres. Payload 1 ton. '*Autocarro*' version of TL37 artillery tractor, produced 1937–48. Four-wheel steering. Similar: AS37 *Autocarro Sahariano*.

Truck, Light, 4×4, Cargo (Fiat/Spa CL39) 4-cyl., 25 bhp, 5F1R×1, wb 2300 mm, 3890×1500×2300 mm, 1590 kg. 7.00–18 solid tyres. Payload 1140 kg. 'Colonial' version had 210–18 pneumatic tyres (payload 1 ton, width 1536 mm, weight 2790 kg). Both produced 1940–45.

ITALY

TRUCKS
MEDIUM and HEAVY
4×2, 4×4, 6×4, 6×6

Makes and Models: *4×2 types:* Alfa Romeo 430 (AUM), 500DR, 800RE (AUP; 6½-ton, 1940-44). Bianchi 'Miles' (AUM). Ceirano 50 C. Fiat 626 BL, NL, N, RNL (AUM; 3-ton, 1940-45); 632NM (5-ton, 1933); 633N (5-ton, 1933-36); 634N (7·6-ton, 1933-39); 633NM (5-ton, 1935); 633GM (3½-ton, 1935-36); 633BM (5-ton, 1936-38); 665NM, NL (5-ton, 1942-44); 666N, 666NM-RA & RE (AUP; 6-ton, 1940-45). Fiat/Spa 36, 38R and 38RA (2½-ton, 1936-44). Isotta Fraschini D65 (AUM), D80 (AUP). Lancia RO NM Mil. (1933), ROB (1935), 3ROB, 3RO(N) (AUP, 1938-44), ESARO (AUM, 1943-44). OM Taurus N/B/C (AUM; 3-ton), Ursus (AUP).
4×4 types: Fiat/Spa T40 (2½-ton).
6×4 types: Breda 41, 52, 53. Fiat 611, 621PN (3·6-ton, 1934-39). Fiat/Spa Dovunque 33 (2-ton, 1933), Dovunque 35 (3-ton, 1936-48).
6×6 types: Fiat/Spa Dovunque 41 (6-ton, 1943-48).

General Data: The trucks listed above indicated with AUM or AUP were standardised by the Italian forces, although not to the extent where one model was produced by various manufacturers to exactly the same specification as was the case in Germany. AUM trucks (*Autocarro Unificato Medio*—standardised medium truck) had a payload capacity of at least 3 tons and a maximum speed of at least 60 km/h. They were fitted with either a petrol or diesel engine. AUP trucks (*Autocarro Unificato Pesante*—standardised heavy truck) carried a maximum load of at least 6 tons at a top speed of at least 45 km/h, or 38 km/h with a 12-ton trailer. These had a diesel engine. Most of these standard vehicles had a general purpose cargo body, others had tanker, workshop (*autofficina*), wrecker or other special bodywork. The heavy type was also used as a light tank carrier or as prime mover for tank carrier trailers. Fiat/Spa vehicles were produced in the Fiat-controlled Spa factory and were usually known as Spa (in 1948 Spa became a division of Fiat). Alfa Romeo 800RE and Fiat/Spa *Dovunque* 41 also appeared as semi-track trucks.

Vehicle shown (typical): Truck, 2½-ton, 4×2, Cargo (Fiat/Spa 38R) (Autocarro Spa 38R)

Technical Data:
Engine: Spa 18R 4-cylinder, I-L-W-F, 4053 cc, 55 bhp @ 2000 rpm.
Transmission: 4F1R.
Brakes: hydraulic.
Tyres: 32×6 (2DT).
Wheelbase: 3500 mm.
Overall l×w×h: 5783×2070×2615 mm.
Weight: 3360 kg.
Note: 38RA differed only in detail. Chassis also with several other body types (*autoambulanza, autofrigoriferi,* etc.).

Truck, 2½-ton, 4×2, Ambulance (Fiat/Spa 38R) 4-cyl., 55 bhp, 4F1R, wb 3500 mm, approx. 5800×2070×2500 mm. Official designation: *Autoambulanza Spa 38R*. Chassis was continued in production after the war, until 1948, as Tipo 38R/45.

Truck, 5-ton, 4×2, Cargo (Fiat 633NM) 4-cyl. diesel, 50 bhp, 4F1R, wb 3750 mm, 6880×2200×2940 mm, 5645 kg. 195×720.5 (solid) tyres. Max. speed 29 km/h. 633GM similar but higher cab, 54-bhp engine, and extended front wings (as 633BM, see below). Note inertia starter unit.

Truck, 6½-ton, 4×2, Cargo (Lancia 3 RO N Mil.) 5-cyl. diesel, 93 bhp, 5F1R×2, wb 4300 mm, 7084×2350×3000 mm, 5500 kg. Tyres 270–20. RO had solid tyres. 3 RO B had petrol engine. *Autocarro Unificato Pesante*. Engine was two-stroke with reciprocating pistons (licence Junkers).

Truck, 5-ton, 4×2, Cargo (Fiat 633 BM) 4-cyl., 80 bhp, 4F1R, wb 3750 mm, 6820×2250×2980 mm, 4660 kg. Tyres 36×8 (2DT). Max. speed 52 km/h. GVW 10,800 kg, payload 5140 kg. 6·6-litre petrol engine. Typical of Italian army trucks of the 1930s. Note auxiliary lamps on sides of scuttle.

Truck, 6-ton, 4×2, Cargo (Alfa Romeo 800 RE) 6-cyl., diesel, 108 bhp, 4F1R×2, wb 3800 mm, 6840×2350×2850 mm, 5500 kg. Alfa's *Autocarro Unificato Pesante*, 1940–44. Medium type (Tipo 430) similar in appearance. Some had 'full' (rather than 'half') doors. 800RE also with half-track conversion.

Truck, 3-ton, 4×2, Cargo (OM Taurus N) 4-cyl. diesel engine. Taurus Tipo B had petrol engine. Tipo C was Colonial (or Tropical) version. Note half-doors which were feature of many Italian military trucks of the period (side-screens with transparent inserts were supplied). IWM photo NA 25024.

Truck, 3-ton, 4×2, Van (Bianchi Miles) Bianchi *Autocarro Unificato Medio* chassis with van-type body. Usually fitted with cargo body and tilt. Picture (IWM photo E12810) shows crew being stopped by British soldier during North African campaign.

Truck, 2½-ton, 4×4, Cargo (Fiat/Spa T40) 6-cyl. diesel, 108 bhp, 5F1R, wb 2600 mm, 5310×2185×2780 mm, 5650 kg. Basically Tipo TM40 tractor chassis with simplified cab (German *Einheits* type) and body for crew, amn and other loads. Four-wheel steering. Chassis produced 1941–48.

Truck, 3-ton, 4×2, Cargo (Fiat 626 BL) 6-cyl., 70 bhp, 5F1R, wb 3320 mm, 6220×2133×2700 mm, 3970 kg. Tipo 626 NL had 65-bhp diesel engine, 626 N ditto but swb. (B = *Benzin*, N = *Nafta*, diesel.) Widely used in Italian Army (RE) and Air Force (RA), also by German forces. 626 RNL was Bus chassis.

Truck, 5-ton, 4×2, Cargo (Fiat 665 NL, NM) 6-cyl. diesel, 110 bhp, 4F1R×2, wb 3760 mm, 7095×2670×3218 mm, 7350 kg. Tipo 665 NM had half-doors. Tipo 666 N and NM (RE and RA) were similar in appearance but had 6-ton payload. Tipo 666 NM had lockable differential. NM = Nafta, Militare.

Truck, 7½-ton, 4×2, Water Tank (Fiat 634N) 6-cyl. diesel, 80 bhp, 4F1R, wb 4300 mm, 7435×2400×3245 mm, GVW 14000 kg. Tyres 42×9. Model 355C 8355-cc engine. 1933–39. Cargo or special bodywork. Shown with water tank, used by Italian Air Force in Libya, 1941.

Van, Bus Type, 4×2 (Alfa Romeo 500 DR) *Autocarri Furgoni* on 6-cyl. Alfa Romeo bus chassis, used by *Genio Militare* (Engineers). Bodywork varied (single or double rear doors, different locations for side doors, etc.). Towed large van-type 4-wh. (4DT) trailers. 1940–43.

Truck, 2–3-ton, 6×4, Cargo (Fiat/Spa Dovunque 35) 4-cyl., 60 bhp, 4F1R×2, wb 3200 (BC 1000) mm, 5030×2070 ×2905 mm, 4400 kg. Model 18D 4053-cc (96×140 mm) SV engine with 4·9:1 CR and magneto ignition. 32×6 tyres. Transversal front springs. Hyd. brakes (air on early prod.). 1936–48.

Truck, 2–3-ton, 6×4, Signals (Fiat/Spa Dovunque 35) House-type van bodywork by Viberti. Chassis as shown on left. *'Dovunque'* means 'anywhere' (cross-country). *Dovunque 33* (1933–35) chassis was similar but had Model 122B 2953-cc 6-cyl. 46-bhp engine, lower payload and other detail differences.

Truck, 5–6-ton, 6×6, Cargo (Fiat/Spa Dovunque 41) 6-cyl. diesel, 115 bhp, 5F1R×2, wb 3900 (BC 1360) mm, 7040 ×2350×3150 mm, 9210 kg. Tyres 11.25–24. 6-ton winch. Lockable central and axle differentials. Produced 1943–48, also with soft-top artillery tractor bodywork.

Gun, 90-mm, Truck-Mounted (Breda 41/Ansaldo 90/55) Produced in 1942. Note heavy outriggers (shown in swung-up position in front of gun, two more at rear). Same armament also mounted on Lancia 3RO chassis. (IWM photos, Breda: NA4760; Lancia: STT7658.) 90/55-mm multi-purpose gun.

ITALY

TRACTORS

Makes and Models: Breda TP 32, 33, 40 (4×4), 61 (semi-track). Fiat 726, 727 SC (semi-track), OCI 708 CM (full-track, 1935). Fiat/Spa TL37 (4×4), TM40 (4×4), T41 (6×6). Lancia (4×4). Pavesi P4, P4-100, P4-110, TL31.

General Data: The Italians constructed a number of interesting four-wheeled artillery tractors, one of their main characteristics being a four-wheel steering system which, in conjunction with their short length, made very short turns possible. A short turning radius was desirable for Italian military tactical vehicles because of the mountainous character of much of their battle grounds. Some of the artillery tractor chassis, notably the TL37 and TM40, were also used with truck bodies. These vehicles, in spite of their comparatively low payload capacity, were of great value in mountain regions, where they supplemented the specially designed but much lighter OM *Autocarretta da Montagna*. The unique Pavesi articulated tractor was also used in fair numbers. It featured four large wheels (usually with single or dual solid rubber tyres), no road springs and an articulating frame. The Pavesi tractor was first introduced in 1924. It was designed by Ing. Pavesi and produced by the SA La Motomeccanica. 45 were produced in 1925 (mod. 25). Fiat started production in 1926 of 1000 units of an improved type (mod. 26) at Spa, which was already controlled by Fiat in the Fiat-Spa-Ceirano Consortium. Fiat/Spa also took part in the production of subsequent versions. They were also produced, under licence, in England, France, Hungary and Sweden, and in addition they were used by the Greek, Polish and Spanish armies. Following the Italian armistice in September, 1943, Germany, on taking over Northern Italy, got possession of large numbers of Italian military vehicles and these were soon pressed into service with *Wehrmacht* units fighting in Italy and elsewhere in Europe. In addition, the Italian industry produced many vehicles directly for the German forces, as the industries of other occupied countries did. It is believed that of the Breda 61 semi-track several hundred were produced for the German *Wehrmacht* in 1944.

Vehicle shown (typical): Tractor, Artillery, Light, 4×4 (Fiat/Spa TL37) (*Trattore Leggero* SPA mod. TL37)

Technical Data:
Engine: Spa 18TL 4-cylinder, I-L-W-F, 4053 cc, 57 bhp @ 2000 rpm.
Transmission: 5F1R×1.
Brakes: mechanical.
Tyres: 9.00–24 (also with solid tyres).
Wheelbase: 2500 mm.
Overall l×w×h: 4250×1830×2150 mm.
Weight: 3300 kg.   *Payload:* 800 kg.
Note: produced 1937–48. Also with truck body (see page 342). Shown with optional oversize tyres.

Tractor, Artillery, Heavy, 4×4 (Breda 40) *Trattore Pesante Campale Breda:* Developed from model 32 (1932), the lwb version of which was model 33 (1933). The 40 (1940) was also produced with open-type bodywork (no cargo body) and some had solid tyres. Towed load 5 tons.

Tractor, Artillery, Medium 4×4 (Fiat/Spa TM40) 6-cyl. diesel, 108 bhp, 5F1R, wb 2600 mm, approx. 4600×2185×2500 mm, 5500 kg. *Trattore Medio*, German nomenclature: *Radschlepper 110 PS Spa (i) TM40*. Also with solid tyres. Produced after the war with larger amn storage compartment.

Tractor, Artillery, Heavy, 4×4 (Fiat/Spa Pavesi P4-110) 4-cyl., 50 bhp, 4F1R, wb 2600 mm, 4170×2050 mm, 4000 kg. Rare version with pneumatic tyres. Originally had solid tyres (wheel diameter 1300 mm). Fitted with 4-ton winch. Towed load $3\frac{1}{4}$ tons. Payload $1\frac{1}{2}$ tons.

Tractor, Artillery, Semi-Track (Fiat 727SC) Designed in 1942/43 (exp.). Weight $3-3\frac{1}{2}$ tons. Towed load 6 tons. Breda *Tipo 61* of 1943/44 had similar German (Richter) bogies but was patterned on German 8-ton *Sd.Kfz.7* (KM m 11) with Breda T14 140-bhp engine and RHD.

ITALY

COMBAT VEHICLES

Principal Types, Makes and Models: *Wheeled, Unarmoured:* Breda 41/Ansaldo 90/55 (6×6 SP gun). Ceirano 50 CMA/75CK (4×2 SP gun). Lancia 3RO/Ansaldo 90/55 (4×2 SP gun).
Wheeled, Armoured: Fiat/Ansaldo 611 (6×4 armoured car). Fiat/Spa AB40, AB41, AB43 (4×4 armoured cars). Also reconnaissance car versions.
Tracked, Armoured: Fiat/Ansaldo L3/35 and L3/38 (mg carriers). Fiat/Ansaldo L6/40 (light tank, three versions). Fiat/Ansaldo/Fossati M11/39, M13/40, M14/41, M15/42 (medium tanks). Fiat/Ansaldo P40 (heavy tank). Fiat/Ansaldo 47/32, 75/18, 75/34, 90/53, 105/25, 149 (SP guns on tank chassis).

General Data: The truck-mounted medium and heavy guns (*Autocannone*) were quite numerous, especially the Ceirano. The Fiat/Ansaldo 611 6×4 armoured car was built in quantity from 1934 but was later replaced by the 4×4 types AB40 to AB43, which were based on a modified version of the Fiat/Spa four-wheel steer artillery tractor chassis, with the engine at the rear. Large open-top reconnaissance cars with various types of armament were also built on this chassis and used in Libya. An interesting Italian production was the 'Lince'. Built only in prototype form it was an almost exact copy of the British Daimler 'Dingo' scout car. On the whole, however, Italian armour in World War II was not very impressive. They took over a good deal of Czech and French armoured and other equipment after these countries were occupied by Germany. Moreover they used many 'vintage' type armoured cars and tanks which were of completely obsolete design. Later, when Italy finished as an Axis force most of what was left of their equipment was taken over by the Germans and the Allies. Large numbers of vehicles had meanwhile been lost in the North African campaign, either destroyed or captured by the Allied forces. The Italian light and medium tanks were transported, if and when required, on full-trailers towed by heavy trucks of the *Autocarro Unificato Pesante* class (Fiat, Lancia, etc.) but the light types could also be carried in the cargo body of these trucks.

Vehicle shown (typical): Carriage, Motor, 75-mm Howitzer, M13/40 (75/18) (Fiat-Ansaldo) (*Carro Semovente 75/18*)

Technical Data:
Engine: Fiat/Spa diesel 8-cylinder, V-I-W-R, 11,140 cc, 125 bhp @ 1900 rpm.
Transmission: 4F1R.
Steering: clutch brake.
Track width: 260 mm.
Overall l×w×h: 4910×2200×1850 mm.
Weight: 13·1 tons.
Note: basically medium 13/40 tank, used as SP artillery. Picture (US official) shows captured specimen at APG.

Carriage, Motor, 47-mm Gun, L40 (Fiat/Spa/Ansaldo)
4-cyl., 76 bhp, 4F1R×2, 3800×1860×1720 mm, 6·7 tons. *Carro Semovente* (SP gun) based on chassis of L6 light tank. 47/32 AT gun. Hull by Ansaldo, automotive components by Fiat/Spa. Crew 3. Introduced in 1941.

Tank, Medium, M15/42 (Fiat/Spa/Ansaldo/Fossati) V-8-cyl., 170 bhp, 4F1R, 5045×2230×2385 mm, 15·5 tons. *Carro Armato* with 47/40 gun and four mgs. Enlarged development of M13/40 and M14/41 medium tanks. Crew 4. Fiat/Spa 15TB 11·98-litre petrol engine. 1942.

Armoured Car, 4×4, AB41 (Fiat/Spa) 6-cyl., 100 bhp, 6F6R (OD top), wb 3245 mm, 5095×1930×2430 mm, 7·5 tons. Tyres 9.74–24. Coil-spring ifs/irs. Propeller shafts to each wheel from central lockable diff. 5-litre OHV engine. 20-mm gun (AB40: three 8-mm mgs; AB43: 47-mm gun).

Reconnaissance Car, 4×4, S43 (Fiat/Spa) *Camionetta 43 Sahariana*, derived from *Autoblinda 41* but with 110-bhp engine of *AB43*. 5200×1800×1490 mm, 4-tons. Armament varied (20-mm AA gun shown). Note large number of fuel cans and neat spare wheel stowage, recessed in front deck.

JAPAN

In 1930 the Japanese motor industry was in its infancy, most cars and trucks being locally assembled American makes. By 1940 the picture had changed drastically and Japan's annual truck output amounted to over 40,000. Today, of course, Japan is one of the world's leading motor vehicle producing countries.

In order to see Japan's automotive production effort during the 1930–45 period in its true perspective, it is necessary to look back at its preceding history.

Motor vehicles first appeared in Japan around the turn of the century and some 600 vehicles were imported during the next decade. Japanese-built vehicles were few and far between. The first Japanese steam car appeared in 1904, followed by the first successful petrol-engined car three years later. The first Japanese motor-car company, Kwaishinsha, was established in 1911, and the origins of the present Isuzu company can be traced back to this enterprise. Apart from these and other relatively early efforts not much progress was made until later, mostly due to the existing backward economic and social conditions.

In military circles, however, attitude and conditions were different. As a result of victory in the Russo-Japanese war in 1905, Japan came to be ranked among the big powers of the world, and caught up in a frenzy of militaristic aggression, set out to conquer Asia and the Pacific during the four decades that followed. The Imperial Army, which was the first official body in Japan to recognise the value of motor vehicles, established a motor-vehicle research and development organisation in 1907 and, during the same year, imported a French truck for tests and evaluation. In 1909 they imported some more European vehicles, namely a Schneider, a Thornycroft, a Laurin-Klement and a Benz-Gaggenau. In 1910 the Army ordered the Osaka Artillery Arsenal to produce some experimental trucks patterned on the original French vehicle. Three prototypes were completed and delivered in the following year, designated A, B and C. The OAA model C, a four-tonner, was approved following endurance tests in Manchuria, and standardised for military use. Series production was taken up by several manufacturers under supervision of the Army Command.

In 1914 Japan participated in World War I and motor transport was used during the attack on Tsingtao. 1918 saw the enactment of the Military Motor-vehicle Subsidy Law which gave aid to manufacturers and operators of vehicles suitable for military use. This law had great influence on the Japanese motor industry, which then developed virtually as an armaments industry.

The great earthquake of 1 September, 1923, gave a big stimulus to the use of motor transport. The Tokyo Electric Tram Bureau placed an order with the Ford Motor Co. in the USA for 1000 bus chassis, and many other American vehicles were imported by private dealers. It was from this time that US cars, because of their low price and quick delivery, came to be more popular than European makes. Henry Ford, in his characteristic style, promptly investigated the possibility of assembling his model T in Japan and, as a result, production started in Yokohama in 1925. General Motors and Chrysler followed in 1927. Thus, by means of foreign capital, mass production was introduced in Japan, and there were great opportunities. The domestic manufacturers, however, suffering from poor business and the earthquake, were forced to take a back seat at least as far as cars were concerned. During 1929 almost 30,000 American cars were assembled. By contrast, only 437 cars were produced by Japanese manufacturers. The Army did not like the American influx either, although they were forced to buy quantities of American vehicles for use in Manchuria (1931). In order to combat the American 'invasion' and the rather poor economic situation in Japan generally, the Ministry of Commerce and Industry decided to establish a motor industry on a sound footing and, in 1931, the Motor Industry Establishment Committee was inaugurated, which drew up a specification for a quasi-military standard medium truck and bus chassis. The eventual results were the Isuzu TX truck and BX bus.

About 1931, the Japanese Army began to feel the need for trucks with better cross-country performance and purchased some 6×4 sample models from Scammell and Thornycroft in England, and Tatra in Czechoslovakia. These vehicles were handed over to the Japanese truck industry with the order to study the basic designs and to develop a Japanese version. The resulting vehicle became known as the Isuzu Type 94. It was produced in various plants, for military and civilian (subsidised) use, and became the most widely used tactical truck of the Imperial Army. 'Type 94', incidentally, indicated the development year: 2594, or 1934 in Western chronology. Likewise, 'Type 1' indicates 2601, which is 1941, etc.

The Army, keen to get rid of the American producers, exercised their authority and, in 1936, this resulted in the enactment of the Motor-car Manufacturing Enterprise Law, which stipulated that the licence system should be applied to the production of vehicles. Also that at least one-half of the capital, company officials and shareholders should be Japanese and that the manufacturers should abide by orders issued by the Army. Isuzu (Jidosha Kogyo), Nissan and Toyota were permitted to operate under this law and the American 'Big Three' were forced to cut down their output. By 1939, as a result of the passing of the Foreign Exchange Control Law, production of US vehicles stopped altogether. The aforementioned companies probably took over the American plants and US-type trucks (Chevrolet, Ford) continued in production with only slight modifications.

By this time Japan's motor industry, by applying the lessons learned from the American 'Big Three', had laid a concrete foundation for its future prosperity. In 1941, the peak year before the war, total production of trucks and buses was more than 45,000, over half of which was for military purposes. Car production amounted to only 1065.

After the Pacific War broke out in December, 1941, the motor industry was placed under direct control of the Government and production for civilian purposes was further restricted. The emphasis was on mass production of standard-type military trucks, but output declined when Toyota and Nissan plants were ordered to increase production of aero engines. Moreover, repeated air raids and relocation of plants as well as shortage of materials accelerated the downward trend. In 1945 (Jan.–Sept.) only 6726 trucks were made; car production had dropped from 19 in 1944 to nil in 1945.

On the other hand, remarkable progress was made in the field of diesel engines, partly because of the need to economise on petrol. A family of standardised diesel engines of various types was developed for use in tanks, trucks and midget submarines.

In addition to their own vehicles, the Japanese Army used large numbers of American and European vehicles captured in their advance southward in 1941–42.

Japanese Ordnance Vehicle Nomenclature

Motorcycle	*Jidojitensha*
Sidecar	*Sokusha*
Motortricycle	*Sanrinsha*
Car (motorcar)	*Jidosha*
Command Car	*Shikisha*
Armoured Car	*Sokosha*
Reconnaissance Car	*Jidoteisha*
Truck	*Jidokasha*
Tracked Truck	*Kidokasha*
Tractor	*Keninsha*
Tracked Tractor	*Mugenkidokeninsha*
Artillery Tractor	*Kahokeninsha*
APC	*Sokoheisha*
Light Tank	*Keisensha*
Medium Tank	*Chusensha*
Heavy Tank	*Jusensha*
Amphibious Tank	*Suirikuryoyosensha*

Who's Who in the Japanese Automotive Industry

Most Japanese motor-vehicle manufacturing companies originally started as branches of companies engaged in other types of engineering, including several shipyards. In most cases they developed separately from the parent firm. Some of these manufacturers merged or found other means of co-operation, especially with regard to the production of military vehicles. This was common particularly during the 1930s when several standardised and/or subsidised types of military trucks and other military equipment had to be produced by more than one factory. These mergers and other agreements, on which little reliable information seems to exist, make Japan's automotive history rather complex.

The following summary is compiled from information supplied by a number of Japanese sources. Only the major military car and truck producers are included, in alphabetical order:

Chiyoda Trade name used by Tokyo Gas & Electric Co. (Tokyo Gas Denki Kogyo or TGE) for their vehicles from 1931. Chiyoda was the name of the residence of the Imperial Household where, in 1931, a TGE truck was taken into use.

DAT producer of small cars, originally known as Kwaishinsha (1912–25), then DAT Motor Co. (1925–26). From 1926 to 1930 only military truck production by DAT Auto Mfg Co. of Osaka. From 1933 Datsun cars were produced by Nissan in Yokohama. By this time the DAT military truck division had merged with the military-vehicle division of Ishikawajima to form Jidosha Kogyo. See also Isuzu.

Hino Originated in 1917 as a branch of Tokyo Gas & Electric Co. and from May 1942 developed separately.

Ishikawajima The Ishikawajima Dockyard & Engineering Co. built its first car in 1916/17 and subsequently produced Wolseley cars and trucks under licence, until 1927. Its military-vehicle division merged with that of DAT to form the Jidosha Kogyo Co., later to be joined by the military vehicle operations of TGE (Chiyoda). Trade names used were Sumida and Isuzu. Also produced armoured fighting vehicles.

Isuzu The name Isuzu (after the sacred river flowing within the grounds of the Grand Shrine of Ise) was originally awarded by the Ministry of Commerce and Industry for use on Ishikawajima-produced Type 94 standardised trucks. These trucks were also produced by TGE (Chiyoda Q). The military-vehicle divisions of Ishikawajima and DAT, which had merged to form Jidosha Kogyo, merged with that of TGE (Chiyoda) in 1937 to form Tokyo Jidosha Kogyo; the Isuzu name was used on all vehicles produced by this group. After the war the group was renamed Isuzu Motors Ltd.

Kurogane Trade name used for the light three- and four-wheeled products of the Nippon Nainenki Seiko Co., later to become the Tokyo Kurogane Motor Co.

Minsei Trade name of Nippon Diesel Engineering Co. From 1935 specialists in diesel engines and diesel-engined vehicles. Became Minsei Diesel, now part of Nissan.

Mitsubishi Started as Mitsubishi Kobe Dockyard. Produced first engine in 1916, first car in 1917/18. Became large-scale producer of military diesel engines and AFVs. Mitsubishi Heavy Industries Ltd was dissolved under the Deconcentration Law enacted after VJ Day and reorganised in a number of smaller companies.

Nissan Formed in Yokohama in December, 1933, continuing production of Datsun cars (taken over from DAT Auto Mfg Co. of Osaka). From 1937 produced US-type cars (Graham) and trucks (Federal).

Sumida Trade name, after river in Tokyo, used by Ishikawajima, from 1923

TGE Tokyo Gas & Electric Co. From 1931, TGE vehicles were known as Chiyoda, later Isuzu. See also Hino. TGE was formed in 1910.

Toyota Toyoda Automatic Loom Works of Kariya City started experiments with motor vehicles in 1933. Toyota Motor Co. was established in 1937 in Toyota City, near Nagoya, to become one of Japan's largest vehicle manufacturers. Probably took over the old GM assembly plant.

JAPAN
CARS, MOTORCYCLES and THREE-WHEELERS

Makes and Models: *Cars, 4×2, Tourer (phaeton):* Chiyoda HF, Nissan 70, Toyota AB and ABR.
Cars, 4×2, Sedan: Chevrolet, Chiyoda H, Ford, Nissan 70, Plymouth, Sumida H (or A) Limousine, Toyota AA, AC, AE and BA.
Cars, 4×4, Scout: Kurogane Type 95.
Cars, 4×4, Command: Isuzu Type 98, KIJI-A and B, Mitsubishi Fuso BX33, Sumida JC.
Cars, 6×4, Command: Chiyoda HS, Isuzu K, Sumida K93.
Motorcycles (solo and with sidecar): Kurogane, Sankyo.
Three-wheelers, 3×2: Iwasaki, Kurogane, etc.

General Data: Most cars used by the Japanese Army were of American origin or patterned on American designs. The main exception was the little Scout Car Type 95, some 4800 of which were built by Kurogane. Toyota and Rikuo produced prototypes for a similar small vehicle. Some four-wheel-drive command cars, petrol and diesel, were developed to replace the earlier 1933 pattern 6×4 types, but production was limited. Daimler-Benz and Tempo 4×4 cars were acquired from Germany for evaluation. Motorcycles were patterned on the American Harley-Davidson and had a high ground clearance. All types could be used with a sidecar, on which a light machine-gun could be mounted. Most common was the Type 97, which had a 24-bhp 1272-cc V-twin engine. The Japanese military police on the main islands used, in addition to motorcycles and other vehicles, a military tourer version of the 1936 Hudson Eight, and captured American 'Jeeps'. A captured Bantam 'Jeep' was copied by Toyota and five pilot models were produced. Too late for active service, this vehicle was eventually put into production after the war, in modified form, as model BJ 'Land-Cruiser'. Motortricycles, known as 'Sanrinsha', had been developed and produced by various manufacturers from about 1930, mainly as freight carriers, and were used by Army and Navy for various purposes. One of the main producers was Kurogane ('New Era' until 1937). There were two basic types, light and heavy.

Vehicle shown (typical): Car, Scout, 4×4, Type 95 (Kurogane 'Black Medal').

Technical Data:
Engine: 2-cylinder, V-I-A-F, 1399 cc, 25 bhp @ 2400 rpm (max. 33 @ 3300).
Transmission: 3F1R×1.
Brakes: mechanical, on rear wheels only.
Tyres: 6.00–18.
Wheelbase: 2000 mm.
Overall l×w×h: 3560×1500×1670 (1575) mm.
Weight: 1000 kg, GVW 1300 kg.
Note: magneto ignition; ifs with coil springs; crew three; late-type body (by Yanase) shown. Earlier production had 2-seater bodywork.

Car, Scout, 4×4, Type 95 (Kurogane) Three-quarter rear view of standardised 'Black Medal' scout car. The fuel consumption was about 4 litres per hour or 0·08 litre per kilometre. Fuel capacity 49 litres. Maximum speed 70 km/h. A folding canvas top was provided. Also with closed cab and truck body.

Motorcycle, Solo, Type 97 (Sankyo) V-2-cyl., 24 bhp, 3F, wb 1600 mm, 2591×915×1168 mm, 280 kg approx. 1196-cc (90×94 mm) engine with cylinders set at 45°. CR 4·8:1. Model 1937. Patterned on US Harley-Davidson. Note high ground clearance. IWM photo STT9514.

Motorcycle, with Sidecar, 3×1, Type 97 V-2-cyl., 24 bhp, 3F1R, wb 1600 mm, 2700×1700×1168 mm, 500 kg. Track 1245 mm. Tyres 4.75-27. Max. speed 70 km/h. Could be armed with light machine gun. 8-in ground clearance.

Motortricycle, 3×2, Cargo and Personnel, Type 1 (Kurogane) V-2-cyl., 3F1R, wb 1900 mm, 2730×1220×1220 mm, 540 kg. Shaft drive to car-type rear axle. Hinged tail gate. Bracket visible on vehicle's right-hand side carried spare wheel. Known as 'Sanrinsha'. Also produced as water-carrier.

Car, 5-seater, 4×2, Tourer (Toyota AB) 6-cyl., 65 bhp, 3F1R, wb 2851 mm, 4737×1734×1740 mm, 1500 kg approx. Max. speed 100 km/h. Special military version (ABR) had 'cut-away' mudguards and other detail modifications. Both based on 1936 Toyota AA sedan.

Car, 5-seater, 4×2, Sedan (Nissan 70) 6-cyl., 80 bhp, 3F1R. In April, 1937, Nissan acquired the tooling of the American Graham-Paige company for the production of this replica of the 1935/36 Graham Sedan. A military 7-seater Tourer version was introduced in 1939. Production ceased in 1940.

Car, 7-seater, 4×4, Command, Type 98 (Isuzu KIJI A) 6-cyl., 70 bhp, 4F1R×2, wb 3300 mm, 4950×1820 mm, 2410 kg. Tyre size 32×6. Model KIJI B had 55-bhp 3·4-litre 4-cyl. diesel engine and weighed 2640 kg. Max. speed 80 and 75 km/h respectively.

Car, 7-seater, 6×4, Command (Sumida K93) This cross-country command car was a Japanese copy of the American Hudson Special, six of which had been supplied to the Japanese Government in 1932 for use in the Manchurian war. Similar cars were produced by TGE (Chiyoda).

JAPAN

TRUCKS
4×2, 4×4, 6×4, 6×6

Makes and Models: *Light (under 1-ton):* Toyota AK 10.
Note: for 3×2 see preceding section.
Medium (1½- to 5-ton): 4×2 *types:* Chevrolet, Dodge, Ford, International, Isuzu Type 94 and 97 (TX 40), Minsei (diesel), Nissan 80 and 180, Toyota GA, KB, KC and BM. 4×4 *types:* Isuzu Type 97 and Type 2 YOKI, Toyota KCY and SUKI (amph.). 6×4 *types:* Chiyoda Q, Isuzu TU 10, Type 94A and B, PCA, TU 23 and Type 1 or 2601, Sumida. 6×6 *types:* Isuzu ROKI.
Heavy (5-ton and over): 4×2 *types:* Isuzu Type 2 (7-ton). 6×4 *types:* Isuzu TH10 (20-ton).

General Data: The above list gives the main types used by the Japanese forces during World War II, including some prototypes. Several types were of pre-war US manufacture or copied from American designs. The 6×4 Types 94A (petrol) and 94B (diesel) were the most common tactical trucks and some 40 variations existed, including artillery prime movers. First designed in 1932 following evaluation of some contemporary European six-wheelers, these trucks were produced in several factories, and for civilian operators there was the attraction of a Government subsidy scheme, not unlike that of some European countries. Certain 4×2 type trucks were also subsidised. The 4×4 and 6×6 types were mainly prototypes; few reached the production stage. Owing to shortage of materials and almost complete lack of technical information from foreign sources during the war, no improvement occurred in the quality of Japanese military trucks and the main emphasis was on quantity of standard types in the 4×2 and 6×4 categories. Nippon Diesel Engineering (later Minsei) introduced a military truck in 1939 powered by a Junkers-Krupp type two-stroke diesel engine. The Isuzu TH10 was a 20-ton dump truck, produced for the Navy and was the heaviest truck produced in Japan before 1945. When the war came to an end some 4×2 type trucks continued in production in modified form and became known as 'lost war' models. Many of them were used by the Japanese Self-Defence Forces which were organised in the early 1950s.

Vehicle shown (typical): Truck, 1½-ton, 4×2, Cargo (Nissan 180).

Technical Data:
Engine: 6-cylinder, I-L-W-F, 3670 cc 82·5×114·3 mm), 69 bhp @ 3000 rpm. CR 5·8:1.
Transmission: 4F1R.
Brakes: hydraulic.
Tyres: 32×6 (2DT).
Wheelbase: 4000 mm.
Overall l×w×h: 6380×2190×2200 mm.
Weight: 2900 kg. *GVW:* 4800 kg.
Note: Max. speed 75 km/h. Load space 3700×2000 mm. Also with other bodywork, incl. fire fighter. Continued in production after 1945 as 2½-tonner ('lost war' model).

Truck, 1½-ton, 4×2, Cargo (Nissan 80) 6-cyl., 80 bhp, 4F1R, wb 3251 mm, 5390×2900×2500 mm, 2880 kg. Patterned on US Federal COE truck. 3·67-litre L-head engine (as in Nissan 70 car). Also with shorter wheelbase. For military use it was superseded by the Nissan 180 (q.v.).

Truck, 2-ton, 4×2, Cargo (Isuzu TX40) 6-cyl., 70 bhp, 4F1R, wb 4000 mm, 6730×2200×2400 mm, 3500 kg. First introduced in 1937/38 (Type 97). 1939 model shown. 4390-cc (90×115 mm) engine. Tyres 34×7. GVW 5900 kg. Also with special bodies, incl. searchlight.

Truck, 1½-ton, 4×2, Chassis (Toyota GB) 6-cyl., 63 bhp, 4F1R. Mechanically very similar to 1936 US Chevrolet truck. Altogether 19,870 produced for commercial and military use during 1938–42, superseding 1936–37 G1 and GA. Appeared mainly with cargo and dump bodywork.

Truck, 1½-ton, 4×2, Cargo (Toyota KB) 6-cyl., 78 bhp, 4F1R, wb 4000 mm, 6470×2200×2200 mm, 2800 kg. Tyres 32×6. 3389-cc (84·1×101·6 mm) OHV engine. 21,130 produced, 1942–44. AK10 (½-ton) had 4-cyl. engine with same bore and stroke. KC had squarer cab and front end.

Truck, 1½-ton, 4×2, Cargo, Type 94 (Isuzu) 6-cyl., 70 bhp, 4F1R. This was the 4×2 version of the more common Type 94 six-wheeled standardised truck, to which it was identical with the exception of the rear bogie. Chassis also with special equipment, incl. aircraft-starting units.

Truck, 7-ton, 4×2, Cargo, Type 2 (Isuzu) 6-cyl. diesel, 100 bhp, 5F1R, wb 4700 mm, 7810×2300 mm, 5500 kg. Tyres 36×8 (2DT). GVW 12900 kg. One of the heaviest Japanese trucks at the time. Engine had 110×150 mm bore and stroke, 8550 cc cubic capacity.

Truck, 2-ton, 4×4, Cargo, Type 2 (Isuzu YOKI) 6-cyl. diesel, 85 bhp, 4F1R×2, wb 4000 mm, 6500×6495×1220 (cab) mm, 4100 kg. Prototype for all-wheel-drive medium truck, designed in 1942. Also 4×2 and civilian versions. Model ROKI was similar but with dual rear axles. Limited production.

Truck, 2-ton, 4×4, Amphibian (Toyota SUKI) 6-cyl., 63 bhp, 4F1R×2, wb 4000 mm, 7620×2220 mm, 4000 kg. GVW 6400 kg. Tyres 32×6 (2DT). Ship-like steel hull with engine, drive train and axles of Toyota model KCY truck. A total of 198 were made during November, 1943–August, 1944.

Truck, 3-ton, 6×4, Cargo (Isuzu TU10) 6-cyl., 34 bhp, 4F1R×2, wb 3040 (BC 1100) mm, 5430×1950 mm, 2950 kg. Tyres 36×6. This was a subsidy-type vehicle, basically similar to the military 1½-ton 6×4 and also made under different trade names.

Truck, 1½-ton, 6×4, Cargo and Personnel, Type 94A/B (Isuzu) *Type 94A:* 70 bhp @ 2800 rpm. *94B:* diesel, 70 bhp @ 2500 rpm. *Both:* 4F1R, wb 3350 (BC 1100) mm, 5430 ×1950×2250 mm. *94A:* 3400 kg. *94B:* 3700 kg. Tyres 34×6. Also used as artillery prime mover.

Truck, 1½-ton, 6×4, Airfield Tanker, Type 94A/B (Isuzu) Chassis similar to cargo truck but shorter wheelbase. Both open and closed cabs were used. Chassis also used for many other special purposes, including searchlight truck and armoured car. Patterned on European-type light six-wheelers.

Truck, 3-ton, 6×4, Cargo, Type 1 or 2601 (Isuzu TU23) 6-cyl. diesel, 85 bhp, 4F1R, wb 4000 (BC 1100) mm, 6840×2200 mm, 4300 kg. Also with petrol engine (70 bhp). Tyres 34×7. Worm-drive axles. GVW 7700 kg. Introduced 1941.

JAPAN

TRACTORS
WHEELED, HALF- and FULL-TRACK

Principal Types: *Wheeled types:* KATO.
Half-track types: Type 98 KO-HI and Isuzu.
Full-track types: KATO 70, Type 94 4-ton, Type 92A and 92B 5-ton, Type 98 6-ton, Type 92A and 92B 8-ton, Type 95A and 95B 13-ton, Medium Prime Mover (V8), Heavy Prime Mover (V12), Armoured Prime Mover, Prime Mover/Recovery vehicle.

General Data: The Japanese Army employed relatively large numbers of full-track artillery prime movers. Most of these were powered by diesel engines. The Type 92 5-ton model was first developed in 1932 and fitted with a Sumida 6-cyl. L-head petrol engine. In 1935 the Automobile Industry Co. (Jidosha Kogyo, predecessor of Isuzu) developed an air-cooled diesel engine for this tractor. Standardised as Isuzu No. 3 power plant, this engine was introduced about 1938 and the diesel version of the tractor was designated Type 92B, the petrol version becoming Type 92A. The same dieselisation programme was applied to the Type 92 8-ton model, for which the diesel power plant was produced by Niigata Iron Works and Kubota Iron Works. Prime Movers powered by Type 100 standardised-type diesel engines (6-, 8- and 12-cyl.) were Type 98 6-ton (6-cyl., 120 bhp), Medium type (V8, 160 bhp), Heavy type (V12, 200 bhp) and Type 98 Half-track (6-cyl., 120–130 bhp). All these Type 100 engines had the same cylinder size (120 × 160 mm bore and stroke) and many parts in common. With the exception of the 4-ton high-speed Type 94 (V8 air-cooled petrol) for field artillery, all other tracked prime movers were completely dieselised. The armoured prime mover was based on a light tank chassis. The engine/transmission unit was mounted at the front on the right-hand side; the driver was placed on the left. A large open-top armoured body, with double doors at the rear, extended over the tracks. The prime mover/recovery vehicle was derived in part from the medium tank. Two swinging booms were provided at the front, traversed by gears and operated independently by two different operators.

Vehicle shown (typical): Tractor, AA, Half-track, Type 98 (Isuzu)

Technical Data:
Engine: 6-cylinder diesel, I-I-W-F, 120 bhp @ 1800 rpm.
Transmission: 4F1R.
Brakes: track brakes.
Tyres: 36 × 6 (front).
Tracks: dry pin, 200 mm.
Overall l × w × h: 5670 × 1900 × 2300 mm.
Weight: 6 tons.
Note: crew 15, max. speed 45 km/h. Track bogies similar to those of light tank.

Tractor, AA, Half-Track, Type 98 (KO-HI) Ikegai 6-cyl. diesel, 110 bhp, 4F1R, 5300×2000×2200 mm, 5700 kg. Prototype for AA prime mover. Crew 15. Same chassis was also used for SP AA gun carriage, which was similar in appearance except for platform-type rear body with gun mount.

Tractor, General Purpose, 4×2 (KATO) K3 4-cyl., 60 bhp, 3F2R, wb 2286 mm, max. width 1880 mm, height 1525 mm. Front wheels 29×5, rear wheels (dual) 40×10, all shod with solid tyres. Basically commercial type, used for general-purpose work. Same engine as in KATO 70 full-track tractor.

Tractor, 4-ton, Artillery, Full-Track, Type 94. V-8-cyl., air-cooled, 73 bhp (max. 88), 4F1R, 3785×1855×2150 mm, 4 tons. Crew six. Clutch-brake steering. Magneto ignition. Fitted with 2·2-ton power winch. Max. speed (when towing) 25 mph. Track width 250 mm.

Tractor, 6-ton, Artillery, Full-Track, Type 98 (Isuzu) 6-cyl. diesel, 88 bhp (max. 110), 4F1R, 4300×2050×1900 mm, 6900 kg. Crew seven. Clutch-brake steering. Fitted with 5-ton power winch. Extensively used as prime mover for 105- and 150-mm howitzers and 105-mm field gun.

JAPAN
COMBAT VEHICLES

Principal Types: *Wheeled types:* Dowa (modified Crossley A/C 4×2). Osaka Type 2592 (A/C 4×2). Chiyoda, Kokusan (Navy) and Sumida, various types (A/C 6×4).
Half-track types: APC Type 1 (HO-HA) (TGE/Hino 2A20).
Full-track types: APC Type 1 (HO-KI) (TGE/Hino CA90). Midget Tanks (Tankette), Type 92, 94 (TK), 97 (TE-KE). Light Tanks, Types 93, 95 (HA-GO/Kyu-Go), 98 (KE-NI), 2 (KE-TO), 3 (KE-RI), 4 (KE-NU). Medium Tanks, Types 89A/B, 97 (CHI-HA), 1 (CHI-HE), 3 (CHI-NU), 4 (CHI-TO), 5 (CHI-RI). Heavy Tanks, Types 92, 95. AA Tank, TA-HA. Gun Tanks (SP), Types 1 and 3 (HO-NI), 2 (HO-I). SP Gun Carriages, KU-SE, NA-TO, KA-TO, HO-RI, HO-RO, HA-TO (SP Mortar). Amphibious Tanks (Navy), SR, Type 2 (KA-MI), 3 (KA-CHI), 4 (KA-TSU), 5 (TO-KU).

Vehicle shown (typical): Tank, Light, Type 95 (HA-GO or Kyu-Go)

Technical Data:
Engine: Niigata 6-cylinder diesel, I-I-A-R, 14,334 cc, 120 bhp @ 1800 rpm (max. 135 bhp).
Transmission: 4F1R.
Steering: clutch brake.
Track width: 250 mm.
Overall l×w×h: 4300×2070×2280 mm.
Weight: 6·7 tons, combat weight 7·4 tons.
Speed: 40 km/h.
Note: designed in 1935, produced by Mitsubishi and other mfrs. One of Japan's principal tanks. Picture (US Official) shows captured specimen at APG.

General Data: In addition to the above types, the Japanese used a variety of specialised tracked vehicles such as armoured recovery vehicles, bridge layers, bulldozers, etc. In 1940/41, following the successes of Hitler's *Blitzkrieg* in Europe, Japan's tank force was reorganised and new field service regulations were introduced for the tactical use of armour. However, for various reasons tanks were mainly used in the old-fashioned infantry support role. Many early Japanese AFVs were either copied from or patterned on Western designs (mainly British and French). The first Japanese tank was designed during 1925–26 and completed at the Osaka Arsenal early in 1927. It featured no fewer than 34 bogie wheels and weighed about 22 tons. By 1940 Japan had produced over 2000 tanks of various types. The peak year was 1942, when just under 1300 tanks were delivered. Most tanks were powered by diesel engines. Relatively few armoured cars were used and all were of pre-war design. Some could be used on railways by fitting flanged wheel-rim attachments. Mitsubishi, in 1941, delivered a diesel-engined 4×2 armoured car for rail use to the South Manchurian Railway, which was used as a 'guide car' when trains were in operation.

Tank, Medium, Type 97 (CHI-HA) V-12-cyl. diesel, 170 bhp, 4F1R×2, 5500×2330×2380 mm, combat wt 15·8 tons. Picture (US Official) shows captured specimen with late model turret (*Shinhoto*) and 47-mm gun. Best Japanese tank of World War II. Early model (1937) had low-velocity 57-mm gun.

Gun tank (SP), Type 1 (HO-NI 1) V-12-cyl. diesel, 170 bhp, 4F1R×2, 5500×2330×2390 mm, combat wt 15·9 tons. Medium tank chassis. 75-mm gun. Only Japanese AFV capable of knocking out any contemporary US tank. US Official picture shows vehicle at APG.

Car, Armoured, 6×4, Type 2593 (Sumida) 6-cyl., 100 bhp, 4F1R, 6570×1900×2950 mm, 7 tons. Railway wheel-rim attachments normally carried on hull sides. Max. speed on roads 40 km/h, on railway 60 km/h. Similar vehicles produced by other mfrs. Crew six.

Carrier, Personnel, Armoured, Half-Track, Type 1 (HO-HA) Hino DB52 6-cyl. diesel, 125 bhp, 6100×2100×2000 mm, 7 tons. Crew 15 (3+12). Air-cooled engine. Speed 50 km/h. Tyres 34×6. Track width 250 mm. Used by mechanised infantry and as AA gun prime mover.

ABBREVIATIONS

Note: for specific German abbreviations see introduction to German section.

AA	Anti-aircraft
A/C	Armoured Car
ACV	Armoured Command Vehicle
AFV	Armoured Fighting Vehicle
AM	Air Ministry (GB)
amn	ammunition
AOP	Armoured Observation Post
APG	Aberdeen Proving Ground
APT	Airportable
ARV	Armoured Recovery Vehicle
AT	Anti-Tank
aux	auxiliary
AVRE	Armoured Vehicle Royal Engineers
BARV	Beach Armoured Recovery Vehicle
BC	bogie centres
B/D	Breakdown (Wrecker)
BEF	British Expeditionary Force
ca	circa (approximate date)
CID	cubic inches displacement
CKD	Completely knocked down
CMP	Canadian Military Pattern
COE	Cab over Engine (FC)
compt	compartment
CR	compression ratio
diff(s)	differential(s)
Dim	dimension(s)
D/S	dropside (body with hinged sides)
DT	dual tyres
FAT	Field Artillery Tractor
FC	forward control (COE)
FFW	fitted for wireless
FVDD	Fighting Vehicle Design Department (GB)
FVPE	Fighting Vehicle Proving Establishment (GB)
Gov't	Government
gpm	gallon per minute
GS	General Service
GVW	gross vehicle weight
HAA	Heavy Anti-Aircraft
HAT	Heavy Artillery Tractor
how	howitzer
ifs	independent front suspension
irs	independent rear suspension
IWM	Imperial War Museum
LAD	Light Aid Detachment
LAT	Light Artillery Tractor
LHD, lhd	left-hand drive
LST	Landing Ship (Tanks)
LVT	Landing Vehicle Tracked
LVT(A)	LVT (Armoured)
LWB, lwb	long wheelbase
MAT	Medium Artillery Tractor
mfr(s)	manufacturer(s)
mg	machine gun
MCC	Morris-Commercial (Cars Ltd)
mod(s)	modification(s)
MoS	Ministry of Supply (GB)
MT	Motor Transport
MWEE	Mechanical Warfare Experimental Establishment (became MEE) (GB)
NA	not available
NC	normal control
N/S	nearside (left-hand side)
OHV	overhead valves
O/S	offside (right-hand side)
pdr	(6-)pounder (gun)
PE	petrol-electic
PTO	power take-off (usually on gearbox)
q.v.	*quod vide* (which see)
RAF	Royal Air Force (GB)
RASC	Royal Army Service Corps (GB)
rds	rounds (amn)
RHD, rhd	right-hand drive
RN	Royal Navy (GB)
SAS	Special Air Services (GB)
SC	Sidecar
SNL	Standard Nomenclature List (US Ord. parts catalogue)
SP	Self-propelled
S-T	Semi-trailer
Std	standard
SV	side valves
SWB, swb	short wheelbase
TEV	Terminal Equipment Vehicle
TT	Tanks and Transport Branches, MoS (GB)
UN	United Nations
US(A)	United States (of America)
USSR	Union of Soviet Socialist Republics
VE Day	Victory in Europe (end of war in Europe)
VJ Day	Victory in Japan (end of war against Japan)
w/	with
wb	wheelbase
WD	War Department
w/o	without
WO	War Office (GB)
Wt	weight
W/T	Wireless Transmitter
WT	Wireless Telegraphy
WVEE	Wheeled Vehicles Experimental Establishment (GB)

INDEX BY TYPE

INDEX BY TYPE	USA	Great Britain	British Commonwealth	USSR	Germany	Italy	Japan
Ambulances	16, 20, 29, 33, 43	149–51	223, 234, 235, 238, 240, 243	265	297, 298	344	
Amphibians	110, 111, 129	200, 201		110, 260	283, 291, 331		360, 364
APCs and ACVs	121, 122	205, 207	249		321, 323		364, 365
Armoured Cars	114–17	202–6	254–6	270, 271	332, 333	350, 351	364, 365
Buses	15	87, 148	236		297, 313		
Carriers	128, 129	205, 208, 209	257				
Cars, Field Cars	13, 14, 17–23	88, 139–47	218, 219	260, 261	280–96, 322	338, 339	355–7
Cranes, Mobile	69, 104, 105	196, 197	240		324, 327		
Fire Fighters	106–9	198, 199	231, 247	262	308, 311, 315		358
Gun Motor Carriages[1]	20, 50, 112, 113, 123, 124, 130, 131	155, 186, 187, 207	235, 245, 253	265, 271	321, 326, 334	347, 350, 351	364, 365
Half-Track Vehicles	118–24	169, 195	253	266, 267	279, 320–9	348, 349	362, 363, 365
Motorcycles[2]	10–12	136–8	218	260, 261	278, 279	338–40	355, 356
Searchlight Trucks	56	164, 165, 172, 174		262	311		359, 361
Snow Fighters	109	89	109	262	303, 324		
Tanks	130, 131	208–11	254	270, 271	334, 335	350, 351	364, 365
Tank Transporters	82, 83, 99–101	92, 93, 100, 182, 183	100, 250		316	343, 350	
Tractors[3]	102, 103, 120, 125–7	89, 184–95	250, 252	268, 269	318, 319, 330, 331	348, 349	362, 363
Trucks	24–93	86–93, 152–81	220–49	259, 262–7	299–317, 328, 329	341	358
Wreckers/Breakdowns	94–8	87, 92, 161, 170, 171, 194, 195	92, 239, 244, 247, 248	264	303, 313, 323	343	

[1] Includes Gun Portees. [2] Includes Motor Scooters and Motortricycles. [3] For Tractor and Prime Mover Trucks, *see* Trucks.

INDEX

Note: 'Canadian types' includes vehicles produced/assembled in other British Commonwealth countries.

A

Adler 277, 284, 286, 291, 297, 299, 300, 320, 322
AEC 135, 166, 168–70, 173–5, 180, 184, 187, 190–2, 194–7, 202, 205, 207, 211, 213
Aero 284
Albion 135, 149, 158, 160, 162, 163, 166, 169, 170, 173, 174, 180–3, 188–92, 194, 196, 246, 258
Alfa Romeo 313, 336–9, 343, 345, 346
Allan Taylor 177, 193
Allied Machinery 112, 113
Allis-Chalmers 9, 102, 118, 120, 125–7
Alvis 135, 202, 204
Alvis-Straussler 135, 188, 192, 193, 202, 204
American Bantam 9, 17–20, 112, 260, 355
American Car & Foundry 128, 130
American LaFrance 45, 106, 108
AMO 258, 262
Amphibian Car Corp. 110, 111
Ansaldo 337, 341, 350, 351
Ariel 135–7
ATZ 268
Audi 277, 284
Austin 135, 139–42, 146, 148, 149, 154, 157, 158, 162, 163, 166, 168, 170, 171, 184, 187, 196–8, 218, 240, 247, 258, 297
Austin-Western 102
Austro-Daimler 277, 302, 305, 318, 330, 331, 333
Autocar 8, 9, 53, 54, 64, 65, 67, 68, 70–5, 80, 94, 98, 118, 119, 121–4, 175, 178, 266
Auto Union 277, 284, 287, 288, 292–4, 297–300, 320, 330, 332
Available 9, 15, 94, 97, 104
Aveling-Barford 208

B

Baldwin 131
Bantam (*see* American Bantam)
Bay City 9, 104, 105, 196
Bedford 134, 135, 148–50, 152, 154, 155, 158–60, 162, 163, 166–9, 175, 178, 179, 184, 186, 187, 192, 194, 195, 198, 199, 202, 206, 211, 213, 226, 236, 238, 258
Benelli 338, 341, 342
Berliet 313
Bernard 316
Bianchi 280, 336–9, 343, 345
Biederman 9, 78, 79, 94, 97, 104, 105
BMW 136, 259, 260, 277–81, 283–5, 320, 321
Bombardier 120, 216, 250
Borgward 277, 299, 300, 306, 308–10, 320, 323–5, 334
Botond 302, 305
Breda 324, 337, 343, 347–50
Brockway 9, 76, 77, 104, 106, 108, 162, 175, 178
Browning 104
BSA 135–7, 218, 260
Buffalo 106
Buick 9, 13, 14, 128, 216, 218, 289
Büssing-NAG 277, 302, 304, 306, 307, 311, 313, 314, 316, 317, 320, 321, 324–6, 332, 333

C

Cadillac 9, 13, 14, 16, 128, 130, 289
Canadian-American 67, 69, 70
Carriers, Universal, etc. 128, 129, 192, 205, 208, 209, 254, 257
Case 102, 192
Caterpillar 9, 125, 126
Ceirano 337, 343, 348, 350
Chevrolet (Canadian types) 144, 149, 152, 154, 155, 158, 162, 166, 185, 187, 188, 212–53
Chevrolet (US types) 9, 13, 15, 16, 22–5, 36, 37, 40–3, 45, 47, 51, 53, 86, 106, 107, 112, 114–17, 141, 149, 158, 162, 170, 198, 258, 289, 306, 353, 355, 358, 359
Chiyoda 354, 355, 357, 358, 364
Chrysler 9, 13, 28, 31, 36, 53, 54, 128, 131, 212, 213, 216–18, 236, 249, 352
Citroën 277, 280, 299, 301, 306, 308, 320
Clark 102, 192
Cletrac (Cleveland) 9, 125, 192
Clydesdale 78
Coleman 9, 104
Coles 196, 197
Commer 105, 135, 151, 152, 154, 155, 158, 159, 162, 175, 177, 178, 202
Cook 84, 99, 112, 113
Corbitt 9, 64, 66, 76, 78, 80, 81, 84, 85, 104, 190
Corgi 138
Coventry 202, 205
Crosley 9, 22, 23, 125 126
Crossley 135, 166, 168, 170, 171, 173, 178, 179, 192, 196–9, 246, 364
Csepel 31
Cushman 9, 10

D

Daimler 135, 202, 204, 205
Daimler-Benz (*see* Mercedes-Benz)
Dart 9, 80, 99, 101, 104
DAT, Datsun 354
David Brown 135, 192
Demag 277, 320, 321
Dennis 135, 158, 160, 162, 164, 192, 194, 209
DeSoto 31, 216, 224
Deuliwag 318
Diamond T 8, 9, 15, 24, 25, 36, 51, 52, 64, 67–70, 72, 82, 83, 89, 94, 95, 97, 99–101, 106, 118, 121, 182, 183, 213, 246, 249
DKW 277, 278, 280
Dodge (British types) 135, 162, 164, 171, 175, 202, 206
Dodge (Canadian types) 28, 31, 152, 154, 162, 219–21, 224, 226, 228, 229, 232, 234, 236, 237, 250
Dodge (US types) 9, 15, 24–34, 36, 40, 43, 49–51, 87, 106, 107, 112, 114, 144, 148, 149, 152, 154, 158, 162, 170, 258, 358
Duplex 55

E

ERF 135, 175, 176
Excelsior 135, 138

F

Famo 273, 277, 318, 324
Farand & Delorme 216, 257
Fargo 9, 26, 31, 213, 216, 219, 224, 232
Faun 277, 316, 318, 319
Federal 9, 15, 51–3, 55, 56, 70–3, 78–80, 94, 97, 99, 100, 106, 182, 198, 213, 354, 359
Fiat 258, 277, 280, 299, 302, 306, 309, 313, 318, 336–9, 341–51
Fisher 128, 131
FMC 9, 128, 129
FN 278
Foden 135, 175, 180, 181
Ford (British types) 88, 135, 139, 141, 142, 148–50, 152–6, 158, 160, 161, 166, 167, 169–72, 175, 177, 178, 192, 193, 195, 198, 199, 205, 206, 208, 258, 265
Ford (Canadian types) 141, 144, 145, 149, 152, 154, 155, 158, 162, 166, 178, 184, 186–8, 205, 212–57
Ford (French types) (see Matford)
Ford (German types) 277, 280, 284, 292, 297, 306, 307, 313, 328, 329
Ford (US types) 9, 13–21, 24–7, 30, 31, 35–40, 42, 44, 45, 47, 55, 88, 106, 107, 110, 112–16, 128, 141, 146, 149, 154, 162, 200, 258, 260, 289, 306, 352, 353, 355, 358
Fordson (see Ford, British types)
Framo 277, 280
Fross-Büssing 277, 306
FWD 9, 64, 65, 74–6, 89, 102–4, 106, 109, 124, 170, 178, 190, 213, 216, 241, 250, 252

G

Garner(-Straussler) 170, 188, 189
GAZ/ZIM 110, 258–67, 270, 271

Gilera 338, 340
GM of Canada 149, 212–56, 258, 264
GMC 9, 15, 24–7, 36–40, 44–6, 51–60, 62–5, 70, 72, 73, 90, 94, 96, 106, 108, 110, 111, 114, 117, 158, 162, 170, 180, 200, 216, 224, 232, 234, 236, 258, 259, 264
Gnome & Rhône 278
Gräf & Stift 277, 330
Graham-Paige 354, 357
Guy 135, 154, 155, 157, 162, 165, 166, 168, 170, 173, 174, 184, 185, 188, 189, 202, 204, 205, 246
Guzzi 338, 340–2

H

Hanomag 277, 280, 281, 283, 318–20, 323, 330
Hansa-Lloyd 277, 306, 320, 323, 324
Harley-Davidson 9, 11, 12, 22, 136, 218, 355, 356
Harnischfeger 104
Hendrickson 9, 104
Henschel 277, 302, 306, 311–15, 335
Hillman 135, 139–42, 202
Hino 70, 354, 364, 365
Horch 277, 284, 285, 287, 288, 292–4, 297, 298, 332
Horex 278
Hotchkiss 304
Hudson 355, 357
Hug 78
Humber 21, 135, 141, 143–5, 149, 150, 152–5, 202–6

I

Indian 9, 11, 12, 23, 136, 218
International (IHC) 8, 9, 15, 26, 28, 30, 31, 34, 36–8, 40, 44, 49, 50, 55–7, 60–4, 66, 72, 73, 80, 81, 91, 106, 121–3, 125–7, 149, 175, 178, 180, 213, 215, 216, 224, 226, 232, 234, 236, 239, 240, 250, 252, 253, 264, 268, 358
Iron Fireman 125, 126
Ishikawajima 354
Isotta-Fraschini 336, 337, 343
Isuzu 353–5, 357–63
Iwasaki 355

J

JAG 258, 262–7
James 136
JAS 258, 262, 263
'Jeep' (see American Bantam, Ford, Willys)
John Deere 9, 102

K

Kaelble 277, 318, 319, 330
Kaiser 22, 23
Karrier 135, 158, 166, 167, 170, 174, 188, 189, 198, 204
KATO 362, 363
Kenworth 8, 9, 70, 72, 82, 83, 94–6, 106, 108
KIM 258, 260
Klöckner(-Humboldt)-Deutz (KHD) 277, 306, 308–10, 312, 313, 315, 328–30
Knuckey 99, 100
Krauss-Maffei 277, 324–6
Krupp 277, 292, 295, 302, 303, 305, 306, 311–13, 318, 324, 332, 335
Kurogane 354–6
Kwaishinsha 352, 354

L

Laffly 277, 284, 302, 304, 316, 318

Lagonda 163, 202
Lancia 313, 336–8, 343, 344, 347, 348, 350
Lanz 277, 318
LaSalle 16, 289
Latil 284, 313, 318, 319
LeTourneau 101
Leyland 135, 148, 158, 162, 164, 170, 173–5, 177, 180, 181, 194, 196–8, 202, 207, 210, 245, 246
Lima 110, 125
Link-belt 104
Linn 9, 16, 118, 119
Lombard 266
Lorain (see Thew)
Loyd Carriers 208, 209

M

Mack 9, 15 51, 55, 56, 72, 74, 76–80, 91, 92, 94, 99, 101, 106, 109, 118, 119, 124, 175, 178, 180, 182, 190, 196, 213, 258
Magirus 277, 302, 304, 306, 308, 311–13, 315, 328, 330, 332
MAN 277, 297, 302, 306, 313, 314, 335
Maple Leaf 216, 232, 233, 236, 238, 239, 250
Marmon-Herrington 8, 9, 13, 25–7, 31, 40, 42, 44, 45, 47, 49, 70, 72, 78, 88, 89, 106, 117, 121, 124, 125, 128, 130, 216, 232, 233, 246, 247, 249, 250, 254, 256, 257, 264
Massey-Harris 9, 128, 130
Matchless 135, 136, 138, 218, 260
Matford 277, 306, 313
Maudslay 135, 175, 176
Maxim 106–8
Maybach 273, 277, 321–7, 330, 331 334, 335

Mercedes-Benz 277, 281, 284, 286, 287, 292, 293, 295–300, 302, 304, 306, 307, 309–16, 319, 320, 324, 325, 327–9, 332–5, 355
Mercury 218, 289
Metropolitan-Cammell 209
Michigan 104, 105
Minneapolis-Moline 9, 102, 103
Minsei 354, 358
Mitsubishi 354, 355, 364
Morris 135, 139, 140, 149, 151, 200–3, 206
Morris-Commercial (MCC) 135, 149, 150, 152–4, 156–8, 161, 162, 170, 184–6, 188, 189, 195, 200–2, 251, 289, 301

N

NAMI 258
Nissan 31, 353–5, 357–9
Normag 318
Norton 135, 136, 138, 218
NSU 273, 277–9, 320, 334
Nuffield 22, 23, 146, 200, 201, 209, 210

O

ÖAF 277, 313, 314
Oldsmobile 13, 306
OM 277, 306, 336, 337, 341, 343, 345
Opel 273, 277, 280, 281, 284, 285, 288, 289, 291, 292, 297–9, 306, 307, 309, 310, 313, 328, 329 341
Opperman 200
Osaka 352, 364
Oshkosh 9, 94, 98, 102, 103, 106, 109

P

Pacific 9, 82, 83, 99, 100, 182
Packard 9, 13–16, 141

Pavesi 318, 337, 348, 349
PAZ 258
Peugeot 277, 280, 297, 299
Phänomen 277, 278, 292, 295, 297–9, 320
Pirsch 106, 108
Plymouth 9, 13, 14, 216, 218, 224, 355
Pontiac 9, 13, 14
Porsche 277, 280, 282, 283, 335
Praga 277, 302, 304, 306, 324, 325, 330, 331, 334
Pressed Steel Car Co. 128, 130
Puch 278

Q

Quick-Way 104

R

Rába 305, 312
Renault 277, 280, 299, 301, 306, 308, 313, 315, 330
Reo 9, 45, 48, 51, 55, 57, 61, 72, 78, 79, 94, 97, 99, 100, 106, 109, 114, 117, 132
Rheinmetall-Borsig 330
Rikuo 355
Riley 141
Roadless 192, 195
Robur 298
Roebling 125
Röhr 280
Rolls-Royce 168, 210
Rover 146
Royal Enfield 135, 136

S

Sachsenberg 330, 331
Sankyo 355, 356
Saurer 277, 313–5, 320, 324, 330
Scammell 135, 170, 178, 179, 180, 182, 183, 190–5, 352
Seagrave 106
Sentinel 208

Sertum 338
Simplex 11
Singer 139, 193
Skoda 272, 277, 280, 282, 284, 290, 292, 294, 295, 313, 316, 318–20, 324
Somua 266, 277, 320, 323
Spa 299, 302, 306, 309, 318, 336, 337, 341–5, 347–51
SS Cars 22, 137, 146, 147
Stalin(ets) 258, 262, 268, 269–71, 330
Standard 22, 135, 139–42, 146, 147, 184, 186, 202–4
Sterling 9, 76, 82, 84, 85, 94, 98, 106, 109
Steyr 277, 278, 280, 284, 290, 292–4, 296, 297, 299, 300, 302, 305, 330
Stoewer 277, 280, 283, 284
Straussler 135, 184, 186, 188, 189, 192, 193, 209, 211, 333
Studebaker 9, 36, 45, 48, 53–7, 60–3, 93, 110–14, 116, 128, 129, 170, 178, 258, 264, 265
STZ 268, 269
Sumida 354, 355, 357, 362, 364, 365

T

Tatra 277, 280, 282, 287, 290–2, 296, 302, 305, 306, 316, 317, 320, 324, 326, 332, 352
Tempo 277, 280, 355
TGE 354, 357, 364
Thew (Lorain) 9, 104, 105, 196
Thornycroft 135, 158, 159, 162, 165, 166, 168, 170, 174, 180, 184, 187, 196, 197, 200, 208, 352
Tilling-Stevens 135, 162, 164, 165
Toyota 31, 353–5, 357–60
Trackless Tank Corp. 114, 117
Trippel 277, 284, 291
Triumph (British) 135–7

Triumph (German) 278
TT (British MoS) 151

U

Unic 277, 320, 323
Ural(-ZIS) 262, 263

V

Vauxhall 132, 135, 141, 142, 195, 211, 260, 324
Velocette 135, 137, 138
Vickers-Armstrong 190, 209, 211
Victoria 278
Volkswagen (KdF) 21, 273, 277–83
Volugrafo 339
Volvo 321
Vomag 277, 313, 335
Vulcan 170

W

Walter 9, 102, 103, 106, 109
Wanderer 277, 284, 286, 300, 320
Ward LaFrance 9, 67, 76, 80, 94–6, 108
Weiss, Manfred (Csepel) 188, 333
Welbike 136, 138
White 8, 9, 15, 53, 64, 67, 70, 71, 76–8, 80, 81, 93, 114, 118, 119, 121–3, 149, 175, 180, 182, 190
Willème 316
Willys 9, 17–23, 30, 112–15, 118, 120, 134, 146, 213, 257, 258, 260
Wolseley 135, 158, 192, 208, 209, 354

Y

Yellow (GMC) 38, 62

Z

ZIS/ZIL 110, 258–60, 262–8, 297
Zündapp 136, 277–9, 334